Item Response Theory

Item Response Theory

R. Darrell Bock
Robert D. Gibbons
University of Chicago

This first edition first published 2021

Registered Office
John Wiley & Sons, Inc., 111 River Street, Hoboken, NJ 07030, USA

Editorial Office
111 River Street, Hoboken, NJ 07030, USA

For details of our global editorial offices, customer services, and more information about Wiley products visit us at www.wiley.com.

Wiley also publishes its books in a variety of electronic formats and by print-on-demand. Some content that appears in standard print versions of this book may not be available in other formats.

Library of Congress Cataloging-in-Publication Data

Names: Bock, R. Darrell, author. | Gibbons, Robert D., 1955- author.
Title: Item response theory / Richard Darrell Bock, Robert David Gibbons, University of Chicago.
Description: First edition. | Hoboken : Wiley, 2021. | Includes bibliographical references and index.
Identifiers: LCCN 2020055709 (print) | LCCN 2020055710 (ebook) | ISBN 9781119716686 (hardback) | ISBN 9781119716679 (adobe pdf) | ISBN 9781119716716 (epub)
Subjects: LCSH: Item response theory. | Psychology–Mathematical models.
Classification: LCC BF39.2.I84 B63 2021 (print) | LCC BF39.2.I84 (ebook) | DDC 150.28/7–dc23
LC record available at https://lccn.loc.gov/2020055709
LC ebook record available at https://lccn.loc.gov/2020055710

Cover Design: Wiley
Cover Image: © Image by Robert Gibbons

Set in 9.5/12.5pt STIXTwoText by Straive, Chennai, India

To Renee, Monica, Paul, and Conrad

R.D.B.

To Carol, Julie, Jason, Ethan, and Michael

R.D.G.

Contents

Preface

Not everything that can be counted counts, and not everything that counts can be counted.

(Albert Einstein)

If we date the origin of modern item response theory from Derrick Lawley's pioneering 1943 paper, "On problems connected with item selection and test construction," or Frederic Lord's 1950 Psychometric Monograph, "A theory of test scores," the field has now enjoyed nearly 75 years of vigorous development. It appears to have reached a level of maturity sufficient to warrant a comprehensive review of the accomplishments up to this point. The previous effort, Lord and Novick's 1968 monograph "Statistical theories of mental test scores," while incorporating the innovative contributions of Allan Birnbaum, was necessarily a report of work in progress in a young field. Only the models for binary-scored items were available at that time, and the estimation theory required to implement them, although intimated, was not yet well developed.

Results in the field are now much richer. Currently, there are models to fit many forms of item response data, and the statistical methods for estimating the parameters of these models exist and are implemented. Procedures for assigning scale scores to respondents are more varied and include efficient adaptive algorithms. Entirely new methods exist for estimating latent distributions of populations without computing scores for individual sample members. Better solutions have been found for the classic problems of test maintenance and forms equating. Connections between item response theory and multilevel sampling models have been clarified. Perhaps most important, the computing facilities now exist to make large-scale applications of these developments practical.

Our aim in the present text is to present a reasonably complete account of this progress with special emphasis on the computer applications. Our discussion therefore includes details on numerical procedures suitable for practical applications of IRT. The most difficult aspect of writing this text has been finding the right level at which to present these topics, which are inherently mathematical and statistical. To make the discussion accessible as possible, without violating the spirit of the subject, we have adopted a level of presentation that assumes a first-year graduate background in the behavioral or social sciences, together with mathematics preparation through calculus and courses in statistics through generalized linear models. To supplement that preparation, we offer in Chapter 2 a review of some of the mathematical and statistical foundations required in the sequel. To motivate the reader and to fix ideas, we have everywhere tried to find real and interesting data with which to illustrate the theory.

Acknowledgments

We thank Bob Mislevy and David Thissen for their contributions to early work on this project; Don Hedeker, Li Cai, and Yanyan Sheng for their contributions, review, and helpful comments; and Cody Brannan for help in preparing the manuscript. We also acknowledge support of the Office of Naval Research, N00014-85-K-0586, and the National Institute of Mental Health R01 MH100155 and R01 MH66302 which funded our work on multidimensional item response theory and computerized adaptive testing. The content is solely the responsibility of the authors.

1

Foundations

To measure is to know.

(Source: Lord Kelvin (William Thompson) 1824–1907)

Most observations of behavior are recorded as distinct qualitative events. For example:

> a student responds correctly to certain specified questions, responds incorrectly to others, and declines to respond to still others;
> on the fifth trial of a learning experiment, the subject recalls six abstract and ten concrete words from a list of thirty;
> a reader rates each paragraph of an essay exercise on a scale of rhetorical effectiveness graded from 1 to 7.
> a participant in a class discussion group speaks up three times on issue A, once on issue C, but not at all on issues B, D or E;
> in response to a social survey, a head of household endorses five out of ten statements concerning a public issue, but disagrees with the others;
> an applicant for a secretarial position makes two spelling errors in transcribing 300 words of dictation;
> a patient in a primary care clinic reports specific problems with mood, cognition and somatic symptoms of depression during the past two weeks.

These types of data have in common the fact that each respondent is reacting qualitatively to multiple stimuli in a specified set. In the present context, we call all such stimuli *items* and define item response theory, or "IRT," as the statistical study of data that arise in this way. That each respondent is responding to more than one item is essential to the definition: if each respondent were presented only one item, an enumeration of the observed qualitative responses would result in a simple contingency table that could be analyzed by conventional chi-square or

Item Response Theory, First Edition. R. Darrell Bock and Robert D. Gibbons.

log-linear methods. Such methods can be extended to perhaps three or four items by assigning respondents to distinct categories generated by all possible combinations of the repeated qualitative response, but they quickly become unworkable as the numbers of items increase. When there are repeated qualitative responses to relatively large numbers of items, the data are the special province of IRT. This form of data must be regarded to arise from two stages of sampling – the sampling of responses within each respondent, and the sampling of respondents from some population.

There are three main uses of IRT methods of data analysis. The first is to summarize information in the responses in a way that is suitable for some practical decision about a given respondent. The IRT reduction of the data either classifies the respondents qualitatively or assigns each a quantitative measure that supports such a classification. This is a traditional treatment of data from multiple-item tests. A score on an educational test, for example, may support the decision to admit a student to a college or university; a profile of performance in a battery of vocational tests may influence the choice of job applicants or military recruits; a self-report on a personality inventory may suggest the best approach to counseling or psychotherapy. Under favorable conditions, these kinds of uses of item response data can substantially improve the chances of successful outcomes of the decision compared to subjective or more arbitrary methods of selection or classification.

The second important use of these methods is to describe various groups to which the respondents may belong. The paradigm of this use is the randomized experiment, in which subjects are assigned with equal probability to control or treatment groups and a multiple-item test is administered in order to evaluate the effects of the treatments. The object of the IRT analysis of the test data is to estimate the distribution of response tendencies in the several groups. Similarly, in survey studies, respondents may be randomly selected from defined subpopulations of some larger population and administered an attitude or opinion questionnaire. The responses to the questionnaire items could in principle be used to classify the individual respondents, as in an employment interview, but this is not the purpose of the typical survey. The aim is rather to compare the subpopulations with respect to the distribution of response tendencies among their members.

The classical approach to analysis of data from either of these sources is to make comparisons among the groups or subpopulations by estimating scores for the respondents and analyzing them as if they were the primary data. One of the important contributions of IRT has been to show that this is not necessarily the best way to proceed. We present methods by which population characteristics can be directly estimated from the original item responses without computing intermediate respondent-level scores.

The third use of IRT analysis is to characterize the items. In some types of study, the items themselves are the objects of interest. For example, the aim may be to construct items with levels of difficulty and discriminating power suitable for a particular population of respondents. These activities will inevitably involve the analysis of empirical item data in order to verify that the methods of item construction are succeeding. Some items may be far wide of the mark and will have to be made easier or harder. Others, especially among multiple choice items, may contain hidden ambiguities that weaken their discriminating power; they can usually be corrected by reworking the response alternatives. In either case, IRT methods can identify and estimate characteristics of the items that are diagnostic of these problems. These methods now extend beyond the traditional multiple-choice item formats to rating scales, nominal categories, and item clusters. They also encompass the empirical study of the cognitive processes involved in the item response. In these studies, interest centers on classes of items distinguished by common stimulus or task features. The objective is to identify such features, connect them to other cognitive theory, and predict their effect on statistical characteristics of the items (e.g., item difficulty or discriminating power). Chapters of this text devoted to this aspect of IRT are Chapter 4 on parameter estimation for binary items, Chapter 5 on multiple-category items, and Chapter 6 on item factor analysis.

1.1 The Logic of Item Response Theory

Classical and IRT methods of measurement differ dramatically in the ways in which items are administered and scored. The difference is clarified by the following analogy. Imagine a track and field meet in which 10 athletes participate in men's 110-m hurdles race and also in men's high jump. Suppose that the hurdles race is not quite conventional in that the hurdles are not all the same height and the score is determined not only by the runner's time but also by the number of hurdles successfully cleared, i.e. not tipped over. On the other hand, the high jump is conducted in the conventional way: The crossbar is raised by, say, 2-cm increments on the uprights, and the athletes try to jump over the bar without dislodging it. The first of these two events is like a traditionally scored objective test: Runners attempting to clear hurdles of varying heights is analogous to questions of varying difficulty that examinees try to answer correctly in the time allowed. In either case, a specific counting operation measures ability to clear the hurdles or answer the questions. On the high jump, ability is measured by a scale in millimeters and centimeters at the highest scale position of the crossbar the athlete can clear. IRT measurement uses the same logic as the high jump. Test items are arranged on a continuum at certain fixed points of increasing difficulty. The examinee attempts to answer items until she can no longer do so correctly. Ability is measured by the location on the continuum of the last item answered correctly.

In IRT, ability is measured by a scale point, not a numerical count. These two methods of scoring the hurdles and the high jump, or their analogues in traditional and IRT scoring of objective tests, contrast sharply: If hurdles are arbitrarily added or removed, the number of hurdles cleared cannot be compared with races run with different hurdles or different numbers of hurdles. Even if percent of hurdles cleared were reported, the varying difficulty of clearing hurdles of different heights would render these figures noncomparable. The same is true of traditional number-right scores of objective tests: Scores lose their comparability if item composition is changed. The same is not true, however, of the high jump or of IRT scoring. If the bar in the high jump were placed between the 2-cm positions, or if one of those positions were omitted, height cleared is unchanged, and only the precision of the measurement at that point on the scale is affected. Indeed, in the standard rules for the high jump, the participants have the option of omitting lower heights they feel they can clear. Similarly, in IRT scoring of tests, a certain number of items can be arbitrarily added, deleted, or replaced without losing comparability of scores on the scale. Only the precision of measurement at some points on the scale is affected. This property of scaled measurement, as opposed to counts of events, is the most salient advantage of IRT over classical methods of educational and psychological measurement.

1.2 Model-Based Data Analysis

The IRT discussed in this text is aptly described as "model-based." There are cogent reasons for taking this approach to item response data rather than relying on enumerative summaries or nonparametric statistical methods. Perhaps most important is the economy of thought and discussion that results from substituting quantitative complexity for voluminous descriptive detail. In this, IRT emulates modern physical science, which attempts to account for a wide range of observable phenomena by a possibly complicated mathematical function depending on relatively few free parameters. IRT achieves this kind of economy by expressing the probability of an observed response to a stimulus in terms of a limited number of characteristics of the stimulus and of the respondent. The mathematical functions used for this purpose, the most important of which we discuss in Chapters 3–6, are called *item response models*. They are a central feature of IRT. They are capable of accounting succinctly for the kinds of data exemplified above, and their parameters concisely describe the operating characteristics of the items.

Another merit of the model-based approach is that, when it leads us to a restricted class of parsimonious models that fit a wide range of data, our confidence in the theory behind the models is strengthened. We are then encouraged to extend the theory to new situations and further test its generality. Apart from fortuitous discoveries, this is the main avenue of progress in scientific work. Purely descriptive methods of data analysis do not give us the same reassurance that we have the right conception of the phenomenon. Nonparametric curve fitting procedures, for example, merely produce smoothed representations of the data, possibly under continuity restrictions. They have no definite limit on the number of free parameters implicitly fitted in the construction of the curve. Having no definite form, they are difficult to compare, discuss, or extend to other domains. Admittedly, they are useful for limited purposes, such as interpolating values between observations when no suitable functional forms can be found. But absence of suitable functions is not generally the case in item response data: most of the response models that have been proposed for both binary and multiple category data account for the observations within the limits of sampling error, and they do so with comparative few free parameters. It is unlikely that worthwhile improvement in fit could be expected by a model-free approach to item response data in the domains typically analyzed.

A main strength of existing IRT is that it provides a coherent and rigorous methodology for the analysis of a very wide range of multiresponse qualitative data. The familiar statistical tools for measured, quantitative variables are not generally suitable for such data. The most widely used procedures for such variables, including linear least-squares regression, univariate and multivariate analysis of variance, discriminant analysis, linear structural analysis, etc., all model the distribution of the observations on a continuous interval scale and assume homogeneous error variation. Except in limiting cases, qualitative response data do not even remotely satisfy these assumptions. They are discrete events: They do not refer to any continuum, do not have interval scale properties or have homogeneous error from one stimulus to another. Even the familiar population descriptors in the classical statistical analysis – means, standard deviations, product-moment correlations, etc. – do not serve these forms of data well.

1.3 Origins

IRT is not primarily a theory in the sense of a putative explanation of some phenomenon. Rather, it is a coherent methodological system, similar to estimation theory or least-squares theory in the field of statistics. The exception in IRT is the concept of an observed qualitative responses arising from underlying quantitative variation through the action of an intervening threshold process. This conception, especially as it applies to sensory discriminations or preference judgments, is implicit in the more psychologically oriented applications of the theory. In other areas of application, such as educational measurement, IRT is viewed merely as a means of relating the response probabilities to a much smaller number of underlying parameters in terms of which respondents can be characterized, populations compared, or items described.

The psychological orientation in IRT had its origins in the nineteenth and early twentieth-century work on scaling of stimuli; the educational measurement orientation is associated with the development of educational tests during the twentieth century and is now referred to as "classical test theory." Running through both of these approaches is a common thread of concepts and methods borrowed from mathematics and mathematical statistics. An understanding of these sources of present theory is a good foundation for study of the topic. In Section 1.3.1, we review briefly the contributions from each and discuss their relationships to the theory in its present form.

1.3.1 Psychometric Scaling

The first attempt to estimate scale values from discrete data was by Fechner (1966) in connection with his study of Weber's Law. Weber had found in a careful series of experiments that the magnitude of errors made by human observers in judging the size or intensity of a physical stimulus tends to be proportional to the intensity of the stimulus. Fechner reasoned that this indicated the existence of a sensory continuum on which the intensities of the stimuli are perceived as the logarithm of their physical measures. In a typical experiment demonstrating the Weber effect, the investigator requires the observer, by a pulley arrangement, to adjust the length of a variable line to match that of a displayed line of fixed length. This procedure is called the "method of adjustment." The general finding by this method, that the average absolute error in reproducing the stimulus is a constant proportion of the stimulus size (usually about 10%), is now known as Weber's Law. To establish the generality of this law and gain support for his theory relating stimulation to sensation, Fechner wanted to extend these studies to stimuli that could not easily be adjusted continuously, such as flavors, odors, or weights. For this purpose, he developed what he called the "method of right-and-wrong cases," but which is now called the "method of constant stimuli," or the "constant method."

In the modern version of the "lifted-weight" experiment discussed by Fechner in 1860 (Guilford 1954), the observer is presented a trial weight, and an identically appearing standard weight, and asked to lift them and state whether the former is heavier than the latter. The trial weights, which are set at several different levels smaller than, equal to, and larger than the standard, are sufficiently close to the standard that the observer makes a certain proportion of errors in repeated attempts at this task. The data from the experiment consist of the number of times the observer chooses the trial weight as heavier in a fixed number of attempts.

To infer a continuous measure of the average absolute error from these frequencies, Fechner invoked the same assumption that Gauss had made earlier for errors in astronomical observations – namely, that the errors are normally distributed with a mean and standard deviation typical of the observer. On this assumption, the expected size of the error associated with each observed proportion of "greater-than" judgments is the deviate at the corresponding percentage point of the normal distribution. If the assumption of normally distributed errors is tenable and Fechner's theory is correct, the plot of the corresponding normal deviates versus the log stimulus intensity (or in this case the difference between the logs of the test weights and the standard weight) should form a straight line, apart from sampling error in the observed proportions. With properly counterbalanced orders of stimulus presentation, the 50 percent point (or zero deviate) should occur at the point of stimulus equality. In that case, the probable error in judgments involving the standard stimulus can be defined as difference between the 75 and 25 percent points read from the fitted line – the so-called "difference limen" or "difference threshold" (Bock and Jones 1968).

Moreover, if Weber's Law holds over a wide range of physical intensities, the difference limen is constant on the sensory continuum and can serve as the unit of the scale that measures the psychological construct "sensation." The origin of the continuum can be set at the log of that value of the stimulus that can be correctly distinguished from the null stimulus 50% of the time (the "absolute threshold"). The logarithmic relationship between stimulation and sensation is now referred to as "Fechner's Psychophysical Law."[1]

In 1928, Louis Leon Thurstone formalized the concept of a sensory scale by introducing the discriminal process construct and a threshold mechanism. We discuss his model in more detail in Chapter 3, but briefly his assumption was that the stimulus gives rise in the observer to an unobservable random variable consisting of a fixed component attributable to the stimulus and a random component due to temporal instability of the sensory system. He called this unobservable variable a "discriminal process." To explain the observed response, he posited the existence of a point, or threshold, on the continuum such that the observer responds in one category if the process is above the threshold, and in another category if not.[2]

In Thurstone's model, it is the *difference* of the discriminal processes that is the relevant variable. If the process for the test stimulus momentarily exceeds that of the standard, the difference is positive and the observer responds that the test

1 In the 1950s, Stevens (1961) disputed the validity of Fechner's Law and proposed in its place a class of power functions relating physical to sensory intensity. But Stevens was discussing the relationship between magnitude judgments and stimulus intensity, which is a quite different phenomenon than the stimulus confusions on which Fechner's logarithmic psychophysical law is based. Although the power-law includes the logarithmic function in the limiting case as the exponent goes to zero, it makes no special contribution to the modeling of sensory discrimination phenomena.

2 Fechner (1860) attributed a similar account of the constant method results to his colleague, Möbius.

stimulus appears greater or more intense; otherwise, not. Thus, for the difference process, the threshold is at zero. Like Fechner, Thurstone assumed that the discriminal processes, and thus the difference process, are normally distributed. The continuous value attributed to the difference between the two stimuli is therefore consistently estimated by the normal deviate corresponding to the proportion of times the trial stimulus is judged greater than the standard.

Thurstone's most important contribution to scaling was the demonstration that a purely psychological scale not dependent on any form of physical measurement can be constructed from a suitable set of interlocking comparisons. If the deviates corresponding to these comparisons determine, or overdetermine, the locations of the stimuli on the sensory continuum, it then becomes possible to assign to the stimuli quantitative values in a well-defined metric. In particular, Thurstone (1928) showed that in the method of paired comparisons, where the observer compares all $n(n-1)/2$ distinct pairs of n stimuli, the equations implied by the discriminal process model overdetermine the locations of the stimuli on the psychological continuum. From the observed qualitative judgments, the locations can be estimated on a continuous scale with arbitrary origin and unit, and $(n-1)(n-2)/2$ degrees of freedom remain to test the fit of the model (see Chapter 5 in Bock and Jones 1968). In principle, the dispersions of the random processes could also vary and require estimation, but in the classical "Case V" analysis of paired-comparisons, they are assumed constant. To emphasize that such a scale could be constructed for any sort of pairwise orderable objects, Thurstone referred to it as a "psychological" continuum rather than a sensory continuum.

IRT borrows heavily from these earlier conceptions of psychological scaling. It accepts the idea that a discrete behavioral response to a set task, object, or proposition (in short, to an *item*) is the expression of a stochastic mechanism that can be modeled by an unobservable random variable and a threshold. The variable is assumed to have some form of distribution on an infinite latent continuum. The items are characterized by their locations, or thresholds, on this continuum and by the dispersion of the corresponding random variable. If an external variable or criterion is correlated with the unobservable random variable, the IRT scaling can proceed on the same basis as psychophysical scaling (see Section 4.1). But in the more typical case where no such external variable exists, the scale depends only on relationships internal to the data, which in IRT are direct responses to multiple items rather than the multiple comparisons of Thurstonian scaling.

As we elaborate in Chapter 3, when Lawley (1943) and Lord (1952) formulated the IRT model, they added a random component associated with the observer (or in our terminology the *respondent*). The paired comparison model does not require this component because, if present, it would subtract out of the difference process (Andrich 1978).[3]

The main objective of IRT is to estimate from the item responses the attribute values of the respondents. We discuss statistical methods for this purpose in Chapters 4–6 and 10. Because the values are on a scale determined by the error process implicit in the item response models, they are often referred to as "scale scores" to distinguish them from the traditional "test score," which is just a count of the number of correct item responses.

There is a precise formal sense in which the psychological scaling model, which does not include an individual difference component, stands in the same relationship to the IRT model, as the analysis-of-variance fixed-effects model stands to the mixed-effects model with one random dimension. In the IRT model, the parameters that characterize items are the fixed effects and the attribute components associated with the respondents are the random effects. The only difference between the models is that the estimators of these quantities are linear in the analysis of variance and nonlinear in IRT. This is a very big difference methodologically; however, the simple noniterative calculations of the mixed-effects analysis of variance give way to the more complex iterative procedures of nonlinear estimation that characterize much of IRT analysis (see Chapters 4–6).

In IRT, the item response model expresses the probability of a specified response to a test item as a function of the quantitative attribute of the respondent and one or more parameters of the items. Such models now exist for many types of item responses. Those for binary (right–wrong) scored items were the first to be developed and are still the most common, but models for ratings and graded or multiple-category item responses are now available and enjoying increasing application. We discuss binary item response models in Chapter 3 and multiple-category items in Chapter 5. (See Thissen and Steinberg (1986) for a taxonomy of item response models.)

1.3.2 Classical Test Theory

Classical test theory is essentially an extension of the Gaussian theory of errors (Gauss 1809) to the measurement of individual differences. Originating in the work of Spearman (1907), Brown, E.L. Thorndike, and others, the classical theory was first applied to scores from cognitive tests in which item responses were

---

3 If each observer judges all stimuli in a paired comparison experiment, random interactions between observers and stimuli need to be included in the *sampling* model for the Case V analysis. Otherwise, the correlation between deviates for comparisons involving common objects are correlated, and the test of fit is grossly biased (Chapman and Bock 1958).

scored "right" or "wrong." The test score of a respondent was the number of right responses. Later the theory was extended to any multiple-item psychological test in which items can be meaningfully scored in a consistent direction. Sometimes the direction cannot be specified *a priori*, and empirical evidence must be invoked. In the personality tests – scales of the Minnesota Multiphasic Personality Inventory (MMPI), for example – the direction for scoring each item is chosen so that the scale will best discriminate between normal and pathological groups identified by expert judgments. A similar method is available in IRT whenever an external variable correlated with the attribute dimension is available (see Section 4.1).

Classical test theory assumes that the test score obtained by counting "right" responses is an additive linear model consisting of two random components – one due to the individual differences in the population of respondents – the other due to error, defined as the item-by-respondent interaction. Both components are assumed normally distributed, although this assumption is not required in many of the results. A main motivation of the theory is to obtain a scale-free index of the precision with which test scores estimate attributes of the respondents. Because the origin and range of the test score depends arbitrarily on the number of items in the test, the scale-dependent measure of precision used by Gauss (the reciprocal of the mean square error) is not applicable in classical test theory. It is replaced by the *reliability coefficient*, defined as the ratio of the variance of the individual difference component (true score variance in classical test theory terms) to the sum of that variance plus the variance of the error component. In statistical terms, reliability is the intraclass correlation of within-to between-individual variation in the population of respondents. The reliability coefficient therefore ranges from 1, indicating error-free measurement, and 0, indicating no variation other than error.

One of the main results of reliability theory, based on the assumption of independent responses within individuals, is that the size of the error component in the test score is constant as the number of items increases, whereas the variance of the individual-difference component increases proportionately with the number of items. The reliability coefficient therefore tends to unity as the number of items increases indefinitely, assuming that the items that are added have parallel content. The effect on reliability of increasing the test length by some arbitrary factor is given in the *Spearman–Brown formula*. In practice, the between-respondent variance component and the item-by-respondent interaction are estimated by analysis of variance methods or related formulas.

1.3.3 Contributions from Statistics

Simultaneous with, but largely independent of, the developments in psychological scaling and reliability theory, similar concepts and methods were elaborated in the field of statistics. Fechner's constant method reappeared in the 1920s in procedures for assaying the potency of drugs and other biological agents. An example is insulin, which had to be tested on laboratory mice in order to determine the dose for human subjects. The trial doses could only be set at a few fixed levels, and the response in mice (convulsion) could only be observed qualitatively. The data are therefore formally identical to those of the constant method, and the same least-squares analysis of the inverse normal transform of the proportion of animals convulsing was used to estimate the log-dose level at the 50 percent point. These biological methods of determining drug potency are referred to as *bioassay*.

Because experiments involving animals are more expensive and time-consuming than sensory discrimination trials, however, the early workers in bioassay were especially concerned with the efficiency of the data reduction. In particular, they were reluctant to discard the information contained in cases where none or all of the animals respond at extreme dose levels and the corresponding normal deviates are infinite. When this problem was presented to R.A. Fisher, he suggested the maximum likelihood method for estimating the dose–response relationship – a procedure that works directly with the observed numbers of animals responding and not responding, and does not require the transformation of empirical proportions to normal deviates (Bliss 1935). This technique, called *probit* analysis, is now a standard part of the bioassay methodology (Finney 1952).

It was, of course, immediately recognized that the maximum likelihood method applied equally well to logistic deviates ("logits") which have some advantages in simplifying the computations. Their use, also suggested by Fisher, led to what is now called *logit* analysis. It was also recognized that the simple linear model implicit in the logit model could be elaborated into a wider class of general linear models with which many forms of binomial response relations could be investigated (Anscombe 1956). From that point, a short step led to the multinomial Logit – a vector quantity capable of modeling multinomial response relations (Bock and Lieberman 1970, Bock 1975). This armamentarium of data analytic methods has become known as "logit-linear" analysis, essentially equivalent to methods embracing also Poisson variables known as "log-linear" analysis (Goodman 1968, Haberman 1978, 1979). We review these various methods of analyzing qualitative data in Section 2.16 and apply them at various points throughout the book.

Much the same history was repeated in reliability theory. The results that psychologists such as Spearman, Brown, and others had obtained in terms of Pearson product-moment correlations, R.A. Fisher and others obtained by analysis of variance and variance-component estimation based on the mixed and random-effects models for analysis of variance. These methods now appear side by side with correlation methods in the texts on classical test theory (Cronbach 1970). We review them briefly in Section 2.17.

1.4 The Population Concept in IRT

Earlier statistical formulations of IRT, especially those of Birnbaum (1958a,b) and Rasch (1960), avoided the concept of a population of respondents or of a distribution of attribute values. Clearly, there is no precedent in physical measurement for taking into account the population distribution of the entities measured. Why should it be introduced in psychological measurement? If the origin and unit of the measuring scale is the problem, why not set them by convention as in physical science? How can we justify classical test theory's assumption of a normally distributed attribute in applications where the respondents are screened or self-selected samples as, for example, college students or military recruits? Difficulties in answering these questions seemed to suggest that population assumptions only complicate the theory.

But abandoning the concept of a population distribution of the respondent attribute presents even greater complications. It implies that respondents to whom the IRT analysis is applied need not be a sample. Any respondents will do; they do not need to be representative of a population. But this flies in the face of common experience that tests can behave very differently in different groups of examinees. Cognitive tests consisting of binary scored items can be highly sensitive to "floor and ceiling" effects – the tendency for some items to be all correct or all incorrect in the sample – if the groups vary widely in ability. Selection tests with good predictive validity developed in one group may not be predictive in another. See Bock and Moore (1986) for examples of validity studies in which vocational tests developed in samples of males that were found to be poorly predictive in samples of females. These experiences show us that the kind of population-free measurement typical of physical science is not all that is required in social science.

There are also technical barriers to dispensing with populations. Without a population distribution, there is no way to express any *average* property of a test. Quantities that reveal the operating characteristics of the test, such as test information, mean square error, or reliability, cannot be summarized in single indices, but must be expressed conditional on location on the latent continuum. Neither is there any way to know whether the test is informative in the region of the continuum where the respondents are located. Classical theory provides such information in the form of population statistics such as test reliability, item difficulty, and item validity (albeit based on the often unrealistic normal assumptions), but IRT theory without population assumptions does not provide such information at all.

To exclude population assumptions is also to preclude the use of Bayes estimation that can be justified for both long and short tests, and thus it throws estimation back on the assumption of indefinitely long tests, just as in classical test theory.

Similarly, it misses the opportunity to strengthen estimation by Bayes methods when the test consists of relatively few items.

Fortunately for IRT, there is a straightforward way out of this dilemma in those situations where there exists a population, however arbitrarily defined, from which a probability sample of respondents can be drawn for purposes of test development. Statistical methods are now available for estimating the latent distribution without restrictive assumptions as to its shape. These procedures can be carried out during item development or in those operational uses of the test where a population is sampled or canvassed. As we point out in Chapters 10 and 11, the logic of estimating latent populations should be considered an integral part of IRT. Like other IRT estimation procedures for IRT item analysis, it is based on large samples of respondents, applies equally well to long and short tests, and uses the data efficiently.

These considerations lead to the conclusion that the population concept in classical test theory should not be ignored in IRT. It requires, however, that the data for item modeling and analysis represent some population, just as in classical theory, and not merely an arbitrary collection of individuals. But the incorporation of a population concept does not mean that IRT comes to resemble the classical theory. IRT has many distinctive features that give it greater power and broader utility. Among the most salient of these are the following:

1. IRT does not use number-right as the estimate of the respondent attribute (see Chapter 10). Several different types of estimators of scale scores are available that have better properties than the conventional test scores. The scale scores are less subject to the "typical distortions of test score distributions" due to arbitrary item difficulty (Lord and Novick 1968), and they exhibit less nonlinearity of relationship between tests of different length and reliability.
2. The IRT item characteristics have an explicit role in the estimation of the scale scores of the respondents and in the calculation of the associated standard error of measurement.
3. Both respondents and items are located on the same IRT latent continuum. This makes possible the interpretation of scale values on the continuum in terms of the typical items that respondents at various points on the continuum have a specified probability of answering correctly. This possibility, which is unique to IRT, has many uses in the communicating of test results to the general public.
4. Characteristics of the population of respondents can be estimated directly without necessarily computing scores for individual respondents. This is not possible by classical methods.
5. IRT models and estimation methods are straightforwardly generalizable to multiple-category responses, including graded scoring of items, as in rating

scales, and nominal response categories as in multiple-choice tests (see Chapter 5). The nominal categories model is even capable of efficiently scoring a multiple-choice test without the use of a fully specified answer key and of recovering information in wrong responses.

6. In large-scale testing programs, IRT methods permit the item content of tests to be updated during operational use without compromising the long-term comparability of the estimated scores. Separate equating studies are not required (see Chapter 12).
7. In cognitive testing, not all items need to be presented to all examinees. This property opens the way to various forms of testing, such as computerized adaptive testing (CAT) and two-stage testing (TST), that reduce test administration time without loss of precision of estimation (see Chapter 8).
8. IRT models handle the effects of guessing on multiple-choice items more effectively than the classical corrections for guessing.

1.5 Generalizability Theory

In educational testing, the items of any particular test are assumed to be a sample of all possible tasks defined by a specified curricular objective. The hypothetical population from which the sample is drawn is called a *content domain*, where content is understood to include facts, concepts, procedures, and skills. For purposes of evaluating a student's mastery of the objective, there is assumed to be a corresponding population of test items called an *item universe*. If the items chosen for the test comprise a probability sample of this universe, the student's percent-correct score on the test is a best estimate of the percentage of items in the universe to which he or she could respond correctly. Thus, to the extent that the item content of the domain is well-defined, the quantity estimated by the student's percent-correct score has a clear meaning – namely, the degree of mastery of the curriculum objective. Measurement in these terms is essentially enumeration; there is no attempt to assign the student a scale value on a defined continuum.

Nor does this conception of measurement necessarily assume the content of the domain to be homogeneous: the scope of the domain is purely a matter of definition. A subject matter such as ninth-grade mathematics, for example, could be considered a single domain covering arithmetic, algebra, and geometry, or each of the latter topics could be regarded as separate domains. Curricular objectives usually have a hierarchical structure that permits aggregation at several levels.

A simple example of a well-defined domain, one which we discuss at various points in this text, is a list of words to be spelled correctly. The word books compiled for secretaries are examples of such lists. Because its universe is exhaustively defined, a spelling test consisting of a random sample of words from such a book has a clearly interpretable domain percent-correct score. If the book is assumed to contain all the words that an administrative assistant will have to transcribe from dictation, a score of 98% on the test is an estimate that, out of every 100 *different* words dictated, the administrative assistant is expected to misspell two. Thus, a score of 98% on the test might be a reasonable requirement for employment.

The domains representing most educational tests are not as clearly defined as this, but in principle, they can be understood as a large set of objectively defined tasks possibly structured in a crossed and/or nested classification. To the extent that two independent test constructors are working from the same domain definition, they should be able to construct tests that estimate the same domain percent-correct score, just as two independent social survey experts working from the same sampling frame should be able to reach consistent estimates of, for example, the number of persons older than age 30 who have never married.

The domain score concept can be brought into IRT, as we have attempted to do in Chapter 10. If the fitted response models for the test items accurately predict the probability of a correct answer from respondents at any point on the continuum, then the expected percent-correct score for any particular respondent is the sum of these probabilities at the point corresponding to the respondent's attribute value. If the items have been randomly sampled from their universe, this IRT expected percent correct is as good an estimate of the domain score of the respondent as is the percent-correct score on the test. Indeed, it may be a better estimate because of the smoothing of the response probabilities by the fitted IRT models.

Generalizability theory is concerned with evaluating the precision of the domain score when it is estimated from the test score. In the development of the theory by Cronbach et al. (1972), the index of precision is the ratio of the estimate variance component between respondents to the sum of that component and the error component or components. In principle, the error components for an educational test could include the:

1. temporal instability of the respondent,
2. item-by-respondent interaction,
3. between-form variation due to sampling of items, and
4. variation among the readers who are scoring the test items.

In practice, however, when the test is used to compare students all tested on the same occasion, temporal variation is excluded; when the scores on forms of the test containing different items are equated, between-form variation is excluded; and when the items are scored objectively, the reader effects are excluded.

This leaves item-by-respondent interaction as the error component, and this is the same source of variation that is assessed by the standard error associated with an IRT estimate of a respondent's attribute value. A measure of precision derived from the IRT error function is therefore comparable to a conventional generalizability coefficient computed from the item-by-respondent variance component. In the IRT context, the corresponding index is computed from the squared error function integrated over the population distribution of the attribute. If the IRT estimated attribute values have been scaled to unit standard deviation in the sample, the index of precision is just the complement of the integrated squared error function.

Incorporating other sources of error variation into the IRT framework would proceed along lines similar to those discussed by Cronbach et al. (1972). To assess reader variation in the scoring of an essay test, for example, a sample of paper would have to be read by more than one reader and the IRT score computed separately for each reading. Similarly, if there were more than one form of unequated test booklets with randomly assigned items, a sample of respondents would have to take more than one booklet, which again would be scored separately by IRT. In either case, the resulting score values would be subjected to a random-effects analysis of variance, just as in a conventional generalizability analysis. But more elaborate treatments of such data are available in methods for *multilevel analysis* (see (Bock 1989a); (Bock 1989b); (Hedeker and Gibbons 2006)). Generally speaking, IRT estimated scores are as good or better than conventional number-correct or percent-correct scores for purposes of these analyses. More modern approaches also allow the differential uncertainty in the IRT score estimates to be incorporated into the analysis (Hedeker and Gibbons 2006).

An interesting extension of the domain score concept is presented in Chapter 10, where the existence of an item-parameter distribution is assumed. If, as generalizability theory asserts, the test items are a sample from a larger domain, then the item parameters or the response functions are random variables and should have a (possibly multivariate) distribution. Suppose that distribution can be estimated from a sufficiently large random sample of items in the domain. In that case, the domain percent correct score for the respondent can be computed from the IRT estimate of his or her attribute value by integrating over the model-parameter distribution. This method of computing the domain score has the advantage that the set of items in the test administered to the respondent *does not have to be a random sample from the domain*. The effects of nonrandom sampling of items are absorbed by the parameters of those particular items. The logic of this use of IRT methods is analogous to that of survey sampling when statistics from an allocation sample are converted to those for a probability sample by use of case weights based on demographic characteristics of each respondent. In the IRT scale score estimation, the item parameters play a role analogous to case weights, such that the test constructed from allocated items estimates the same domain score as a test constructed from items randomly sampled from the domain. Indeed, the item response model need not be unidimensional as we have assumed up to now. The integration to compute the domain score could be carried out with respect to attributes in more than one dimension. The necessary multidimensional item response models and methods of estimating their parameters already exist (see Chapter 6). We pursue a number of promising implications of this conception of IRT-estimated domain scores in Chapter 10.

As our experience with IRT applications continues to grow, we have a clearer view of the theory's limitations as well as its advantages. Some uses that first appeared attractive now seem dubious. Others that seemed ordinary have become surprisingly important. An example of the former is the concept of creating "item banks," which pools of precalibrated items for educational tests or opinion scales would be assembled so that researchers or teachers could select items for a local test or scale. Because comparable IRT scale scores can be estimated from subsets of the original normative calibration, it seemed that the locally constructed test could be scored on a normed scale with a minimum of effort.

But more recently two of the assumptions on which a calibrated item bank would depend have become questionable. The first is the absence of context, position, or administration effects that might alter item difficulties or other properties in the local test. The second is unbiasedness of the local selection of items with respect to operating characteristics of items in the pool. Neither of these assumptions now seems tenable. Experience shows that a correct response to a given item can be facilitated or hindered by the items that precede it in a test, and endorsement of opinion scale items can be influenced by the tone set by earlier items. Moreover, in the cognitive test domain, local conditions of administration – the number of items in the test, the instructions to the respondents, the time allowed – could affect item characteristics. The existence of any or all of these effects implies that rigorous comparability of scores can be obtained only by the administration and calibration of intact test forms under the same conditions in the same population of respondents.

As for bias in the local selection of items, the effect of the test constructor's choosing items that appear "appropriate" for his or her local population would have an even greater impact on comparability of scores. It would be especially evident in educational tests constructed from items tailored to the local curriculum. Such a test would undoubtedly produce higher scores in the local schools than another test constructed elsewhere to conform to some other curriculum. Generally speaking, students study and learn the topics they are taught, and so will perform better on items pertaining to those particular topics. The only way that a school could create a local test comparable with another local test would be to draw items randomly from the same subject-matter categories within the item pool. But this would defeat the purpose of creating local tests, which is precisely to avoid testing students on topics they have not had an opportunity to learn. The conclusion from these considerations is that, for fair comparisons among local subpopulations, items must be selected at a higher level, and the test must be calibrated in a probability sample of respondents from the larger population to which the local subpopulations belong. This is a job best done by an agency external to the subpopulations.

The proper use of IRT belongs rather to those realms where comparisons must be kept consistent from one place to another and one time to another. Some of these situations are illustrated in the examples at the beginning of this chapter. They include any use of tests to guide decisions affecting the fates of individual persons, as in educational selection or certification, employment or military selection, and classification, and psychological or psychiatric evaluation. They include research settings where accurate, objective, and well-defined measures are needed to provide consistent and reproducible results among different sites and over extended periods of time. In these settings, IRT-based descriptions of behavioral response tendencies have important advantages over traditional methods based on arbitrary weighting schemes. The IRT method of item factor analysis helps define the dimensions to be measured, and scale-score estimation provides most precise measurement on these dimensions. The models and procedures that apply to multiple-category scoring, which is typical in individually administered cognitive tests, utilize this form of information with greater efficiency than traditional methods. Indeed, IRT methods also enable different types of items – multiple-choice, graded items with varying numbers of categories, nominal category items – to be mixed together consistently and efficiently in one scale. These methods cut through the many arguments in the classical test literature on how to assign points or weights in computing a total test score from multiple-category responses. They are especially valuable in educational testing when an achievement test consists of part multiple-choice and part open-ended items. The multiple-choice tests are scored "right–wrong," but the open-ended responses can be rated on graded scales according to their quality or completeness. A suitable IRT model will allow such a test to be scored on one scale while optimally weighting the contributions of each type of item to the scale. No comparable classical scoring procedures exist for these types of tests.

Another form of IRT measurement impossible within classical theory is adaptive testing. Presently, it exists in two main forms – computerized adaptive testing (CAT) and two-stage testing (TST). In CAT, the items are administered to individual respondents by computer, notebook, laptop, or smart-phone and a provisional scale score for the respondent is computed immediately after each response. On the basis of this provisional score, the next item is chosen at the level of difficulty that is most discriminating for that individual. Context effects are not a problem in this form of administration because items are chosen randomly from those available at each level of difficulty and any local effects of context are averaged out as the item presentations continue. The item presentations are continued until the standard error of the estimated scale score for the respondent becomes acceptably small. To attain a given standard error, a fully adaptive computerized test requires only about one third as many items as a conventional test, and we have developed applications of CAT for mental health testing based on multidimensional IRT which only require adaptive administration of an average of 3% of the items from the bank (12 of 400), yet maintain a correlation of $r = 0.95$ with the total 400 item bank score (Gibbons et al. 2012).

In TST, the respondents are first administered a relatively brief locator test. Then at a later time, each respondent is administered a form of the second-stage test that is tailored to his or her level of performance on the first-stage test. The second-stage data are then analyzed and scored by special, multiple-group IRT procedures that link the second-stage forms into a single scale. Studies by Lord (1980) have shown that a TST consisting of four second-stage forms is almost as efficient as a fully adaptive computerized test of approximately the same number of items. But TST is better suited to educational testing where relatively large numbers of students must be tested in groups. CAT is at present limited to vocational, medical, or military testing, where the respondents are interviewed individually. Both CAT and TST not only save time, but each is a more satisfactory experience for the respondent, who then confronts items that are not so easy as to lack challenge nor so difficult as to be frustrating. The IRT procedures that make adaptive testing possible are discussed in Chapter 8. In addition, CAT eliminates response set bias which is produced when the same items are repeatedly administered to the same respondents within relatively short periods of time. In CAT, different items are administered across different testing sessions for a given individual and the next testing session can begin at the ability or impairment level that was estimated in the last session. Even high-frequency measurements such as ecological momentary assessments or the benefits of short-term pharmacologic (e.g. ketamine) or surgical (deep brain stimulation) interventions can be assessed using CAT. Remarkably, test–retest reliability was shown to be higher for CAT relative to similarly sized traditional fixed length tests for the assessment of depression ($r = 0.92$ versus $r = 0.84$), despite the administration of different questions on repeat administrations (Beiser et al. 2016).

The model-based methods of IRT also solve other problems of measurement and test construction that are difficult in classical test theory. These problems arise because of nonlinear effects that are inherent in the classical number-right score. As Lord (1953) has shown, the distribution of number-right scores of a sample of respondents depends not only on the distribution of true scores in the population to which the respondents belong, but also on the distributions of item operating characteristics in that particular test. Only test forms that are matched item by item in their characteristics will tend to the same distributions as the number of respondents increases. Moreover, the regression of the number-right scores from one test on those of others will exhibit nonlinearities dependent on the distribution of item characteristics in each. These typical nonlinearities of number-right scores make the analysis of test data and the equating of alternate forms extremely difficult. They force the test constructor who is attempting to maintain the equivalence of test forms as the items are updated to resort to the equipercentile method of equating number-right scores. But this method has no good property other than to insure that the same percent of respondents in the population will exceed the equated score levels. The two forms of the test will appear fair in the sense that the overall rates of "passing" will be the same, but the equated scores from the two forms are not necessarily best estimates of the same true score for any given respondent.

IRT avoids these intractable problems by directly estimating an invariant latent distribution from tests whose items may have different distributions of parameters. Groups can therefore be compared in terms of their latent distributions without artifacts due to differing distribution of item characteristics within the alternative forms. This means that the two forms can be equated in the equipercentile sense by simple linear methods such as setting the mean and standard deviation of their latent distributions to the same values. This happens automatically in the calibration procedure when the two forms have been administered in the same population and the undetermined location and scale parameters of the scale scores of the two forms are set, respectively, to the same population values.

These properties of the IRT methods make it relatively easy to update test forms by replacing a certain proportion of items with parallel items on a regular basis. For example, successive forms of a test that are updated each year can be equated during calibration merely by setting the latent distributions estimated from the common items in the two forms so that their means and standard deviations are identical in the previous year's sample and the current year's sample. To obtain such equivalence by classical methods would require either an external equating study or the replacing of each of the retired items by a new item with the same difficulty and discrimination parameters – an obviously difficult task. This invariance of the measurement scale as tests are updated and recalibrated is one of the most attractive properties of IRT procedures. It simplifies and economizes the often difficult job of maintaining consistent performance of standardized tests that must be regularly changed in order to release a certain proportion of the items publicly and to protect the test from unauthorized disclosures of its contents. These advantages are especially important in educational measurement and assessment, where the scales on which results are reported must be kept consistent and uncompromised over extended periods of time. It is perhaps in these large-scale and highly visible applications of testing that IRT has made its greatest contribution. We discuss these uses of IRT in Chapter 12.

2

Selected Mathematical and Statistical Results

If it can't be expressed in figures, it is not science. It is opinion.
(Source: Robert Heinlein)

In this chapter we present, mostly without proof, various definitions, results, and algorithms needed in the sequel. They serve to introduce the notation and concepts of item response theory in the remaining chapters.

2.1 Points, Point Sets, and Set Operations

1. A *point* represents any entity that can be labeled. Usually, we will label a point by a lower-case Roman letter, such as x or y. Two distinct points cannot have the same label, and the same point cannot have two different labels unless the second label is declared equivalent to (*aliased* to) the first.
2. A *point set* or, briefly, *set*, is a collection of points. Usually, we will represent a set by an upper-case Roman letter, such as S. For example, the set consisting of the points x and y will be represented as

$$S = \{x, y\}. \tag{2.1}$$

   The points x and y are called *members* of the set S.
3. The points of a set may themselves be sets, in which case they are called *subsets* of the set of which they are members. For example,

$$S = \{A, B\} \tag{2.2}$$

   is a set containing the subsets A and B, where $A = \{x\}$ and $B = \{y\}$.

Item Response Theory, First Edition. R. Darrell Bock and Robert D. Gibbons.

4. A set may be devoid of points, in which case it is called the *null set* and designated by the special symbol

$$S = \varnothing. \tag{2.3}$$

5. A set is an (improper) subset of itself:

$$S = \{S\}. \tag{2.4}$$

6. Every set contains the null set:

$$S = \{S, \varnothing\}. \tag{2.5}$$

7. If A is a subset of S, the set of all points in S not contained in A is called the *complement* of A with respect to S and written A'. Then,

$$S = \{A, A'\}. \tag{2.6}$$

8. If C is the set of all points contained in *either* A or B, we say C is the *union* of A and B. Expressed in terms of the set operator $\cup$,

$$C = A \cup B. \tag{2.7}$$

For example, if $A = \{u, v, w, x\}$ and $B = \{w, x, y, z\}$,

$$A \cup B = \{u, v, w, x, y, z\}. \tag{2.8}$$

If D is the set of all points contained in *both* A and B, we say that D is the *intersection* of A and B. Expressed in terms of the operator $\cap$,

$$D = A \cap B. \tag{2.9}$$

For example, if A and B contain the above points,

$$A \cap B = \{w, x\}. \tag{2.10}$$

9. If $A \cap B = \varnothing$, the subsets A and B are called *mutually exclusive* or *disjoint*.
If $A \cup B = S$, the subsets are called *exhaustive*.
Note that the points in S, if regarded as subsets each containing one point, are mutually exclusive and exhaustive.
10. If the number of its points is sufficiently small, a subset may be defined by listing the labels of the points exhaustively. If the number is too large for such a listing, one way to define the subset is to state a rule that identifies the points unambiguously. This type of set definition may be written in the form,

$$A = \{\boldsymbol{x} = x \mid x \text{ is a member of } S \text{ satisfying } y\}, \tag{2.11}$$

in which the symbol | is read as "where," and y is a member of a set of conditions. For example, let A be those 12th grade public school students in a certain school district who have successfully completed two courses in algebra. Then $\boldsymbol{x}$ in the above definition is a *variable* label taking on values equal to the names of those students in the district roster who meet the condition $y =$ {successful-completion-of-two-courses-in-algebra}.

11. Another way to define the points of a large set is to associate each with a unique member of an already defined set. The set of natural numbers, which generate an indefinitely large number of distinct labels according to rules, is often used for this purpose. For example,

$$A = \{\mathbf{x} = x_i \mid i = 1, 2, \ldots, N\} \tag{2.12}$$

defines the subset in terms of a variable label that takes on the values x_i, where i runs over the natural numbers from 1 to N. Points that can be so labeled are called *discrete*. Even when N is indefinitely large, the points in A remain countable in the sense that they are in one-to-one correspondence with the natural numbers.

12. The set S may be a line or line segment, in which case it may be divided into a finite number of subsets consisting of nonoverlapping intervals, each of which contains a non-countable infinity of points. Let the points in each subset be *continuous*, in a sense to be defined later. Then S is continuous except perhaps at a finite number of points at the boundaries of the intervals. A set of continuous points is called a *continuum*.

An important non-countable infinite point set is the real-number line, designated R_1. The real plane, R_2, is a doubly infinite non-countable point set. The real n-space, R_n, is a similar n-fold infinite set.

Intervals on the real line are subsets of R_1. For example,

$$A = \{\mathbf{x} = x \mid 0.0 \leq x \leq 1.0\}, \tag{2.13}$$

which includes the boundary points 0.0 and 1.0, is a *closed* interval. In contrast,

$$A = \{\mathbf{x} = x \mid 0.0 < x < 1.0\}, \tag{2.14}$$

which excludes the boundary points 0.0 and 1.0, is an *open* interval.

Regions in the real plane are subsets of R_2. For example, the subset A could be defined as

$$A = \{\mathbf{x} = x, \mathbf{y} = y \mid x^2 + y^2 \leq c\}, \tag{2.15}$$

for some real c.

2.2 Probability

1. Probability is a real-valued measure associated with each of the points of a set, S. If the set contains N discrete points, we express the probability of the points x_i as

$$P(\mathbf{x} = x_i) = f(x_i), \quad \text{for } i = 1, 2, \ldots, N, \tag{2.16}$$

and require that

$$f(x_i) \geq 0 \tag{2.17}$$

and

$$\sum_{i=1}^{N} f(x_i) = 1. \tag{2.18}$$

Then,

$$P(S) = 1, \tag{2.19}$$

and

$$P(\emptyset) = 0. \tag{2.20}$$

If the set has a countable infinity of points, we require the *limit* of the sum to be 1.0. For example, if $f(x_i) = (\frac{1}{2})^i$, for $i = 1, 2, \ldots$, the limit of the sum of the infinite series

$$\frac{1}{2} + \frac{1}{4} + \frac{1}{8} + \frac{1}{16} + \cdots, \tag{2.21}$$

equals 1, and the requirement is satisfied.

2. The probability of a subset of $n \leq N$ discrete points is the sum of the probabilities of the points. For example, the probability of the subset corresponding to the first three terms of the above series is

$$\sum_{i=1}^{3} f(x_i) = \frac{7}{8}. \tag{2.22}$$

3. If the points x of the set S are continuous, the measure is called a *probability density* and defined by a *probability density function*, $f(x)$, for a real x. For example, we define the probability of the subset

$$A = \{\boldsymbol{x} = x \mid a < x < b\} \tag{2.23}$$

as

$$P(A) = \int_a^b f(x)dx \tag{2.24}$$

and require as before that $f(x)$ is nonnegative and

$$P(S) = \int_{-\infty}^{\infty} f(x)dx = 1. \tag{2.25}$$

It does not matter whether the subset is closed or open because for continuous x any finite number of points in S has probability measure zero.

4. Probability measures on subsets obey the following rules:
   (a) *Addition rule for disjoint subsets.*

$$\text{For } C = \{A, B\} \text{ and } \{A \cap B\} = \emptyset, \tag{2.26}$$

$$P(C) = P(A \cup B) = P(A) + P(B). \tag{2.27}$$

   (b) *Complement rule.*

$$P(A') = 1 - P(A). \tag{2.28}$$

   (c) *Addition rule for non-disjoint subsets.*

$$\text{For } C = \{A, B\} \text{ and } A \cap B \neq \emptyset, \tag{2.29}$$

$$P(C) = P(A \cup B) = P(A) + P(B) - P(A \cap B), \tag{2.30}$$

2.3 Sampling

1. Described abstractly, sampling is equivalent to an experiment in which points of the set S are observed and classified according to the subsets of S to which they belong. In this context, S is usually called the *sample space*, the selection of a point is called a *trial*, and the observation of a point belonging to a particular subset such as A is called an occurrence of the *event A*.
2. We employ probability measure to describe the relative frequency of such events in an indefinitely large number of sampling trials. By the use of suitable physical devices, we can choose the points to be observed in such a way that, as the trials continue, the relative frequency with which each point is observed, in a sample space of N points, tends to $1/N$. We then say that the sampling is *equiprobable*, or *random*, and assign the points the probability measure $1/N$. With the measure of the points assigned, we are in a position to use the addition rule for discrete sets to compute the relative frequencies to which the event subsets will tend. For example, if the event A corresponds to a subset of size n, the probability of A in the sample space S is

$$P(A) = \sum_{i=1}^{n} \frac{1}{N} = n/N. \tag{2.31}$$

2.4 Joint, Conditional, and Marginal Probability

1. Sometimes the sample points are classified as more than one non-disjoint event. To fix ideas, suppose that respondents to a social survey of a large

population are classified as (M)ale or (F)emale, and as (E)ver-married or (N)ever-married. Assuming that all respondents within the scope of the sample can be so classified, their numbers in each category of the joint classification could be represented in a two-way table:

	1. E	2. N	
1. M	n_{11}	n_{12}	$n_{1\cdot}$
2. F	n_{21}	n_{22}	$n_{2\cdot}$
	$n_{\cdot 1}$	$n_{\cdot 2}$	$n_{\cdot\cdot}$

In the margins of the table, n_{ij} is the number in row category i and column category j; $n_{i\cdot}$ and $n_{\cdot j}$ are the respective row and column totals; and N is the grand total for the population.

Represented in the same way, the probabilities of the corresponding events relative to the population as a whole are:

$P(M \cap E) = n_{11}/n_{\cdot\cdot}$	$P(M \cap N) = n_{12}/n_{\cdot\cdot}$	$P(M) = n_{1\cdot}/n_{\cdot\cdot}$
$P(F \cap E) = n_{21}/n_{\cdot\cdot}$	$P(F \cap N) = n_{22}/n_{\cdot\cdot}$	$P(F) = n_{2\cdot}/n_{\cdot\cdot}$
$P(E) = n_{\cdot 1}/n_{\cdot\cdot}$	$P(N) = n_{\cdot 2}/n_{\cdot\cdot}$	1

2. The measures on the intersections of M and E are called *joint* probabilities; those on subclasses M and F are called the row *marginal* probabilities; those on subclasses E and N, are called the column marginal probabilities.
3. For certain purposes, it is often useful to limit the joint events under consideration to a subset of the sample space. The measure in the restricted space is called a *conditional* probability. It is found by dividing the probability of the intersection by the marginal probability of the event conditioned on.
   For example, conditional on the respondent's being female (and assuming $P(F) \neq 0$), the probability that she has ever married is
$$P(E \mid F) = \frac{P(E \cap F)}{P(F)}; \tag{2.32}$$
   here the symbol "|" is read as "given." Similarly, given that the respondent is male and assuming $P(M) \neq 0$, the probability that he has ever married is
$$P(E \mid M) = \frac{P(E \cap M)}{P(M)}. \tag{2.33}$$
4. The definition of conditional probability implies that every joint probability can be expressed as the product of a conditional and a marginal probability:
$$P(A \cap B) = P(A) \cdot P(B \mid A). \tag{2.34}$$

5. If $P(A \mid B) = P(A)$ and $P(B \mid A) = P(B)$, the conditioning has no effect and the events A and B are said to be *independent*. Independence of events implies and is implied by

$$P(A \cap B) = P(A) \cdot P(B). \tag{2.35}$$

6. Random sampling trials are assumed to be independent. Thus, in the above example, the probability of observing respondents of sex M, F, M, F in successive trials is

$$P(MFMF) = P(M) \cdot P(F) \cdot P(M) \cdot P(F) = P^2(M) \cdot P^2(F). \tag{2.36}$$

7. When events occur at different times and are nonindependent, conditioning a later event upon a previous event is in effect a statistical prediction of the later event. For example, having observed A, the space of future events is restricted such that

$$P(B \mid A) = \frac{P(B \cap A)}{P(A)} \tag{2.37}$$

is in general greater than $P(B)$ in the unrestricted space prior to the occurrence of A. But the relationship is symmetric, so that

$$P(A \cap B) = P(A) \cdot P(B \mid A) = P(B) \cdot P(A \mid B), \tag{2.38}$$

or

$$P(A \mid B) = \frac{P(A) \cdot P(B \mid A)}{P(B)}. \tag{2.39}$$

This result is *Bayes Theorem*; it implies, in situations where A is an unobservable "cause" of B, that the occurrence of A can be *postdicted* statistically after having observed B. In this context,
$P(A)$ is the *prior* probability of A,
$P(B \mid A)$, the *likelihood* of B, and
$P(A \mid B)$ is the *posterior* probability of A, given B.

2.5 Probability Distributions and Densities

1. In applications of probability theory, there may be other real values associated with the points of the sample space in addition to the probability measure. Any real-valued function that defines these extrinsic values for every point is called a *random variable* or, briefly, a *variate*.
2. If the points of a sample space are discrete, the random variable associated with them is discrete. A mathematical function that assigns probabilities to the points is called a *probability function* and written as

$$P(\mathbf{x} = x) = f(x), \tag{2.40}$$

where $\boldsymbol{x}$ is the random variable and x is the value it takes on. If the space is finite,

$$\sum_{i=1}^{s} f(x_i) = 1 \tag{2.41}$$

over $i = 1, 2, \dots, s$, the *range* of $\boldsymbol{x}$. The values of the probability function over the range of the random variable comprise a *probability distribution*.

3. The simplest nontrivial probability distribution is that of a *Bernoulli* variable, which describes the outcome of a so-called "Bernoulli trial." The variable takes on only two values, 1 or 0 according to the outcome of the trial being a "success" or "failure." Its distribution function is

$$P(\boldsymbol{x} = 1) = P,$$
$$P(\boldsymbol{x} = 0) = Q,$$

for $P + Q = 1$.

4. Bernoulli trials are required to be independent, so that a particular pattern of r successes and $n - r$ failures in a sequence of n trials has probability $P^r Q^{n-r}$.
5. A random variable that gives the *number* of successes in n Bernoulli trials is called a *binomial* variable. Because the number of ways that r successes can appear in a sequence of n trials is $n!/r!(n-r)!$, the distribution function of a binomial variable is

$$P(\boldsymbol{r} = r) = \frac{n!}{r!(n-r)!} P^r Q^{n-r}. \tag{2.42}$$

We recognize the values of this function as the terms of the binomial expansion

$$(P + Q)^n = \sum_{r=0}^{n} \frac{n!}{r!(n-r)!} P^r Q^{(n-r)} \tag{2.43}$$

showing, as required, that the sum of the probability measure over the range of the binomial variable is 1.

6. A Bernoulli trial that takes on more than two values is called a *multivariate* Bernoulli variable. For the case of three values, the distribution function may be expressed as

$$P\left(\boldsymbol{x}_1 = \begin{bmatrix} 1 \\ 0 \\ 0 \end{bmatrix}\right) = P_1,$$

$$P\left(\boldsymbol{x}_2 = \begin{bmatrix} 0 \\ 1 \\ 0 \end{bmatrix}\right) = P_2,$$

$$P\left(\boldsymbol{x}_3 = \begin{bmatrix} 0 \\ 0 \\ 1 \end{bmatrix}\right) = P_3,$$

for $P_1 + P_2 + P_3$. The values of the random variable correspond to the assignment of the outcomes of the trial to three mutually and exhaustive categories labeled 1, 2, and 3. The probability of a particular pattern of outcomes in the trials in which the numbers of occurrences of the categories is r_1, r_2, and r_3, respectively, is

$$P_1^{r_1} P_2^{r_2} P_3^{r_3}, \tag{2.44}$$

where $n = r_1 + \cdots + r_m$.

7. The number of occurrences of the categories indicated by a multivariate (in this case, trivariate) Bernoulli variable in n trials is called a *multinomial* variable. Its distribution function for m categories is

$$P(\boldsymbol{r}_1 = r_1, \boldsymbol{r}_2 = r_2, \ldots, \boldsymbol{r}_m = r_m) = \frac{n!}{r_1! r_2! \cdots r_m!} P_1^{r_1} P_2^{r_2} \cdots P_m^{r_m}. \tag{2.45}$$

These values are terms of the multinomial expansion

$$(P_1 + P_2 + \cdots + P_m)^n = 1. \tag{2.46}$$

8. If the sample space is non-countably infinite, a continuous random variable, $\boldsymbol{x}$, may be defined on the sample points. Then a *probability density function*, $f(x)$, where x is a value of $\boldsymbol{x}$, may be defined that assigns a finite probability measure to intervals of the continuum as follows:

$$P(a < \boldsymbol{x} < b) = \int_a^b f(x)dx, \tag{2.47}$$

where $f(x) \geq 0$ for all x and

$$\int_{-\infty}^{\infty} f(x)dx = 1. \tag{2.48}$$

9. The function

$$F(x) = P(\boldsymbol{x} < x) = \int_{-\infty}^{x} f(t)dt, \tag{2.49}$$

which gives the cumulative probability up to the point x is called the distribution function of $\boldsymbol{x}$. If $F(x)$ is differentiable,

$$f(x) = \frac{dF(x)}{dx}. \tag{2.50}$$

Example 1 A distribution is called uniform on the interval (a, b) if the density is

$$f(x) = \frac{1}{b - a} \quad \text{for } a < \boldsymbol{x} < b. \tag{2.51}$$

The corresponding distribution function is

$$F(x) = \begin{cases} 0 & \text{for } \boldsymbol{x} < a \\ x & \text{for } a < \boldsymbol{x} < b \\ 0 & \text{for } \boldsymbol{x} > b. \end{cases}$$

Example 2 The density function of the normal distribution is

$$h(x) = \frac{1}{\sqrt{2\pi}\sigma} \exp\left[-\frac{1}{2}(x-\mu)^2/\sigma^2\right],$$

where the parameter μ, which determines the location of the distribution on the $\boldsymbol{x}$-continuum, is called the *mean*, and the parameter σ, which determines the dispersion of the distribution on the continuum, is called the *standard deviation*.

The distribution function,

$$P(\boldsymbol{x} < x) = \Phi(x) = \int_{-\infty}^{x} \phi(t)dt,$$

cannot be expressed in closed form. (See (Hastings, 1955), for computing approximations.)

10. If more than one random variable are associated with the points of the sample space, we say that the probability, density, or distribution function is *multivariate*, in contrast to the one-dimensional, or *univariate* case. When there are two random variables, the distribution is called *bivariate*. A bivariate probability function or density, may be expressed as

$$P(\boldsymbol{x} = x, \boldsymbol{y} = y) = f(x, y). \tag{2.52}$$

Then the corresponding distribution function for discrete variates is

$$P(\boldsymbol{x} \le x, \boldsymbol{y} \le y) = \sum_{s \le x} \sum_{t \le y} f(s, t), \tag{2.53}$$

and, for continuous variates,

$$P(\boldsymbol{x} \le x, \boldsymbol{y} \le y) = \int_{-\infty}^{x} \int_{-\infty}^{y} f(s, t) ds\, dt, \tag{2.54}$$

Example 3 The bivariate normal density is

$$f(x, y) = \frac{1}{2\pi\sqrt{1-\rho^2}\sigma_x\sigma_y} \exp\left[-\frac{1}{2(1-\rho^2)}\left(\frac{(x-\mu_x)^2}{\sigma_x^2} - 2\rho\frac{(x-\mu_x)(y-\mu_y)}{\sigma_x\sigma_y} + \frac{(y-\mu_y)^2}{\sigma_y^2}\right)\right],$$

where the parameter ρ, called the *correlation coefficient*, measures the dependency between $\boldsymbol{x}$ and $\boldsymbol{y}$.

11. A *marginal* distribution of one or more multivariate random variables is the sum or integral over the remaining variables. For example, the $\boldsymbol{x}$-margin of a discrete bivariate distribution may be expressed as

$$P(\boldsymbol{x} = x) = \sum_{y} f(x, y), \tag{2.55}$$

and that of a continuous bivariate distribution as

$$P(\boldsymbol{x} < x) = \int_{-\infty}^{x} \int_{-\infty}^{\infty} f(s, t) dt\, ds. \tag{2.56}$$

Example 4 The substitution $\mu_y^* = \mu_y + \rho \frac{\sigma_y}{\sigma_x}(x - \mu_x)$ makes it possible to express a bivariate normal density in the factored form,

$$\frac{1}{\sqrt{2\pi}\sigma_y\sqrt{1-\rho^2}} \exp\left[-\frac{1}{2}\frac{(y-\mu_y^*)^2}{\sigma_y^2(1-\rho^2)}\right] \cdot \frac{1}{\sqrt{2\pi}\sigma_x} \exp\left[-\frac{1}{2}\frac{(x-\mu_x)^2}{\sigma_x^2}\right].$$

Since the left-hand factor is a proper density, it integrates to 1 with respect to y, and the marginal distribution of $\boldsymbol{x}$ is just the univariate normal distribution of $\boldsymbol{x}$; that is, the distribution of $\boldsymbol{x}$ ignoring $\boldsymbol{y}$ (see (Bock, 1975), p. 121).

12. The *conditional* probability function of a multivariate random variable, given specified values of one or more of the variates, is the joint probability function divided by the marginal probability of the given value or values. For example, the probability function of $\boldsymbol{y}$, given x, for the bivariate probability, $f(x, y)$, is

$$p(\boldsymbol{y}|x) = \frac{f(x, y)}{h(x)}, \tag{2.57}$$

where $h(x)$ is the marginal probability function of $\boldsymbol{x}$ evaluated at x. If $\boldsymbol{x}$ and $\boldsymbol{y}$ are discrete, $p(y|x)$ is a probability; if they are continuous, it is a density.

Example 5 We saw in the previous example that the left-most two terms were the joint density and a marginal density of bivariate normal variables. Dividing through by the marginal density of $\boldsymbol{x}$, we see that the right-most term is the conditional density of $\boldsymbol{y}$ given x. It is apparent that the conditional distribution is also normal, and that its mean,

$$\mu_y^* = \mu_y + \rho \frac{\sigma_y}{\sigma_x}(x - \mu_x),$$

is a function of x. This function is called the *regression* of $\mathbf{y}$ on x. The conditional standard deviation, which in this case does not depend on x, equals

$$\sigma_y^* = \sigma_y \sqrt{(1 - \rho^2)}.$$

2.6 Describing Distributions

Various indices have been proposed for describing probability distributions in one or more dimensions. Borrowed from physics, where it is used to describe the distribution of mass, is the concept of *moments* of a distribution.

1. The univariate first moment, which is mean or *expected value* of the distribution, is defined as

$$\mu = \mathcal{E}(x) = \sum_x x f(x) \tag{2.58}$$

   in the discrete case and as

$$\mu = \int_{-\infty}^{\infty} x f(x) dx \tag{2.59}$$

   in the continuous case.
2. The univariate second moment about the mean, called the *variance*, is defined as the expected value of the square of deviations of the variable from the mean of the distribution – that is, as

$$\sigma^2 = \mathcal{V}(x) = \mathcal{E}\left[(x - \mu)^2\right] = \mathcal{E}(x) - \mathcal{E}^2(x) = \sum_x (x - \mu)^2 f(x) \tag{2.60}$$

   and

$$\sigma^2 = \int_{-\infty}^{\infty} (x - \mu)^2 f(x) dx \tag{2.61}$$

   in the discrete and continuous cases, respectively. The square root of the variance is the standard deviation.
3. In general, the rth moment about the mean, called the rth central moment and designated μ_r, is defined as the expected value of the r power of deviations about the mean $\mu_r = \mathcal{E}\,[(x - \mu)^r]$. The ratio of the third central moment to the cube of the standard deviation,

$$\beta = \frac{\mu_3}{\sigma^3}, \tag{2.62}$$

   is a measure of asymmetry or *skewness* of the distribution.
   The ratio of the fourth central moment to the square of the variance,

$$\gamma = \frac{\mu_4}{\sigma^4}, \tag{2.63}$$

   measures the peakness or *kurtosis* of the distribution.

4. The joint distribution of probability for two random variables is characterized by the *product* moment. In particular, the expected value of the product of two variates deviated from their respective means,

$$\sigma_{xy} = \mathcal{E}(xy) - \mathcal{E}(x)\mathcal{E}(y) = \sum_x \sum_y (x - \mu_x)(y - \mu_y) f(x, y) \tag{2.64}$$

in the discrete case, or

$$\sigma_{xy} = \int_{-\infty}^{\infty} \int_{-\infty}^{\infty} (x - \mu_x)(y - \mu_y) f(x, y) dy\, dx \tag{2.65}$$

in the continuous case, is called their *covariance.* A covariance divided by the standard deviations of the respective variates,

$$\rho_{xy} = \frac{\sigma_{xy}}{\sigma_x \sigma_y}, \tag{2.66}$$

is called the *correlation* of the two variates.

Example 1 The expected value of a Bernoulli variable is

$$\mu = 1 \cdot P + 0 \cdot (1 - P) = P.$$

The variance is

$$\sigma^2 = (1 - P)^2 \cdot P + (0 - P)^2 \cdot (1 - P) = P(1 - P).$$

Example 2 The expected value of a trivariate Bernoulli variable (see Section 2.5) is

$$\begin{aligned}\mathcal{E}(\mathbf{x}) &= x_1 P_1 + x_2 P_2 + x_3 P_3 \\ &= P_1 \begin{bmatrix} 1 \\ 0 \\ 0 \end{bmatrix} + P_2 \begin{bmatrix} 0 \\ 1 \\ 0 \end{bmatrix} + P_3 \begin{bmatrix} 0 \\ 0 \\ 1 \end{bmatrix} = \begin{bmatrix} P_1 \\ P_2 \\ P_3 \end{bmatrix}.\end{aligned}$$

The covariances of the variables are

$$\begin{aligned}\mathcal{V}(\mathbf{x}) &= \mathcal{V} \begin{bmatrix} x_1 \\ x_2 \\ x_3 \end{bmatrix} \\ &= P_1 \begin{bmatrix} 1 & 0 & 0 \\ 0 & 0 & 0 \\ 0 & 0 & 0 \end{bmatrix} + P_2 \begin{bmatrix} 0 & 0 & 0 \\ 0 & 1 & 0 \\ 0 & 0 & 0 \end{bmatrix} + P_3 \begin{bmatrix} 0 & 0 & 0 \\ 0 & 0 & 0 \\ 0 & 0 & 1 \end{bmatrix} \\ &\quad - \begin{bmatrix} P_1^2 & P_{12} & P_{13} \\ P_1 P_2 & P_2^2 & P_{12} \\ P_1 P_3 & P_{23} & P_3^2 \end{bmatrix}\end{aligned}$$

$$= \begin{bmatrix} P_1(1-P_1) & -P_1P_2 & -P_1P_3 \\ -P_1P_2 & P_2(1-P_2) & -P_2P_3 \\ -P_1P_3 & -P_2P_3 & P_3(1-P_3) \end{bmatrix}.$$

Example 3 The normal distribution.

Since the normal density function (see Section 2.5) is symmetric about μ,

$$\int_{-\infty}^{\infty} (x-\mu)h(x)dx = \int_{-\infty}^{\infty} xh(x)dx - \mu \int_{-\infty}^{\infty} h(x)dx = 0.$$

The expected value of a normal variate is therefore

$$\int_{-\infty}^{\infty} xh(x)dx = \mu,$$

and it is also the *median* of the distribution. Similarly, the variance of a normal variate is

$$\int_{-\infty}^{\infty} (x-\mu)^2 h(x) = \sigma^2.$$

The covariance of bivariate normal variates

$$\int_{-\infty}^{\infty}\int_{-\infty}^{\infty} (x-\mu)(y-\mu)h(x,y) = \rho\sigma_x\sigma_y = \sigma_{xy}.$$

2.7 Functions of Random Variables

Frequently we need to obtain the distribution of a function of random variables for which the joint distribution is known. Most estimators of population parameters are examples of such functions. We consider in this section the distributions of linear and nonlinear functions of random variables.

2.7.1 Linear Functions

The expected values, variances, and covariances of linear functions of similarly distributed variables are easily found:

1. In Section 2.6, we expressed the expected value of a random variable x as the expectation operation $\mathcal{E}(\boldsymbol{x})$. For any real constant c, the *expectation operator* has the properties

$$\mathcal{E}(c\boldsymbol{x}) = c\mathcal{E}(\boldsymbol{x}), \tag{2.67}$$

and

$$\mathcal{E}(\boldsymbol{x}+c)=\mathcal{E}(\boldsymbol{x})+c. \tag{2.68}$$

For any two random variables $\boldsymbol{x}$ and $\boldsymbol{y}$,

$$\mathcal{E}(\boldsymbol{x}+\boldsymbol{y})=\mathcal{E}(\boldsymbol{x})+\mathcal{E}(\boldsymbol{y}). \tag{2.69}$$

Example 4 As a binomial variable is the sum of similarly distributed Bernoulli variables, the expected value of a binomial variate is

$$\mathcal{E}\left(\sum_{i=1}^{n}\boldsymbol{x}_i\right)=\sum_{i=1}^{n}\mathcal{E}(\boldsymbol{x}_i)=nP.$$

2. We denote the variance operator of a random variable by $\mathcal{V}(\mathbf{x})$. It has the properties

$$\mathcal{V}(c\boldsymbol{x})=c^2\mathcal{V}(\boldsymbol{x}), \tag{2.70}$$

$$\mathcal{V}(\boldsymbol{x}+c)=\mathcal{V}(\boldsymbol{x}), \tag{2.71}$$

and, if and only if $\boldsymbol{x}$ and $\boldsymbol{y}$ are uncorrelated,

$$\mathcal{V}(c\boldsymbol{x}+\boldsymbol{y})=\mathcal{V}(\boldsymbol{x})+\mathcal{V}(\boldsymbol{y}). \tag{2.72}$$

Example 5 Because Bernoulli trials are independent, the variate values from different trials are uncorrelated. Thus, the variance of a binomial variate is

$$\mathcal{V}\left(\sum_{i=1}^{n}\boldsymbol{x}_i\right)=\sum_{i=1}^{n}\mathcal{V}(\boldsymbol{x}_i)=nP(1-P).$$

3. The variance operator with two arguments denotes a covariance: $\mathcal{V}(\boldsymbol{x},\boldsymbol{y})$. For any real constants c and d, the covariance operator has the properties

$$\mathcal{V}(c\boldsymbol{x},d\boldsymbol{y})=cd\mathcal{V}(\boldsymbol{x},\boldsymbol{y}) \tag{2.73}$$

and

$$\mathcal{V}(\boldsymbol{x}+c,\boldsymbol{y}+d)=\mathcal{V}(\boldsymbol{x},\boldsymbol{y}). \tag{2.74}$$

Whether or not $\boldsymbol{x}$ and $\boldsymbol{y}$ are uncorrelated,

$$\mathcal{V}(c\boldsymbol{x}+\boldsymbol{y})=\mathcal{V}(\boldsymbol{x})+\mathcal{V}(\boldsymbol{y})+2\mathcal{V}(\boldsymbol{x},\boldsymbol{y}). \tag{2.75}$$

Example 6 The multinomial frequencies for categories j and k are the sums of the corresponding multivariate Bernoulli variables. The covariance of these frequencies is therefore

$$\mathcal{V}\left(\sum_{i=1}^{n} \boldsymbol{x}_{ij}, \sum_{i=1}^{n} \boldsymbol{x}_{ik}\right) = \sum_{i=1}^{n} \mathcal{V}(\boldsymbol{x}_{ij}, \boldsymbol{x}_{ik}) = nP_j P_k.$$

4. The variance of a linear combination of random variables with coefficients $(c_1, c_2, \ldots, c_n)$ is

$$\sigma^2 = \sum_{i=1}^{n} \sum_{j+1}^{n} c_i c_j \sigma_{ij}, \tag{2.76}$$

where σ_{ij} for $i = j$ is the variance of variable j and σ_{ij}, $i \neq j$, is the covariance of variables i and j.

5. By a suitable linear transformation, any variate with finite mean and variance can be *standardized*; that is, its mean can be set to 0 and its standard deviation to 1. Thus, if the mean and variance of $\boldsymbol{x}$ are μ_x and σ_x^2, respectively, the variate

$$\boldsymbol{z} = \frac{\boldsymbol{x} - \mu_x}{\sqrt{\sigma_x^2}} \tag{2.77}$$

is in standard form.

Only the location and scale of a probability or density function is affected by linear transformation of the random variable; the shape of the function remains unchanged.

6. By suitable linear transformations, any two variates with finite means and variances can be made uncorrelated and standardized. Thus, if the means of $\boldsymbol{x}$ and $\boldsymbol{y}$ are a and b, respectively, their variances c and d, and their covariance e, the variates

$$\boldsymbol{u} = (\boldsymbol{x} - a)/\sqrt{c}, \tag{2.78}$$

$$\boldsymbol{v} = \left[(\boldsymbol{y} - b) - \frac{e}{c}(\boldsymbol{x} - a)\right] \Big/ \sqrt{d - \frac{e^2}{c}} \tag{2.79}$$

have means 0, standard deviations 1, and correlation 0.

7. Although it is not true of any other random variables, normally distributed variates have special property that linear functions of normal variates are normal. This property and those of the expectation and variance operators imply that the mean of n independent normal variables with means μ and variances σ^2 is normally distributed with mean μ and variance σ^2/n.

8. *Central limit theorem*: The mean of n independent random variables, having any forms of distribution with finite means and variances, tends to a normal distribution with mean equal to the mean of the variables and variance proportional to $1/n$ of their variances as n increases indefinitely. This theorem generally justifies the treating of sample statistics as normally distributed when the number of independent observations is large. In many cases, sample sizes in excess of 30 can be considered "large" in this respect.

2.7.2 Nonlinear Functions

Nonlinear transformations of random variables change the shape of the distributions as well as location and scale. If the variable is continuous and the transformation single valued, the form of the distribution function under the transformation can be found by changing the variable of integration. Suppose the density function of $\boldsymbol{x}$ is $f(x)$ and the transformation is $y = w(x)$. Then, if the inverse function is $x = w^{-1}(y)$, $dx = dy/w'(x)$, and

$$P(\boldsymbol{y} \leq y) = \int_{-\infty}^{y} f(y)\frac{1}{w'(x)}dy. \tag{2.80}$$

The corresponding density function is

$$dP(y) = f(y)\frac{1}{w'(x)}. \tag{2.81}$$

1. As n increases indefinitely, the mean and variance of a nonlinear transform $y = s(\boldsymbol{u})$ of a statistic $\boldsymbol{u}$ with mean μ and variance σ^2/n approaches

$$\mathcal{E}(\boldsymbol{y}) = s(u), \tag{2.82}$$

$$\mathcal{V}(\boldsymbol{y}) = \left[\frac{ds(u)}{du}\right]^2 \mathcal{V}(\boldsymbol{x}). \tag{2.83}$$

   This method of computing the large-sample standard errors of nonlinear transforms of statistics is referred to as the "delta" method.

2. Similarly, the large-sample covariance of two nonlinear transforms of two statistics,

$$y = s(u) \tag{2.84}$$

   and

$$z = t(v), \tag{2.85}$$

   is

$$\mathcal{V}(\boldsymbol{y}, \boldsymbol{z}) = \left[\frac{ds(u)}{du}\right] \mathcal{V}(\boldsymbol{u}, \boldsymbol{v}) \left[\frac{dt(v)}{dv}\right]. \tag{2.86}$$

2.8 Elements of Matrix Algebra

Matrix algebra deals with relations and operations defined for rectangular arrays of elements called *matrices*. An example of a matrix is[1]

$$\underset{m\times n}{A} = \begin{bmatrix} a_{11} & a_{12} & \cdots & a_{1n} \\ a_{21} & a_{22} & \cdots & a_{2n} \\ \cdots & \cdots & \cdots & \cdots \\ a_{m1} & a_{m2} & \cdots & a_{mn} \end{bmatrix}.$$

Although the elements of a matrix may be any quantities that satisfy the postulates of an abstract field, e.g. rational numbers, complex numbers, and polynomials, for our purposes they will be restricted to real numbers.

The number of rows in a matrix is called the *row order*; the number of columns, the *column order*. If the number of rows is m and the number of columns n, we say that the matrix is of order m by n; if there is possible ambiguity as to order, we insert $m \times n$ under the letter which designates the matrix. A matrix may also be designated by exhibiting a typical element in square brackets, possibly with an indication of the range of subscripts; e.g.

$$A = [a_{ij}], \quad i = 1, 2, \ldots, m, \; j = 1, 2, \ldots, n. \tag{2.87}$$

In data analysis, we often refer to the *data matrix*, any one row of which consists of observations obtained from one subject. Thus, if p responses of each of N subjects have been observed, the data matrix is of order $N \times p$.

Definitions and postulates of matrix algebra are as follows:

1. *Equality*. Two matrices are equal if and only if corresponding elements are equal:

$$A = B \text{ implies and is implied by } a_{ij} = b_{ij} \quad \begin{matrix} i = 1, 2, \ldots, m \\ j = 1, 2, \ldots, n \end{matrix}.$$

2. *Addition*. The sum of two matrices of the same order is obtained by summing corresponding elements:

$$A + B = [a_{ij}] + [b_{ij}] = [a_{ij} + b_{ij}] \quad \begin{matrix} i = 1, 2, \ldots, m \\ j = 1, 2, \ldots, n \end{matrix}.$$

Matrix addition is associative

$$A + (B + C) = (A + B) + C, \tag{2.88}$$

and commutative

$$A + B = B + A. \tag{2.89}$$

1 Adapted from Bock (1975).

The additive identity is the *null matrix* O consisting of zero elements:

$$\underset{m\times n}{A} + \underset{m\times n}{O} = \underset{m\times n}{A}. \tag{2.90}$$

3. *Scalar multiplication.* The product of a scalar number and a matrix is obtained by multiplying each element of the matrix by the scalar number:

$$cA = c[a_{ij}] = [ca_{ij}] \quad \begin{array}{l} i = 1, 2, \dots, m \\ j = 1, 2, \dots, n \end{array}. \tag{2.91}$$

Scalar multiplication is associative

$$(cd)A = c(dA), \tag{2.92}$$

and distributes with respect to scalar addition

$$(c + d)A = cA + dA, \tag{2.93}$$

and with respect to matrix addition

$$c(A + B) = cA + cB. \tag{2.94}$$

The additive inverse for matrix addition is the scalar product of -1 and the matrix:

$$A + (-1)A = A - A = O. \tag{2.95}$$

4. *Matrix multiplication.* Matrix multiplication is defined for matrices in which the column order of the left-hand factor equals the row order of the right-hand factor. The (i, k)th element of the product matrix is the sum of products of corresponding elements from the ith row of the left-hand factor and the kth column of the right-hand factor:

$$\begin{aligned} AB = \underset{m\times q}{A} \cdot \underset{q\times n}{B} &= \left[\sum_{j=1}^{q} a_{ij} b_{jk}\right] \\ &= [a_{i1}b_{1k} + a_{i2}b_{2k} + \cdots + a_{iq}b_{qk}] = [c_{ik}] = \underset{m\times n}{C}. \end{aligned} \tag{2.96}$$

Matrix multiplication is associative

$$(AB)C = A(BC), \tag{2.97}$$

but in general it is not commutative

$$AB \neq BA. \tag{2.98}$$

Thus, it is necessary to refer to AB as *pre*multiplication of B by A, or *post*multiplication of A by B. Matrix multiplication, pre- or post-, distributes with respect to matrix addition:

$$A(B + C) = AB + AC,$$

$$(A + B)C = AC + BC.$$

A square matrix with unities in the diagonal and zeros elsewhere is called the *unit matrix*:

$$I = \begin{bmatrix} 1 & 0 & \cdots & 0 \\ 0 & 1 & \cdots & 0 \\ \cdots & \cdots & \cdots & \cdots \\ 0 & 0 & \cdots & 1 \end{bmatrix}.$$

Because the unit matrix is the identify for matrix multiplication, it is also called the *identity matrix*:

$$\underset{m\times n}{A} \cdot \underset{n\times n}{I} = \underset{m\times n}{A},$$

$$\underset{m\times m}{I} \cdot \underset{m\times n}{A} = \underset{m\times n}{A}.$$

Any square matrix in which the off-diagonal elements are identically equal to zero is called a *diagonal matrix*:

$$D = \text{diag}[d_1, d_2, \ldots, d_n] = \begin{bmatrix} d_1 & 0 & \cdots & 0 \\ 0 & d_2 & \cdots & 0 \\ \cdots & \cdots & \cdots & \cdots \\ 0 & 0 & \cdots & d_n \end{bmatrix}. \tag{2.99}$$

Premultiplication by a diagonal matrix rescales the rows of the postfactor. Postmultiplication by a diagonal matrix rescales the columns of the prefactor.

5. *The inverse matrix*. If an $n \times n$ square matrix B exists such that $AB = I$ and $BA = I$, then B is called the inverse of A, in which case B is usually denoted by A^{-1}. If a unique inverse of A exists, it is an inverse with respect to multiplication both on the left and on the right. For, suppose B is a right inverse of A, and C is a left inverse of A. Then

$$CA = AB = I. \tag{2.100}$$

But

$$CAB = C = B = A^{-1} \tag{2.101}$$

or

$$A^{-1}A = AA^{-1} = I. \tag{2.102}$$

That is, multiplication of a matrix and its inverse is commutative. If a matrix has an inverse, the matrix is called *nonsingular*; otherwise, it is called *singular*. Note that the inverse of a matrix product is the product of the inverses in the reverse order:

$$(AB)^{-1} = B^{-1}A^{-1} \tag{2.103}$$

for

$$B^{-1}A^{-1}AB = B^{-1}B = I, \tag{2.104}$$

and

$$ABB^{-1}A^{-1} = AA^{-1} = I. \tag{2.105}$$

Note that the inverse of a diagonal matrix contains the reciprocals of the diagonal elements; that is, $D^{-1} = \text{diag}[1/d_1, 1/d_2, \dots, 1/d_n]$.

6. *The transpose matrix.* A matrix obtained by interchange of the rows and columns of a given matrix is called the *transpose* of the original matrix. It is usually designated by a prime. That is, if $\underset{n\times m}{A} = [a_{ij}]$, then

$$\underset{m\times n}{A}{}' = [a_{ij}]' = [a_{ji}] \tag{2.106}$$

is the transpose. The transpose of a matrix sum is the sum of the transposes:

$$(A + B)' = A' + B'. \tag{2.107}$$

The transpose of a matrix product is the product of the transposes in the reverse order:

$$(AB)' = B'A'. \tag{2.108}$$

Note also that $(A^{-1})' = (A')^{-1}$, since $(A^{-1}A)' = A'(A^{-1})' = I' = I = A'(A')^{-1}$.

2.8.1 Partitioned Matrices

Let the initial r rows of an $m \times n$ matrix A be partitioned from the remaining $s = m - r$ rows, and the initial p columns partitioned from the remaining $q = n - p$ columns. Then A may be represented in submatrices as follows:

$$A = \begin{array}{c} \left[\begin{array}{cc} A_{11} & A_{12} \\ A_{21} & A_{22} \end{array}\right] \begin{array}{l} r \\ s\,. \end{array} \\ \begin{array}{cc} p & q \end{array} \end{array} \tag{2.109}$$

If a matrix B is similarly partitioned, we readily verify that

$$A + B = \begin{bmatrix} A_{11} + B_{11} & A_{12} + B_{12} \\ A_{21} + B_{21} & A_{22} + B_{22} \end{bmatrix}. \tag{2.110}$$

Suppose now that B is partitioned into p and q rows and t and u columns. Then,

$$\begin{aligned} A = {} & \begin{array}{l} r \\ s \end{array} \begin{array}{c} \left[\begin{array}{cc} A_{11} & A_{12} \\ A_{21} & A_{22} \end{array}\right] \\ \begin{array}{cc} p & q \end{array} \end{array} \begin{array}{c} \left[\begin{array}{cc} B_{11} & B_{12} \\ B_{21} & B_{22} \end{array}\right] \\ \begin{array}{cc} t & u \end{array} \end{array} \begin{array}{l} p \\ q \end{array} \\ = {} & \begin{array}{c} \left[\begin{array}{cc} A_{11}B_{11} + A_{12}B_{21} & A_{11}B_{12} + A_{12}B_{22} \\ A_{21}B_{11} + A_{22}B_{21} & A_{21}B_{12} + A_{22}B_{22} \end{array}\right] \\ \begin{array}{cc} t & u \end{array} \end{array} \begin{array}{l} r \\ s\,. \end{array} \end{aligned} \tag{2.111}$$

We may similarly verify that

$$\begin{array}{c} p \\ q \end{array} \begin{bmatrix} A & B \\ C & D \end{bmatrix}^{-1}_{\quad p \quad q}$$

$$= \begin{bmatrix} A^{-1} + A^{-1}B(D - CA^{-1}B)^{-1}CA^{-1} & -A^{-1}B(D - CA^{-1}B)^{-1} \\ -(D - CA^{-1}B)^{-1}CA^{-1} & (D - CA^{-1}B)^{-1} \end{bmatrix} \begin{array}{c} p \\ q \end{array} \tag{2.112}$$

(column orders p, q)

for A and $D - CA^{-1}B$ nonsingular.

Corollary: For

$$\begin{array}{c} p \\ q \end{array} \begin{bmatrix} A & B \\ C & D \end{bmatrix}$$

with A, D, and $D - CA^{-1}B$ nonsingular,

$$(A - BD^{-1}C)^{-1} = A^{-1} + A^{-1}B(D - CA^{-1}B)^{-1}CA^{-1}. \tag{2.113}$$

2.8.2 The Kronecker Product

When working with linear statistical models, we have frequent occasion to employ the Kronecker product of matrices:

$$\underset{mp\times nq}{C} = \underset{m\times n}{A} \otimes \underset{p\times q}{B} = \left[a_{ij}B\right]. \tag{2.114}$$

In the product, the position occupied by each element in A is replaced by the matrix B times that element. Thus the row and column order of the resulting matrix is the product of the corresponding orders of the factors.

Example

$$\begin{bmatrix} a_{11} \\ a_{21} \end{bmatrix} \otimes \begin{bmatrix} b_{11} & b_{12} \\ b_{21} & b_{22} \end{bmatrix} = \begin{bmatrix} a_{11}b_{11} & a_{11}b_{12} \\ a_{11}b_{21} & a_{11}b_{22} \\ a_{21}b_{11} & a_{21}b_{12} \\ a_{21}b_{21} & a_{21}b_{22} \end{bmatrix}. \tag{2.115}$$

The Kronecker product has a number of useful properties:

$$(A \otimes B) \otimes C = A \otimes (B \otimes C),$$

$$(A + B) \otimes C = A \otimes C + B \otimes C,$$

$$(A \otimes B)' = A' \otimes B',$$

$$(A \otimes B)^{-1} = A^{-1} \otimes B^{-1},$$

$$(A \otimes C)(B \otimes D) = AB \times CD.$$

2.8.3 Row and Column Matrices

A position vector in its coordinate representation may be considered an $n \times 1$ column matrix, e.g.

$$x = \begin{bmatrix} x_1 \\ x_2 \\ \vdots \\ x_n \end{bmatrix}. \tag{2.116}$$

Column matrices are therefore often referred to as vectors, or more specifically as *column vectors*, and are written in vector notation. The transpose of a column matrix is called a row vector ($1 \times n$) and indicated by the transpose sign:

$$x' = [x_1, x_2, \dots, x_n]. \tag{2.117}$$

2.8.3.1 Rank and Nullity

The vector space generated by the rows of A is called the *row space* of A. Similarly, the vector space generated by the columns of A is called the *column space*. The row rank of A is the dimensionality of the row space, and the column rank is the dimensionality of the column space. These two dimensionalities are equal and therefore the row and column rank are one and the same number $r = \text{rank}(A)$ (see (Bock, 1975), Section 2.3.2).

Suppose A is an $m \times n$ matrix. If $m \leq n$, the rank of A cannot exceed m. If $r = m$, we say that A is of *full* row rank. If $r < m$, then A is not of full rank or, equivalently, the *row nullity* of A is $m - r$. Similarly, if $r = n$, A is of full column rank. If $r < n$, A is not of full column rank, and the column nullity is $n - r$. For an $n \times n$ square matrix, the unique nullity is $n - r$.

2.8.4 Matrix Inversion

For computational purposes, there are several approaches to matrix-inversion, many of which make use of a compact version of Gauss' method called the *Gauss–Jordan* method (Householder, 1953, 1964). The calculations are recursive, elements of the matrix at stage $i + 1$ being replaced by the result of a so-called *pivoting* operation on elements at stage i:

$$a_{kl}^{(i+1)} = a_{kl}^{(i)} - \frac{a_{kj}^{(i)} a_{jl}^{(i)}}{a_{jj}^{(i)}}, \quad k \neq l \neq j,$$

$$a_{jl}^{(i+1)} = \frac{a_{jl}^{(i)}}{a_{jj}^{(i)}}, \quad l \neq j,$$

$$a_{kj}^{(i+1)} = \frac{a_{kj}^{(i)}}{a_{jj}^{(i)}}, \quad k \neq j,$$

$$a_{jj}^{(i+1)} = \frac{1}{a_{jj}^{(i)}}.$$

The element $a_{jj}^{(i)}$ is called the *pivot*, and its row and column are called the *pivotal row* and *pivotal column*. After n pivoting operations, the original matrix is replaced by its inverse, provided that each row and column is pivoted only once.

Example Matrix inversion by the Gauss–Jordan method. Let us illustrate the Gauss–Jordan method by inverting the order-4 Vandermonde matrix

$$V_4^{(0)} = \left[\begin{array}{ccc|c} 1 & 1 & 1 & 1 \\ 1 & 2 & 4 & 8 \\ 1 & 3 & 9 & 27 \\ \hline 1 & 4 & 16 & 64 \end{array}\right].$$

Stage 1. Choosing 64 as the pivot, we have, after the first pivoting operation,

$$V_4^{(1)} = \left[\begin{array}{cc|c|c} \frac{63}{64} & \frac{15}{16} & \frac{3}{4} & -\frac{1}{64} \\ \frac{7}{8} & \frac{3}{2} & 2 & -\frac{1}{8} \\ \hline \frac{37}{64} & \frac{21}{16} & \frac{9}{4} & -\frac{27}{64} \\ \hline \frac{1}{64} & \frac{1}{16} & \frac{1}{4} & \frac{1}{64} \end{array}\right].$$

Stage 2. The largest element and next pivot is $\frac{9}{4}$:

$$V_4^{(2)} = \left[\begin{array}{c|ccc} \frac{19}{24} & \frac{1}{2} & -\frac{1}{3} & \frac{1}{8} \\ \hline \frac{13}{36} & \frac{1}{3} & -\frac{8}{9} & \frac{1}{4} \\ \frac{37}{144} & \frac{7}{12} & \frac{4}{9} & -\frac{3}{16} \\ -\frac{7}{144} & -\frac{1}{12} & -\frac{1}{9} & \frac{1}{16} \end{array}\right].$$

Stage 3. Now the largest element in a row and column not already pivoted is $\frac{19}{24}$, and that is the next pivot:

$$V_4^{(3)} = \left[\begin{array}{c|c|cc} \frac{24}{19} & \frac{12}{19} & -\frac{8}{19} & \frac{3}{19} \\ \hline -\frac{26}{57} & \frac{2}{19} & -\frac{14}{19} & \frac{11}{57} \\ \hline -\frac{37}{114} & \frac{8}{19} & \frac{21}{38} & -\frac{13}{57} \\ \frac{7}{114} & -\frac{1}{19} & -\frac{5}{38} & \frac{4}{57} \end{array}\right].$$

Stage 4. Now $\frac{2}{19}$ is chosen as the last pivot, since it is the element from the only row and the only column not yet pivoted:

$$V_4^{(4)} = V_4^{-1} = \begin{bmatrix} 4 & -6 & 4 & -1 \\ -\frac{13}{3} & \frac{19}{2} & -7 & \frac{11}{6} \\ \frac{3}{2} & -4 & \frac{7}{2} & -1 \\ -\frac{1}{6} & \frac{1}{2} & -\frac{1}{2} & \frac{1}{6} \end{bmatrix}.$$

2.9 Determinants

An important scalar function of an $n \times n$ square matrix A is the *determinant* of order n. The determinant is denoted by $|A|$ and defined as

$$|A| = \sum^{n!} (-1)^s a_{1\alpha} a_{1\beta} \cdots a_{nv}, \tag{2.118}$$

where $\alpha, \beta, \dots, v$ represents one of the permutations of the natural numbers 1 through n. The number of terms in this sum is $n!$, that is, the number of ordered arrangements (permutations) of n distinct things taken n at a time. The exponent s, which determines the sign of each term in the expansion, is 1 or 0 according as the *class* of the permutation is *odd* or *even*. A permutation is called *odd* if the number of pairs of numbers which are out of natural order is odd; otherwise, it is called *even*. For the numbers 1 through 4, three are out of natural order, namely, 32, 42, and 43. The permutation 1342, on the other hand, is even because two pairs are out of order, namely, 32 and 42.

Since each of the numbers $1, 2, \dots, n$ appears once as a row subscript and once as a column subscript, it is apparent that each term contains as a factor exactly one element from each row and column of A. Because each pair of numbers appears in natural order in one of these terms and in reverse order in another, exactly half the terms are positive and half negative (for $n > 1$).

Example

$$\left| a_{11} \right| = a_{11},$$

$$\begin{vmatrix} a_{11} & a_{12} \\ a_{21} & a_{22} \end{vmatrix} = a_{11}a_{22} - a_{12}a_{21},$$

$$\begin{vmatrix} a_{11} & a_{12} & a_{13} \\ a_{21} & a_{22} & a_{23} \\ a_{31} & a_{32} & a_{33} \end{vmatrix} = a_{11}a_{22}a_{33} - a_{11}a_{23}a_{32} + a_{12}a_{23}a_{31} - a_{12}a_{21}a_{33}$$

$$+ a_{13}a_{21}a_{32} - a_{13}a_{22}a_{31}.$$

2.10 Matrix Differentiation

In the following chapters, we will have frequent occasion to differentiate functions that depend on many variables. When the functions are expressed in terms of matrices, we can obtain the required results more easily by vector and matrix differentiation than by the more familiar methods of scalar calculus. Although matrix differentiation is merely a symbolic representation of the results of multiple scalar differentiations, the compactness of expression and the ease of converting the results to matrix operations for numerical evaluation are highly advantageous.

How this symbolism should be defined is to some extent arbitrary, and numerous schemes have been been proposed. In presenting here some results needed in the sequel, we follow the conventions of Henderson and Searle (1979) and Magnus and Neudecker (1988).

2.10.1 Scalar Functions of Vector Variables

The first-order derivative of a scalar function of vector variable is a vector consisting of the partial derivatives of the function with respect to each component. Magnus and Neudecker prefer the row orientation of the result. If $f(x)$ is a scalar function of an n-vector, x, we have

$$\frac{\partial f(x)}{\partial x} = \left[\frac{\partial f(x)}{\partial x_1}, \frac{\partial f(x)}{\partial x_2}, \ldots, \frac{\partial f(x)}{\partial x_n}\right]. \tag{2.119}$$

Example 1 Linear forms.

$$y = a'x,$$
$$\frac{\partial y}{\partial x} = \frac{\partial xa'}{\partial x} = a'.$$

Example 2 Quadratic forms.

$$\begin{aligned} y &= x'Ax \\ \frac{\partial y}{\partial x} &= x'\frac{\partial Ax}{\partial x} + \frac{\partial Ax}{\partial x}x \\ &= x'(A + A') \end{aligned}$$

or, if A is symmetric, $= 2x'A$.

Geometrically, the vector of partial derivatives of a continuous function at the point $x = c$, interior to the space of x, describes the slopes, in the direction of the x-axes, of the plane tangent to the the function at c. This vector is called the *gradient* of the function; as a column, we will represent it by $G(x)$, and as a row, by $G'(x)$.

With the aid of the gradient, the increment of the function from the point c to $c + u$ is approximated to the first order of accuracy by the first two terms of a Taylor series,

$$f(c + u) \approx f(c) + G'(c)u,$$

where c and u are n-vectors. If the function is linear in x, the relation is exact.

Example *Minimizing quadratic forms.* In statistical procedures such as principal factor analysis (see Chapter 6), we seek the linear contribution $c = x'y$ that maximizes the variance of the n-vector random variable y with a known covariance matrix A. Since the variance of the combination,

$$c^2 = x'Ax$$

is arbitrarily large when elements of x are arbitrarily large, we choose to maximize the ratio

$$\lambda = \frac{x'Ax}{x'x}.$$

Seeking the maximum of λ, we differentiate with respect to x and set the derivatives to zero. The result is

$$\begin{aligned}\frac{\partial \lambda}{\partial x} &= \frac{2Ax(x'x) - 2(x'Ax)x}{(x'x)^2} \\ &= (A - \lambda I)x = 0,\end{aligned}$$

for $x'x \neq 0$. The necessary and sufficient condition for this set of n homogeneous equations to have a nontrivial solution in x is for the matrix $A - \lambda I$ to be singular, which implies

$$|A - \lambda I| = 0.$$

Expansion of the determinant results in a polynomial equation in λ. It is referred to as the *characteristic equation* of the matrix A; its degree is equal to the rank of A.

If A is a given real symmetric matrix, the solutions, or roots, of the characteristic equation are real. They are referred to variously as the *characteristic roots, latent roots, principal values*, or *eigenvalues* of A. Similarly, the corresponding solutions of the system of homogeneous equations $(A - \lambda I)x = 0$ are called *latent* or *principal vectors*, or *eigenvectors*, of A.

If A is of order n and rank $r \leq n$, the number of nonzero roots is r and the number of zero roots is $n - r$. If A is positive-semidefinite, the r nonzero roots are all positive. If A is of full rank ($r = n$), all roots are nonzero. This is another test of the definiteness of A. If all roots are positive, A is positive-definite. If all roots are negative, A is negative-definite. If some roots are positive and others negative, A is called "indefinite."

2.10.2 Vector Functions of a Vector Variable

Suppose m functions of an $n \times 1$ vector x are the components of an $m \times n$ vector function, $\mathbf{f}(x)$. In other words, $\mathbf{f}(x)$ is a transformation of x. Then the derivatives of the transform with respect to x take the form of an $m \times n$ matrix:

$$\frac{\partial \mathbf{f}(x)}{\partial x} = \begin{bmatrix} \partial f_1(x)/\partial x_1 & \partial f_1(x)/\partial x_2 & \dots & \partial f_1(x)/\partial x_n \\ \partial f_2(x)/\partial x_1 & \partial f_2(x)/\partial x_2 & \dots & \partial f_2(x)/\partial x_n \\ \cdot & \cdot & \cdot & \cdot \\ \partial f_m(x)/\partial x_1 & \partial f_m(x)/\partial x_2 & \dots & \partial f_m(x)/\partial x_n \end{bmatrix}. \tag{2.120}$$

This matrix is called the *Jacobian matrix* of the transformation. When $m = n$, its determinant is called the *Jacobian* of the transformation.

Example *Linear transformation.* Let A be the $m \times n$ matrix of the transformation,

$$y = Ax.$$

Then,

$$\frac{\partial y}{\partial x} = A,$$

and A is the Jacobian matrix.

2.10.3 Scalar Functions of a Matrix Variable

The derivative of a scalar function of an $m \times n$ matrix X consists of partial derivatives with respect to each element of X. Although it may seem natural to arrange the partials in the corresponding positions of an $m \times n$ matrix, this may not be convenient in subsequent operations, such as taking the second derivatives of the function. A better convention is to arrange the elements of X into a column vector by stacking its columns upon one another. The result of the stacking operation is called the "vec." Thus,

$$\text{vec } X = \text{vec}[x_1, x_2, \dots, x_n] = \begin{bmatrix} x_1 \\ x_2 \\ \vdots \\ x_n \end{bmatrix}. \tag{2.121}$$

Then the partial of $f(X)$ with respect to vec X is a $mn \times 1$ vector, and with respect to $\text{vec}'X = (\text{vec } X)'$, which is a $1 \times mn$ vector.

Example 1 *Determinants.* Differentiating a determinant with respect to the vec of the matrix yields the vec of the transposed adjoint matrix:

$$\frac{\partial \mid X \mid}{\partial \text{vec } X} = \text{vec}'(\text{adj } X').$$

Differentiating the *log* determinant divides the elements of the transposed adjoint matrix by the determinant and thus yields the elements of the transposed inverse matrix:

$$\frac{\partial \log \mid X \mid}{\partial \text{vec } X} = \mid X \mid^{-1} \text{vec}'(\text{adj } X') = \text{vec}'(X^{-1})'.$$

If a real symmetric matrix S is decomposed into the product of a square matrix and its transpose $S = TT'$ (e.g. the Cholesky decomposition), the derivative of the log determinant with respect to T is

$$\frac{\partial \log \mid S \mid}{\partial \text{vec} T} = 2\text{vec}' S^{-1} T.$$

Example 2 *Traces.* Differentiating a trace with respect to the vec of the matrix merely yields the vec of the identity matrix of the same order. Although the trace is defined only for square matrices, it applies to the products of non-square matrices if the row order of the right-most factor and the column-order of the left most factor are equal. Some derivatives in this case are the following:

$$\frac{\partial \text{tr} XA}{\partial \text{vec } X} = \text{vec}' A',$$

$$\frac{\partial \text{tr } X^{-1}A}{\partial \text{vec } X} = -\text{vec}' X^{-1} A X^{-1},$$

$$\frac{\partial \text{tr} XX'}{\partial \text{vec } X} = 2 \text{vec}' X.$$

These results make use of the identity

$$\text{tr} AB = \text{tr} BA = \text{vec}' A \text{ vec } B.$$

2.10.4 Chain Rule for Scalar Functions of a Matrix Variable

Suppose a scalar function of X is a function of another function that depends on X, say,

$$h(X) = f(g(X)). \tag{2.122}$$

Then the chain rule for differentiation of compound functions applies as follows:

$$\frac{\partial h(X)}{\partial \text{vec } X} = \frac{\partial h(X)}{\partial g(X)} \cdot \frac{\partial g(X)}{\partial \text{vec } X}. \tag{2.123}$$

Example Let $f(\xi) = e'De$, where ξ is an n-vector, D is an $n \times n$ diagonal "weight" matrix, and $e = (y - A\xi)$ is an m-vector. Then,

$$\frac{\partial f(\xi)}{\partial \text{vec } \xi} = \frac{\partial f(\xi)}{\partial \text{vec } e} \cdot \frac{\partial \text{vec } e}{\partial \text{vec } \xi}$$
$$= (y - A\xi)'DA.$$

2.10.5 Matrix Functions of a Matrix Variable

Let $F(X)$ represent an $m \times n$ matrix of scalar functions $f_{ij}(X)$, where X is a $p \times q$ matrix. Then the derivatives of $F(X)$ with respect to each of the elements $x_{k\ell}$ of X will consist of the $mnpq$ elements $\partial \mathbf{f}_{ij}/\partial x_{k\ell}$. The only question is in what order to write them down. Magnus and Neudecker recommend the $mn \times pq$ Jacobian matrix,

$$\frac{\partial \text{vec } F(X)}{\partial \text{vec } X}. \tag{2.124}$$

On this definition, $\partial \text{vec } X/\partial \text{vec } X$ equals the $pq \times pq$ identity matrix, but

$$\frac{\partial \text{vec } X}{\partial \text{vec } X'} = K_{p,q}, \tag{2.125}$$

where $K_{p,q}$ is the $pq \times pq$ *commutation* matrix that carries the vec of a matrix into the vec of its transpose; i.e.

$$\text{vec } X' = K_{p,q} \text{ vec } X. \tag{2.126}$$

A commutation matrix K is a matrix of 1's and 0's that reorders the rows of a matrix of the same order as K. It is necessarily an orthogonal matrix and therefore has the property $K^{-1} = K'$.

For A constant,

$$\frac{\partial \text{vec } AX}{\partial \text{vec } X} = I_q \otimes A, \tag{2.127}$$

where $\otimes$ designates the Kronecker product described in Section 2.8.2.

If A and B are matrix functions of X conformable for addition, the derivative of their sum is

$$\frac{\partial \text{vec}(A + B)}{\partial \text{vec } X} = \frac{\partial \text{vec } A}{\partial \text{vec } X} + \frac{\partial \text{vec } B}{\partial \text{vec } X}. \tag{2.128}$$

To express the result for the product, we specialize the identity

$$\text{vec } ABC = (C' \otimes A)\text{vec } B \tag{2.129}$$

to obtain, for $A(m \times n)$ and $B(n \times r)$,

$$\text{vec } AB = (B' \otimes I_m)\text{vec } A = (I_r \otimes A)\text{vec } B. \tag{2.130}$$

Then,

$$\frac{\partial AB}{\partial \text{vec } X} = (B' \otimes I_m)\frac{\partial \text{vec } A}{\partial \text{vec } X} + (I_r \otimes A)\frac{\partial \text{vec } B}{\partial \text{vec } X}. \tag{2.131}$$

Applying this result to $XX^{-1} = I_p$ and using the identities for Kronecker products given in Section 2.8.2, we obtain

$$\begin{aligned}\frac{\partial \text{vec } X^{-1}}{\partial \text{vec } X} &= -[I_p \otimes X^{-1}][(X^{-1})' \otimes I_p]\frac{\partial \text{vec } X}{\partial \text{vec } X} \\ &= -(X^{-1})' \otimes X^{-1}.\end{aligned}$$

Example For A symmetric and constant, we make use of the derivative

$$\frac{\partial X^{-1}A(X^{-1})'}{\partial \text{vec } X} = [X^{-1}A(X^{-1})' \otimes X^{-1}] + [(X^{-1})' \otimes (X^{-1})'AX^{-1}].$$

2.10.6 Derivatives of a Scalar Function with Respect to a Symmetric Matrix

When X is a $p \times p$ real symmetric matrix, say S, $F(X)$is a function of only $p(p+1)/2$ distinct variables. To express the corresponding elements as a column vector, Henderson and Searle (1979) introduced the operator

$$\text{vech}(S) = [s_{11}, s_{12}, s_{22}, s_{13}, s_{23}, s_{33}, \dots, s_{pp}]', \tag{2.132}$$

which stacks elements in the columns of S down to the main diagonal.

To transform the vech of a matrix into the vec, one multiplies by the $p^2 \times p(p+1)/2$ full-rank matrix G:

$$\text{vec } S = G \text{ vech } S. \tag{2.133}$$

Example For $p = 3$,

$$G = \begin{bmatrix} 1 & 0 & 0 & 0 & 0 & 0 \\ 0 & 1 & 0 & 0 & 0 & 0 \\ 0 & 0 & 0 & 1 & 0 & 0 \\ 0 & 1 & 0 & 0 & 0 & 0 \\ 0 & 0 & 1 & 0 & 0 & 0 \\ 0 & 0 & 0 & 0 & 1 & 0 \\ 0 & 0 & 0 & 1 & 0 & 0 \\ 0 & 0 & 0 & 0 & 1 & 0 \\ 0 & 0 & 0 & 0 & 0 & 1 \end{bmatrix}.$$

To transform the vec into the vech, one multiplies by the pseudo-inverse $(G'G)^{-1}G'$.

Example For $p = 3$,

$$(G'G)^{-1}G' = \begin{bmatrix} 1 & 0 & 0 & 0 & 0 & 0 & 0 & 0 & 0 \\ 0 & 0.5 & 0 & 0.5 & 0 & 0 & 0 & 0 & 0 \\ 0 & 0 & 0 & 0 & 1 & 0 & 0 & 0 & 0 \\ 0 & 0 & 0.5 & 0 & 0 & 0 & 0.5 & 0 & 0 \\ 0 & 0 & 0 & 0 & 0 & 0.5 & 0 & 0.5 & 0 \\ 0 & 0 & 0 & 0 & 0 & 0 & 0 & 0 & 1 \end{bmatrix}.$$

An important example of the use of the vech notation in multivariate statistics is the differentiation of tr $S^{-1}A$ with respect to the elements of S. The matrices S and A are real-symmetric, positive-definite, and A is constant. On the one hand, when we ignore the duplication of elements in vec S, we have the p^2 elements of the Jacobian matrix,

$$\frac{\partial \text{tr } S^{-1}A}{\partial \text{vec } S} = -\text{vec}' \, S^{-1}AS^{-1}\frac{\partial \text{vec } S}{\partial \text{vec } S}, \tag{2.134}$$

where, as we saw above, the latter derivative is the $p^2 \times p^2$ identity matrix.

Differentiating with respect to the functionally independent elements of S, on the other hand, we have the $p(p+1)/2$-element gradient vector,

$$\begin{aligned} \frac{\partial \text{tr } S^{-1}A}{\partial \text{vech} S} &= -\text{vec}' S^{-1}AS^{-1}G\frac{\partial \text{vech} S}{\partial \text{vech} S} \\ &= -\text{vec}' S^{-1}AS^{-1}G\frac{\partial \text{vech} S}{\partial \text{vech} S} \\ &= -\text{vec}' S^{-1}AS^{-1}G. \end{aligned} \tag{2.135}$$

2.10.7 Second-Order Differentiation

If the function is twice differentiable in some subspace of x, we may obtain the second-order derivatives of the function by differentiation of the first-order derivatives. We will discuss here only the case required in the sequel, namely, the second derivative of a scalar function of an $m \times n$ matrix variable, say X. As we have seen, the first derivative of such a function can be expressed, with the aid of vec notation, as the column vector,

$$\text{vec}\frac{\partial f(X)}{\partial \text{vec } X} = \begin{bmatrix} \partial f(X)/\partial x_{11} \\ \partial f(X)/\partial x_{21} \\ \vdots \\ \partial f(X)/\partial x_{mn} \end{bmatrix}. \tag{2.136}$$

The first derivative of *this* function with respect to the vec of X requires the differentiation of a vector function with respect to a vector. As above, we have:

$$\frac{\partial \text{vec } \partial f(X)/\partial \text{vec } X}{\partial \text{vec } X}$$

$$= \begin{bmatrix} \partial^2 f(X)/\partial x_1 \partial x_1, & \partial^2 f(X)/\partial x_1 \partial x_2, & \dots, & \partial^2 f(X)/\partial x_1 \partial x_{mn} \\ \partial^2 f(X)/\partial x_2 \partial x_1, & \partial^2 f(X)/\partial x_2 \partial x_2, & \dots, & \partial^2 f(X)/\partial x_2 \partial x_{mn} \\ \dots & \dots & \dots & \dots \\ \partial^2 f(X)/\partial x_{mn} \partial x_1, & \partial^2 f(X)/\partial x_{mn} \partial x_2, & \dots, & \partial^2 f(X)/\partial x_{mn} \partial x_{mn} \end{bmatrix}.$$

For points at which $f(X)$ is twice differentiable,

$$\partial^2 f(X)/\partial x_i \partial x_j = \partial^2 f(X)/\partial x_j \partial x_i, \tag{2.137}$$

and the matrix of second derivatives is symmetric. It is referred to as the *Hessian matrix* of the function and is conventionally represented by $H(X)$. Its determinant is called the *Hessian* of the function.

Geometrically, $H(c)$ describes the curvature of the function at the point c. If $H(c)$ is positive semi-definite, the function is concave up at c; if it is negative semi-definite, the function is concave down; if it is indefinite, there is an inflection point or a saddle point at c. The definiteness of the Hessian matrix therefore determines whether a zero of the gradient at c is, respectively, a minimum, a maximum, or an inflection or saddle point.

With the aid of the gradient and the Hessian matrix, the increment of the function from the point c to $c + u$ is approximated to the second order of accuracy by the first three terms of a Taylor series,

$$f(c + u) \approx f(c) + G'(c)u + \frac{1}{2}u'H(c)u. \tag{2.138}$$

If the function is quadratic in x, the relation is exact.

Example We have seen that the derivative of the function $f(\xi) = e'De$ defined in Section 2.10.4 is

$$\frac{\partial f(\xi)}{\partial \text{vec } \xi} = (y - A\xi)'DA. \tag{2.139}$$

The second derivative is

$$\frac{\partial \text{vec } \partial f(\xi)/\partial \text{vec } \xi}{\partial \text{vec } \xi} = A'DA. \tag{2.140}$$

The $n \times n$ Hessian matrix is therefore at least positive semi-definite, and any extremum of the function is a minimum.

2.11 Theory of Estimation

Models describing population distributions generally contain unknown parameters that have to be estimated in samples drawn from the population. Estimation theory provides criteria for evaluating the performance of various procedures that might be proposed for this purpose. The following are some of the important "good" properties that are desired of an estimator.

1. *Consistency*. A minimal requirement of a good estimator is that it converges in probability to the true parameter as the sample size increases. A statistic with this property is called *consistent*.

For example, the sample mean,

$$\bar{x} = \sum_{i=1}^{N} x_i / N, \tag{2.141}$$

is a consistent estimator of the population mean, provided the population distribution has a finite mean and variance. This is established by Tchebychev's theorem, which states that for a distribution with mean μ and standard deviation σ, a probability bound on the magnitude of the mean $\bar{x}$ of n random observations is given by

$$P\left(|x - \mu| < \frac{k\sigma}{\sqrt{n}} \right) \geq \frac{1}{k^2}, \tag{2.142}$$

where k is any positive number. Taking n large enough, we can make arbitrarily small the probability that the mean of the sample deviates from the population mean by more than a specified fraction of the population standard deviations. A sample proportion, for example, is a consistent estimator of the mean of a binomial distribution.

2. *Unbiasedness*. An estimator is called *unbiased* if, at any sample size, the expected value of the statistic is equal to the population value of the estimated parameter. Thus, x is an unbiased estimator of θ if, and only if,

$$\mathcal{E}(x) = \theta. \tag{2.143}$$

An unbiased estimator is obviously consistent, but a consistent estimator is not necessarily unbiased.

It is important to realize that only when an estimator is unbiased does the mean of many small samples estimate consistently the corresponding population parameter. For example, the mean of sample variances computed with the biased statistic

$$s_j^2 = \frac{1}{n} \sum_{i=1}^{n} (x_{ij} - \bar{x}_j)^2 \tag{2.144}$$

estimates $(n-1)\sigma^2/n$ rather than σ. Dividing by the degrees of freedom $(n-1)$ instead of n corrects the bias in the jth estimate and makes the mean of the estimates consistent for σ.

The foregoing is an example of inconsistency of estimation caused by a wrongly specified model for the observations. It neglects the fact that deviations from the sample mean are negatively correlated (the correlation is $-1/n$), so that the expected value of the sum of squared deviations is $(n-1)\sigma^2$ rather than σ.

3. *Efficiency*. Among competing estimators, we would generally prefer the one that is in some sense closest to the population value on average. A traditional measure of goodness of approximation of an estimator $\hat{\theta}$ is the *root-mean-square* (RMS) error:

$$\text{RMS} = \left[\sum_{j=1}^{N} (\hat{\theta}_j - \theta)^2 \right]^{1/2}. \tag{2.145}$$

In most applications, the RMS cannot be computed because θ is unknown. If x is unbiased, however, the sample variance estimates the RMS. Therefore, among unbiased estimators, the RMS criterion indicates that we should prefer a *minimum variance* estimator if it exists.

The so-called *Cramer–Rao inequality* tells us that, if $\hat{\theta}$ is based upon a random sample of n observations of x from a population with probability $f(x)$, the variance of the estimator satisfies

$$\mathcal{V}(\hat{\theta}) \geq \frac{1}{n}\mathcal{E}\left[\left(\frac{\partial \ln f((x)}{\partial \theta} \right)^2 \right]. \tag{2.146}$$

Any unbiased estimator that attains the equality is therefore a minimum variance estimator.

The ratio of the variance of two unbiased estimators is their relative *efficiency*. For example, as an estimate of the mean of a normal distribution, the efficiency of the sample median relative to the sample mean is

$$\frac{\sigma^2/(2n+1)}{\pi\sigma^2/4n} = \frac{\sigma^2/n}{\pi\sigma^2/4n} = \frac{4n}{(2n+1)\pi}, \tag{2.147}$$

which equals $2/\pi$, or approximately 64%, for large n.

As n becomes large, the Cramer–Rao inequality becomes the so-called *information* limit of the variance of a consistent estimator. A consistent large-sample estimator that attains the equality is called *efficient*. We shall make extensive use of Fisher information when comparing item response theory estimators (IRT) estimators.

4. *Sufficiency*. An estimator is called *sufficient* if and only if the conditional distribution of the sample is independent of the parameter that it estimates. Sufficiency of $\hat{\theta}$ for θ implies and is implied by the likelihood factoring as

$$f(x \mid 0) = g(\hat{0} \mid \theta)h(x), \tag{2.148}$$

where x contains the sample values. It is easy to verify, for example, that the likelihood function of a random sample of size n from a population $N(\mu, \sigma)$ can be expressed as

$$f(x \mid \mu, \sigma) = \frac{1}{(\sqrt{2\pi})^{n/2}} \prod_{i=1}^{n} \exp\left\{ -\frac{1}{2}[(\bar{x}_i - \mu)^2/\sigma^2] \right\} \exp\left\{ -[(x_i - \bar{x}_i)^2/\sigma^2] \right\}. \tag{2.149}$$

The sample mean is therefore a sufficient statistic for the mean of a normal distribution.

It is especially convenient if a *simple* sufficient statistic exists for the parameter in question, where "simple" means easily computed and represented. For in that case, a compact table of the sufficient statistics will contain all of the information of the original data, and the number of operations in analysis may be greatly economized. Prior to the introduction of electronic computers, there was a great premium on finding sufficient statistics. With the advent of high-speed computing, it is often possible to work directly from the original data, and the existence of simple sufficient statistics is less important.

5. *Robustness*. An estimator that is resistant to untoward effects of failures of assumptions is called *robust*. In applications of statistics to large-scale data, as in social surveys and educational testing, it is especially important that an estimator is robust to outliers caused by spurious observations and clerical errors. Various procedures have been proposed to limit the influence of outliers on statistics. A well-known example is the Tukey–Mostellor *biweight* applied to estimation of the population mean μ from a sample of size n. It takes the form,

$$\tilde{x} = \Sigma w_i x_i / \Sigma w_i,$$

$$W_i = \begin{cases} (1 - u_i^2)^2 & \text{if } |u_i| \leq | \\ 0 & \text{otherwise,} \end{cases}$$

where $u_i = (x_i - \tilde{x})/sC$ with s equal to the sample standard deviation. C is an arbitrary constant that adjusts the severity of the biweight trimming. $C = 2$ is a typical value.

The biweight is applied to scale-score estimation.

2.11.1 Analysis of Variance

As mentioned earlier in this section, the extremal equations of linear least squares are referred to as normal equations because they correspond to the resolution of the observational vector y into the vector $X\beta$, in the space of the model, and a vector of residuals, $y - X\hat{\beta}$, orthogonal to (or normal to) the plane of the model. By the Pythagorean theorem for n-space, the least-squares solution therefore partitions the square length of y into, respectively, a component attributable to the model and an orthogonal component attributable to the residual:

$$y'y = \hat{\beta}' X' X \hat{\beta} + (y - X\hat{\beta})'(y - X\hat{\beta}). \tag{2.150}$$

This orthogonal partition of the sum of squares of the observed values is the foundation of R.A. Fisher's analysis of variance (ANOVA) for linear models. He showed that, if the error is normally distributed with mean 0 and variance σ_ϵ^2, the model and residual sums of squares divided by σ_ϵ^2 are independently distributed

Table 2.1 Analysis of variance for a linear model of rank m.

Source of variation	Degrees of freedom	Sums of squares	Expected sums of squares
Model	m	$y'X(X'X)^{-1}C'y$	$m\sigma_\epsilon^2 + \beta'X'X\beta$
Residual	$n - m$	$y'[I - X(X'X)^{-1}X']y$	$(n - m)\sigma_\epsilon^2$
Total	n	$y'y$	

as chi-square variates with degrees of freedom equal to the dimensionality of the corresponding subspaces, namely, m and $n - m$.

On the null hypothesis $\beta = 0$, the chi-square statistic associated with the model is centrally distributed (i.e. as the sum of squares of m unit normal deviates), whereas on the alternative hypothesis it is noncentrally distributed with noncentrality parameter $\beta'X'X\beta$. The residual chi-square, on the other hand, follows a central distribution unconditionally.

These results provide an exact (small sample) test of the null hypothesis. Fisher showed that, in terms of mean squares (the sums of squares divided by their respective degrees of freedom), the ratio of the model mean square to the residual mean square is distributed as an F-statistic on m and $n - m$ degrees of freedom. On the null hypothesis, its distribution is central; on the alternative hypothesis, it is noncentral with noncentrality parameter $\beta'X'X\beta/m$. Significance of this statistic when referred to the central F-distribution implies that the noncentrality parameter is positive, which, since $X'X$ is positive-definite, can only happen when β is non-null.

These statistical tests are summarized in Table 2.1.

Moreover, by partitioning the model vector into further orthogonal components, the nullity of separate parameters in β can be successively tested by quasi-independent F-statistics. This provides step-wise testing of parameters corresponding to additional x-variables added successively to the model (see (Bock, 1975), Chapter 4).

2.11.2 Estimating Variance Components

ANOVA can also be applied to designs in which all ways of classification are random (a random-effects ANOVA versus a fixed-effects or mixed-effects ANOVA). The purpose of the analysis is to estimate components of variation in the observations that can be attributed to the various ways of classification and their interactions. This type of analysis has many useful applications in measurement.

Table 2.2 Data scheme for the Respondents × Methods × Occasions random model.

B:	Methods	1		2		
C:	Occasions	1	2	1	2	Mean
A:	1	y_{111}	y_{112}	y_{121}	y_{122}	$y_{1\cdot\cdot}$
Respondents	2	y_{211}	y_{212}	y_{221}	y_{222}	$y_{2\cdot\cdot}$
	.	.	.	.	.	.
	N	y_{N11}	y_{N12}	y_{N21}	y_{N22}	$y_{N\cdot\cdot}$
	Mean	$y_{\cdot 11}$	$y_{\cdot 12}$	$y_{\cdot 21}$	$y_{\cdot 22}$	$y_{\cdot\cdot\cdot}$

Suppose each of a respondents is measured by each of b methods on each of c occasions. Table 2.2 shows the arrangement of the resulting measurements when there are two methods and two occasions. It is of interest to know to what extent the corresponding ways of classification, of respondents A, methods B, and occasions C, are responsible for random variation in the scores.

The model for the score of respondent i on method j on occasion k is

$$y_{ijk} = \mu + \alpha_i + \beta_j + \gamma_k + (\alpha\beta)_{ij} + (\alpha\gamma)_{ik} + \beta\gamma_{jk} + \epsilon_{ijk}. \tag{2.151}$$

The random terms are assumed independently distributed as:

$$\begin{aligned}
\alpha &\sim N(0, \sigma_\alpha^2),\\
\beta &\sim N(0, \sigma_\beta^2),\\
\gamma &\sim N(0, \sigma_\gamma^2),\\
(\alpha\beta) &\sim N(0, \sigma_{\alpha\beta}^2),\\
(\beta\gamma) &\sim N(0, \sigma_{\beta\gamma}^2),\\
(\alpha\gamma) &\sim N(0, \sigma_{\alpha\gamma}^2),\\
\epsilon &\sim N(0, \sigma_\epsilon^2).
\end{aligned} \tag{2.152}$$

Under this model, the ANOVA may be expressed in terms of the marginal means:

$$\begin{aligned}
y_{ij\cdot} &= \textstyle\sum_k^c y_{ijk}/c & y_{i\cdot\cdot} &= \textstyle\sum_j^b \sum_k^c y_{ijk}/bc\\
y_{\cdot jk} &= \textstyle\sum_i^a y_{ijk}/a & y_{\cdot j\cdot} &= \textstyle\sum_i^a \sum_k^c y_{ijk}/ac\\
y_{i\cdot k} &= \textstyle\sum_j^b y_{ijk}/b & y_{\cdot\cdot k} &= \textstyle\sum_i^a \sum_i^b y_{ijk}/ab
\end{aligned} \tag{2.153}$$

$$y_{\cdots} = \textstyle\sum_i^a \sum_j^b \sum_k^c y_{ijk}/abc.$$

The separate components of variance are estimated by equating the expected sums of squares to the observed sums of squares and solving the resulting system of linear equations.

Example Walker and Lev (1953) present scores from an archery contest (taken from (Schroeder, 1945)). Eleven archers shot in three trials at distances 30, 40, and 50 yards with the results shown in Table 2.3. The analysis of variance including the estimated variance components rounded to whole numbers, as computed from this table, is shown in Table 2.4.

The large variance component for distance (2602) is not very meaningful in these data because the distances, although potentially random, have been deliberately set with rather wide spacing. The component for archer-by-distance interaction (152) shows, however, that the effect of distance is not all systematic: some archers are better than others at different distances.

Trials have relatively less influence on individual scores than distance ($\hat{\sigma}^2_{\alpha\beta} = 72$ versus $\hat{\sigma}^2_{\alpha\beta} = 152$). Average scores for distance and trial interact appreciably ($\hat{\sigma}^2_{\beta\gamma} = 208$), presumably because the archers fatigue more on repeated trials at longer distances.

Table 2.3 Scores from an archery contest.

Distance (yards)	30			40			50		
Trial	1	2	3	1	2	3	1	2	3
Archer									
1	305	255	286	248	241	157	114	182	123
2	203	188	225	101	160	153	99	121	119
3	201	184	238	153	149	148	51	100	35
4	163	167	219	114	142	132	78	144	72
5	267	182	259	246	214	237	134	125	201
6	146	120	157	105	117	84	71	38	49
7	195	235	181	122	125	133	66	67	107
8	162	166	213	141	164	113	33	89	51
9	275	240	227	181	222	187	146	159	157
10	170	110	126	109	120	106	80	101	83
11	232	166	221	186	185	188	105	113	113

Table 2.4 Analysis of variance of the archery scores.

Source of variation	d.f.	Sums of squares	Mean square	Variance components
Mean	1	2 331 588		
Archers	10	134 395	13 440	1 524
Distances	2	171 491	85 746	2 602
Trials	2	178	89	9
Archers × Distances	20	19 492	975	152
Archers × Trials	20	14 639	732	72
Distances × Trials	4	11 196	2 799	208
Residual	40	20 242	506	506
Total	99	2 688 582		

The inferred error variance of 506 for a single distance and single trial shows that the archery test is a fairly accurate measure of the archerś skill. The standard error of measurement (SEM) of $\sqrt{506} = 22.5$ is relatively small compared to the standard deviation of the scores with the distance and trial effects eliminated, namely, $\sqrt{1524 + 152 + 72 + 506} = 47.5$. (See the continuation of this example in Section 2.17.)

2.12 Maximum Likelihood Estimation

If the systematic and the stochastic part of a statistical model are not additive, the method of least squares is not generally applicable. Fortunately, in situations where sample sizes are large, as they are in most applications of item response theory, the method of maximum likelihood estimation is a quite satisfactory and widely used alternative.

Although the maximum likelihood concept was intimated by Gauss, its modern development is due to Fisher (1922). Applied to estimation, it requires a specified model for the distribution of observations in question. The principle of MLE is that, given a random sample of observations, we should choose as estimates of unknown parameters in the model those values that make the corresponding probability or probability density function of the sample a maximum. To make clear that probability functions used in this way describe mathematical rather than statistical objects, Fisher called them *likelihood* functions. Briefly, the essential methods and properties of maximum likelihood estimation are as follows.

2.12.1 Likelihood Functions

If the data are discrete, the likelihood of a parameter θ given N independent observations $y_i, i = 1, 2, \ldots, N$, is the product of the corresponding probability functions:

$$L(\theta) = \prod_{i=1}^{N} P(y_i \mid \theta). \tag{2.154}$$

If the data are continuous, the likelihood is the product of density functions:

$$L(\theta) = \prod_{i=1}^{N} f(y_i \mid \theta). \tag{2.155}$$

2.12.2 The Likelihood Equations

In most applications, the likelihood is maximized by differentiating the log likelihood function with respect to the parameters to be estimated. Set equal to zero, these derivatives become the so-called *likelihood equations*, which are necessary conditions on the maximum. For an m-vector parameter θ, the m likelihood equations may be represented symbolically as,

$$\frac{\partial \log L}{\partial \theta} = 0. \tag{2.156}$$

The value of the derivatives at θ, written as $G(\theta)$, is the *gradient* of the likelihood surface at that point. We designate the solution of these equations, if it exists, as $\hat{\theta}$.

To check that the solution corresponds to a maximum of the likelihood, we examine the definiteness of the $(m \times m)$ Hessian matrix at $\hat{\theta}$ (see Section 2.10):

$$H(\hat{\theta}) = \left[\frac{\partial^2 \log L}{\partial \theta \partial \theta'} \right]_{\hat{\theta}}. \tag{2.157}$$

If and only if the Hessian matrix is of full rank and negative definite, the system of equations has a solution that corresponds to a maximum of the likelihood. We then say that the parameters are *identified* by the data.

We also need to verify, however, that the solution corresponds to the proper, or global maximum. For likelihood functions arising from many common distributions, $H(\hat{\theta})$ is independent of the data and always negative definite if the model is identified. Thus the maximum of the likelihood is always a global maximum. If $H(\theta)$ depends on the data, one may have to search the likelihood surface more widely to establish that the solution is proper. Fortunately, improper solutions often lead to implausible values for some of the parameters and are easily recognized.

2.12.3 Examples of Maximum Likelihood Estimation

The following are simple examples of maximum likelihood estimators (MLEs).

Example 1 Suppose that a random sample of N observations from a binomial distribution yields r successes. Then the log likelihood is

$$\log L(P) = \log \frac{N!}{r!(N-r)!} + r \log P + (N-r)\log(1-P).$$

The latter two terms in this expression, which are the only ones that will be involved in the maximization, comprise the *kernel* of the log likelihood. To obtain a necessary condition on the maximum, we set to zero the derivative of the kernel with respect to P to obtain the likelihood equation:

$$\frac{r}{P} - \frac{N-r}{1-P} = 0.$$

The solution of this equation is the maximum likelihood estimate,

$$\hat{P} = r/N.$$

Any value of this estimator is admissible, and the second derivative of the log likelihood is negative in the open interval $0 < P < 1$.

$$\frac{\partial^2 \log L(P)}{\partial P^2} = \frac{-NP(1-P) - (r-NP)(1-2P)}{[P(1-P)]^2}.$$

Example 2 In the continuous case, consider the estimation of the mean μ and variance σ^2 from N random observations y from a normal distribution. The kernel of the log likelihood is

$$-\frac{1}{2}N \log \sigma^2 - \frac{1}{2}\sum_{i=1}^{N}(y_i - \mu)^2/\sigma^2.$$

The likelihood equations for μ and σ^2 are, respectively:

$$\sum_{i=1}^{N}(y_i - \mu)/\sigma^2 = 0$$

and

$$-\frac{1}{2}N\sigma^{-2} + \frac{1}{2}\sum_{i=1}^{N}(y_i - \mu)^2/\sigma^{-4} = 0.$$

The simultaneous solution of these equations gives

$$\hat{\mu} = y. = \sum_{i=1}^{N} y_i/N$$

and

$$\hat{\sigma}^2 = \sum_{i=1}^{N} (y_i - y.)^2/N.$$

Notice that the MLE of σ^2 is not unbiased: the divisor is N rather than $N - 1$. In general, MLEs are neither unbiased nor minimum variance. As we show below, however, they are consistent and efficient; that is, *asymptotically* (as $N \to \infty$) unbiased and of minimum variance.

2.12.4 Sampling Distribution of the Estimator

The large-sample distribution of the MLE may be obtained by expanding the likelihood equation in a Taylor expansion about $\hat{\theta}$. Let $\partial \log L/\partial\theta$ be the log likelihood for a random sample from a population specified except for a parameter θ. Then, at the solution point, $\hat{\theta}$, the likelihood equation expands as,

$$0 = \left[\frac{\partial \log L}{\partial\theta}\right]_{\hat{\theta}} \left[\frac{\partial^2 \log L}{\partial\theta\partial\theta'}\right]_{\hat{\theta}} (\theta - \hat{\theta}) + \mathcal{O}(\theta - \hat{\theta})^2. \tag{2.158}$$

Thus,

$$\theta - \hat{\theta} = \left[\frac{\partial^2 \log L}{\partial\theta\partial\theta'}\right]_{\hat{\theta}}^{-1} \left[\frac{\partial \log L}{\partial\theta}\right]_{\hat{\theta}} + \mathcal{O}(\theta - \hat{\theta})^2, \tag{2.159}$$

and the expectation of $\theta - \hat{\theta}$ in indefinitely large samples is zero; that is, the asymptotic expected value of the estimator is the population value of the parameter.

To obtain asymptotic variance of the MLE, we need the following lemma: Because the likelihood, L, is the same as the density function for the sample, the expected value of the likelihood equation may be expressed as,

$$\mathcal{E}\left(\frac{\partial \log L}{\partial\theta}\right) = \int \frac{\partial \log L}{\partial\theta} L\, dy = 0. \tag{2.160}$$

Differentiating with respect to θ under the integral sign, we have

$$\int \left[\frac{\partial^2 \log L}{\partial\theta\partial\theta'} L + \frac{\partial \log L}{\partial\theta}\frac{\partial \log L}{\partial\theta'}\right] dy = 0,$$

$$\int \left[\frac{\partial^2 \log L}{\partial\theta\partial\theta'} + \frac{\partial \log L}{\partial\theta}\frac{\partial\theta}{\partial\theta'}\right] L\, dy = 0$$

or,

$$\mathcal{E}\left(\frac{\partial^2 \log L}{\partial\theta\partial\theta'}\right) = -\mathcal{E}\left(\frac{\partial \log L}{\partial\theta}\frac{\partial \log L}{\partial\theta'}\right). \tag{2.161}$$

In other words, the expected value of the matrix of second derivatives of the log likelihood is equal to the negative of the expected value of the column-by-row products of the first derivatives. This lemma is used in the following result for the large-sample variance of the likelihood estimator:

From the above Taylor expansion, the asymptotic variance of $\hat{\theta}$ is

$$\mathcal{E}(\theta - \hat{\theta})^2 = \left[\frac{\partial^2 \log L'}{\partial\theta\partial\theta}\right]_{\hat{\theta}}^{-1} \mathcal{E}\left[\frac{\partial \log L}{\partial\theta}\frac{\partial \log L}{\partial\theta'}\right]_{\hat{\theta}} \left[\frac{\partial^2 \log L}{\partial\theta\partial\theta'}\right]_{\hat{\theta}}^{-1}; \tag{2.162}$$

whence, from the lemma, we have

$$\mathcal{V}(\hat{\theta}) = -\left[\mathcal{E}\left(\frac{\partial^2 \log L}{\partial\theta\partial\theta'}\right)_{\hat{\theta}}\right]^{-1} = \left[\mathcal{E}\left(\frac{\partial \log L}{\partial\theta}\cdot\frac{\partial \log L}{\partial\theta'}\right)_{\hat{\theta}}\right]^{-1}. \tag{2.163}$$

In the context of maximum likelihood estimation, the negative of the expected matrix of second derivatives is called the (Fisher) *information matrix* and represented by $I(\theta)$. Thus, the asymptotic covariance matrix of the estimator is the inverse information matrix,

$$\mathcal{V}(\hat{\theta}) = I^{-1}(\hat{\theta}), \tag{2.164}$$

computed from either of the above forms.

2.12.5 The Fisher-Scoring Solution of the Likelihood Equations

In the above examples, the unknowns in the likelihood equations can be transposed to the right-hand side and the solution expressed explicitly. In many other cases, the equations can be solved only in their implicit form. Ordinarily the Newton–Raphson solution would be the method of choice, but in maximum likelihood estimation, the Hessian matrix (second derivatives of the log likelihood) is often difficult to derive and calculate. Fortunately, an almost equally efficient method in large sample is the so-called *Fisher-scoring* method – a Newton method in which the information matrix replaces the Hessian matrix. The Fisher-scoring iterations may then be expressed as

$$\hat{\theta}_{i+1} = \hat{\theta}_i + I^{-1}(\hat{\theta}_i)G(\hat{\theta}_i), \tag{2.165}$$

where $G(\hat{\theta}_i)$ is the gradient vector (the first derivatives evaluated at θ_i). Some inefficient estimate of θ is usually available as a starting value for the Fisher-scoring iterations.

2.12.6 Properties of the Maximum Likelihood Estimator (MLE)

A number of important properties of the MLE follow from the above results.

1. *Consistency*. The MLE is a consistent estimator of the corresponding population parameter.
2. *Efficiency*. When the MLE exists, it is large-sample efficient; it attains the Cramer–Rao information limit of the variance.
3. *Sufficiency*. If a minimum sufficient statistic exists, it will be the MLE.
4. *Uniqueness*. The MLE is large-sample unique.
5. *Transformational invariance*. The MLE of any nonsingular transformation of the parameters is the same transform of the MLE of the original parameters.
6. *Asymptotic normality*. As the sample size increases the sampling distribution of the MLE approaches a multivariate normal distribution with mean equal to the parameter and covariance matrix equal to the inverse information matrix.

In some cases, it is possible to investigate the small-sample bias and variance of the MLE by evaluating additional terms in the Taylor expansion of the likelihood equations (see (Shenton and Bowman, 1977)). There are several examples of this approach in IRT applications (Lord, 1983; Warm, 1989).

2.12.7 Constrained Estimation

A Fisher-scoring solution of the likelihood equations requires that the information matrix be positive-definite. A necessary condition for definiteness is that the likelihood equations be functionally independent. In particular, the model must not be overparameterized; i.e. it must not have more undetermined parameters than the number of independent likelihood equations. If the model is overparameterized, we have the same options as in least-squares estimation – namely, either to augment the likelihood by independent functional constraints on the parameters, or to reparameterize the model in terms of functionally independent parameters.

2.12.8 Admissibility

It is not always the case that MLE will yield admissible values of the quantity estimated. An estimate of a variance might be negative, for example, or an estimated proportion greater than unity. Although it is generally possible to impose inequality constraints on an MLE to keep the value obtained in an admissible region, the computations involved are often difficult (see (Gill and Murray, 1974)). A better strategy often is to reparameterize in a one-to-one transformation to an unrestricted space, which makes a conventional Fisher-scoring solution possible. Because of the transformational invariance of MLE, estimates of the constrained original parameters may then be obtained by applying the inverse transform to the estimate of the transformed parameters.

Example To be admissible, an estimate of a population covariance matrix must be at least positive semi-definite. In general, the sample covariance matrix has this property, but the MLE may not in advanced applications. A convenient way to impose positive semi-definiteness on an $(n \times n)$ covariance matrix is to express it in the Gaussian decomposition,

$$\Sigma = LDL',$$

where L is an $(n \times n)$unit lower triangular matrix and $D = \text{diag}[e^{\pi_i}]$, $i = 1, 2, \ldots, n$. Σ is expressed as a one-to-one function of the unrestricted $n(n-1)/2$ free elements of L and the n values of π_i. The covariance matrix computed from the maximum likelihood estimates of these parameters cannot be negative definite or indefinite and can be semi-definite only in the limit as one or more of the π_i go to $-\infty$. (See (Hedeker, 1989).)

2.13 Bayes Estimation

When a statistical model, contains random components to be estimated, we may wish to constrain the estimator of the components to plausible values from a specified distribution. In other words, we will want to impose a stochastic constraint on the estimator rather than an equality or inequality constraint. Bayes estimation serves this purpose[2]:

1. Suppose a vector observation y of a sample unit has the density function $f(y, \theta|\zeta)$, where θ is a random parameter characterizing a unit of the sampling distribution, and ζ is a fixed parameter characterizing the method or condition of observation. Suppose further that θ has a specified density, $g(\theta; \eta)$, independent of $f(y|\theta; \zeta)$. The joint sampling distribution of y and θ is then $f(y|\theta; \zeta)g(\theta; \eta)$, and, according to Bayes theorem, the conditional probability density of θ, given y, is

$$p(\theta|y) = \frac{f(y|\theta; \zeta)g(\theta; \eta)}{h(y)}, \tag{2.166}$$

2 A consistent estimation theory can also be built upon the concept that the parameter distribution describes the investigatorś subjective belief about the true value of the parameter (see (de Finetti, 1972; Savage, 1954)). Bayes estimation then shows us how he should modify her belief in light of the data. Inasmuch as the parameters to be estimated in the IRT applications have real (though not directly observable) distributions, we will have no need of this conception of Bayes estimation.

where

$$h(y) = \int f(y|\theta;\zeta)g(\theta;\eta)d\theta \qquad (2.167)$$

is the marginal density of y; that is, $h(y)$ is the normalizing constant that makes $p(\theta|y)$ integrate to unity over the range of θ. In the terminology of Bayes estimation, $g(\theta;\eta)$ is the *prior* density of θ, $f(y|\theta;\zeta)$, is the *likelihood*, $h(y)$ is the marginal density of y, and $p(\theta|y)$ is the *posterior* density of θ.
The Bayes estimate of θ for a given sampling unit is the mean of the posterior distribution,

$$\bar{\theta} = \int \theta p(\theta|y)d\theta. \qquad (2.168)$$

We will refer to it also as the *expected a posteriori* or EAP estimator.
A measure of the uncertainty about θ, given y, is the variance of the posterior distribution,

$$\sigma^2_{\theta|y} = \int (\theta - \hat{\theta})^2 p(\theta|y)d\theta. \qquad (2.169)$$

The square root of this quantity, the posterior standard deviation (PSD), plays a role in Bayes estimation similar to the role of the standard error in least-squares estimation.

Example Consider n replicate observations, y, each fallibly measuring the value τ. The $(n \times 1)$ vector observation may then be expressed as

$$y = \tau\mathbf{1} + \epsilon,$$

where τ is the true score for this observation, $\mathbf{1}$ is an n-vector of unities, and ϵ is a vector of independent errors. Assume that τ and ϵ are normally distributed and mutually independent, so that the joint distribution of y and τ are $(n+1)$-variate normal:

$$\begin{bmatrix} y \\ \tau \end{bmatrix} \sim N\left(\begin{bmatrix} \mu\mathbf{1} \\ \mu \end{bmatrix}, \begin{bmatrix} \sigma^2_\alpha\mathbf{1}\mathbf{1}' + \sigma^2_\epsilon I & \sigma^2\alpha\mathbf{1} \\ \sigma^2_\alpha\mathbf{1}' & \sigma^2\alpha \end{bmatrix} \right).$$

The posterior (conditional) distribution of τ, given y, is normal with the following mean:

$$\begin{aligned} \bar{\tau} &= \sigma^2_\alpha\mathbf{1}'[\sigma^2_\alpha\mathbf{1}\mathbf{1}' + \sigma^2_\epsilon I]^{-1}(y - \mu\mathbf{1}) + \mu \\ &= \sigma_\epsilon^{-2}[\sigma_\epsilon^{-2}\mathbf{1}'\mathbf{1} + \sigma_\alpha^{-2}]^{-1}\mathbf{1}'(y - \mu\mathbf{1}) + \mu \\ &= \frac{n\sigma_\epsilon^{-2}}{n\sigma_\epsilon^{-2} + \sigma_\alpha^{-2}}(y_{.} - \mu) + \mu \\ &= \rho_n y_{.} + (1 - \rho_n)\mu, \end{aligned}$$

where $\rho_n = \sigma_\alpha^2/(\sigma_\alpha^2 + \sigma_\epsilon^2/n)$ is the reliability coefficient for the mean of n replicate measures and $ny_. = \mathbf{1}'y = \sum_i^n y_i$.

Similarly, the posterior variance is

$$\begin{aligned}\sigma_{\tau|y}^2 &= [\sigma_\epsilon^{-2}\mathbf{1}'\mathbf{1} + \sigma_\alpha^{-2}]^{-1}\\ &= (n/\sigma_\epsilon^2 + 1/\sigma^2)^{-1}\\ &= \sigma_\alpha^2(1 - \rho_n).\end{aligned}$$

The Bayes estimator of τ is therefore a weighted sum of the mean of the replicate observations and the mean of the population from which they are drawn. If each observation is perfectly reliable, or there are indefinitely many replicates, $\rho_n = 1$ and the estimate of the true scores is equal to the observed mean. If the observed score is completely unreliable, or there are no observations, $\rho = 0$ and we have no better information about the location of the true score than the population mean. For any intermediate situation, the Bayes estimator is, in the sense of minimum RMS error, the optimal weighted average of the prior and observed information about the true score.

2. *Properties of Bayes estimation*. The Bayes estimator has a number of favorable properties:
   (a) Although the Bayes estimator is not unbiased, it is consistent as the amount of information in the observation increases without limit (e.g. as n goes to infinity in the preceding example).
   (b) With respect to the population as a whole, the Bayes estimator has the minimum RMS error. In this sense, the investigatorś long-term risk in taking some action based on the estimator is minimized for random observations from the population. For this reason Bayes estimation has a central role in decision theory.
   (c) Unlike the properties of the MLE, those of the Bayes estimator are not asymptotic; they are exact in samples of any size.
   (d) Only for models such as in the preceding example, where both the likelihood and the prior are normal and the measurement model is linear, is the posterior distribution also normal. This is referred to as the "normal–normal" case. When the model is nonlinear, the posterior will not be normal in general even when the prior and likelihood are normal. Nevertheless, the Bayes estimator, like mean of a sampling distribution, will tend to normality as the number of independent observations increase without limit, and the PSD will tend to the standard error of the Bayes estimator. This justifies treating the Bayes estimator like other large-sample statistics.

(e) For a given likelihood, a prior distribution can often be chosen so that the posterior has the same functional form as the prior. Such a prior is called the "conjugate" of the likelihood. In this situation, Bayes estimation can be applied recursively to accumulate information about the quantity being measured (in adaptive testing for example).

(f) For many choices of likelihood and prior, the integral in the Bayes estimator will not exist in closed form. In these cases, numerical integration will be required to evaluate the estimator and the posterior variance. Most of the applications of Bayes estimation in item response theory will be this type.

2.14 The Maximum A Posteriori (MAP) Estimator

If the number of parameters to be jointly estimated is large and the integral in the Bayes estimator is analytically intractable, the high dimensionality of the estimation space may frustrate any attempt to obtain the Bayes estimate by quadrature. In this case, a good alternative is to substitute the mode of the posterior distribution in place of the mean. The resulting estimator is referred to as *maximum a posteriori*, or MAP, to distinguish it from the Bayes estimator as such.

To obtain an MAP estimate, we maximize the log posterior density with respect to the parameter of interest. Because the posterior density, apart from the arbitrary normalizing constant, is the product of the likelihood and the prior, the function to be maximized is the sum of the log likelihood and the log population density. Thus, the MAP equation for θ is

$$\frac{\partial \log L}{\partial \theta} + \frac{\partial \log g(\theta;\eta)}{\partial \theta} = 0; \tag{2.170}$$

in other words, the MAP equation is the likelihood equations plus the derivative of the log population density of the random parameter. In the context of constrained estimation, the latter is called the *penalty* function.

MAP equations can be solved by EM or Fisher-scoring procedures, just as MLE equations, except that the information matrix contains an additional term due to the prior:

$$J(\theta) = I(\theta) + \frac{\partial^2 \log g(\theta;\eta)}{\partial \theta \partial \theta'} \tag{2.171}$$

called $J(\theta)$, which is the *posterior information*.

The EAP and MAP estimators are both consistent for θ as the amount of information in the observation increases. But to the same extent that the sample mode has a larger sampling variance than the mean, the MAP estimator is in general less efficient than the Bayes estimator.

Example For the same model as in the previous example, the MAP equation for estimating τ from n replicate observations is

$$\begin{aligned}
\partial &\left[\sum_i^n \log f(y_i|\tau,\sigma^2) + \log g(\tau;\mu,\sigma_\alpha^2)\right] \Bigg/ \partial\tau \\
&= -\sum_i^n (y_i - \tau)/\sigma^2 + (\tau - \mu)/\sigma_\tau^2 \\
&= \frac{-n\sigma_\alpha^2 y_. + (n\sigma_\alpha^2 + \sigma^2)\tau - \sigma^2\mu}{\sigma^2\sigma_\alpha^2} \\
&= 0.
\end{aligned}$$

Expressing the solution of the MAP equation in terms of the reliability coefficient for the mean of n replicate observations, $\rho_n = \sigma_\alpha^2/(\sigma_\alpha^2 + \sigma^2/n)$, we therefore have

$$\tilde{\tau} = \rho_n y_. + (1 - \rho_n)\mu,$$

which in this special case is the same as the Bayes estimator above. In general, the MAP estimator will *not* be equal to the EAP estimator.

The posterior information is

$$\begin{aligned}
J(\tau) &= \frac{n}{\sigma^2} + \frac{1}{\sigma_\alpha^2} \\
&= \frac{n\sigma_\alpha^2 + \sigma^2}{\sigma^2\sigma_\alpha^2}.
\end{aligned}$$

The limiting (in n) posterior variance is the reciprocal of this information:

$$\begin{aligned}
\mathcal{V}(\tau) &= \frac{\sigma^2\sigma_\alpha^2}{\sigma_\alpha^2 + \sigma^2/n} \\
&= \sigma_\alpha^2(1 - \rho_n),
\end{aligned}$$

which in this case is the same as the Bayes posterior variance.

2.15 Marginal Maximum Likelihood Estimation (MMLE)

In Section 2.13 on Bayes estimation, the likelihood and the prior were assumed fully specified. But in most real-world applications, not all of the parameters of the likelihood or prior are known *a priori*. If a random sample of observations from the population is available, however, it may be possible to estimate these parameters. In large samples, a method with many good properties for this purpose is marginal maximum likelihood estimation (MMLE). In this section, we discuss MMLE in the context of measurement (see also (Bock, 1989a)).

In measurement applications, the data typically result from a two-stage sampling scheme. The investigator samples units from some population and measures each a number of times in order to insure good precision. If we regard the repeated measures as a sample from a hypothetical population of measurements made under the same conditions, there are two stages of sampling. The measurements are the first-stage sample; the units being measured are the second-stage sample.

Models for two-stage samples (such as the true-score model of Section 2.17) typically contain both fixed and random parameters, where some value of the random parameter is associated with each second-stage unit. We cannot then estimate both sets of these parameters by ordinary maximum likelihood estimation. Because the dimensionality of the parameter space increases along with the number of units observed at the second stage, the asymptotic properties of maximum likelihood estimation would not hold. (The problem of maximum likelihood estimation in the presence of infinitely many of these so-called "incidental" parameters has been discussed by Neyman and Scott (1948), Kiefer and Wolfowitz (1956), and Andersen (1980), among others.)

In specifying the model, however, it is only the fixed parameters that must be estimated, and not the random components. The fixed parameters are of two kinds – those of the likelihood function, which we refer to as *structural* parameters, and those of the population density, which we refer to as the *population* parameters.[3] These two types of parameters are treated in essentially the same way in MMLE, but for clarity, we distinguish them in the following discussion.

2.15.1 The Marginal Likelihood Equations

The principle of MMLE is that the fixed parameters should be estimated jointly in the marginal distribution obtained by integrating over the distribution of the random parameters. Let us suppose that the sample consists of vector observations, y_i, for $i = 1, 2, \ldots, N$, and let observation i have n_i measurements; i.e. the number of replicate measurements may vary from one second-stage unit to another. Then the marginal density for observation i is

$$h(y_i) = \int f(y_i|\theta;\zeta)g(\theta;\eta)d\theta, \tag{2.172}$$

or, briefly,

$$h_i = \int_\theta f_i \cdot g\ d\theta, \tag{2.173}$$

and the marginal log likelihood for the sample is

$$\log L_{\mathrm{M}} = \sum_{i=1}^{N} \log h_i. \tag{2.174}$$

3 The population parameters are also called "second-stage" or "hyper-" parameters.

Taking symbolic vector derivative of the log likelihood with respect to the ζ vector and equating to zero, we obtain

$$\begin{aligned}\frac{\partial \log L_M}{\partial \zeta} &= \sum_i^N \frac{1}{h_i}\left[\frac{\partial (h_i)}{\partial \zeta}\right] \\ &= \sum_i^N \frac{1}{h_i}\int_\theta \frac{\partial f_i}{\partial \zeta} g \, d\theta \\ &= \sum_i^N \int_\theta \frac{f_i \cdot g}{h_i}\frac{\partial \log f_i}{\partial \zeta} g \, d\theta \\ &= \sum_i^N \int_\theta p_i \frac{\partial \log f_i}{\partial \zeta} g \, d\theta = 0.\end{aligned} \tag{2.175}$$

With respect to the η vector, we obtain similarly

$$\begin{aligned}\frac{\partial \log L_M}{\partial \eta} &= \sum_i^N \int_\theta \frac{f_i \cdot g}{h_i}\frac{\partial \log g}{\partial \eta} g \, d\theta \\ &= \sum_i^N \int_\theta p_i \frac{\partial \log g}{\partial \eta} g \, d\theta = 0.\end{aligned} \tag{2.176}$$

Thus we see that the marginal likelihood equations are the *posterior expectations* of the derivatives of the same log likelihoods that we encounter in ordinary one-stage estimation of ζ and η. Our first problem, therefore, is how to compute these expectations. In some cases, the integral will have a known closed-form that will enable the likelihood equation to be evaluated directly. In other cases, including almost all of those found in item response theory, there will be no closed-form, and we will have to resort to numerical integration, or "quadrature." Fortunately, the wide array of computer procedures available for quadratures are so convenient and fast that the lack of closed-forms is not an insurmountable barrier. Notice these integrals have the dimensionality of the random parameter, and not that of the structural and population parameters. Models with large numbers of fixed parameters are therefore quite suitable for quadratures, provided the dimensionality of the random parameters is not greater than four or five. This is not restrictive in many applications, especially in IRT, where the number of latent variables is typically small.

Our second problem is to solve the system of marginal likelihood equations. These systems are typically nonlinear, but in general they can be solved iteratively by Fisher scoring or by repeated substitution methods such as the *EM algorithm* (Dempster et al., 1977). Although the EM algorithm is a first-order process and is slow to converge, it is highly stable even when the starting values are far from the solution point. In contrast, Fisher scoring attains stable second-order convergence only in the immediate neighborhood of the solution point. A good strategy therefore is to begin the solution with EM iterations and switch to Fisher-scoring iterations as the solution point is approached. There are many applications of this technique in this text.

2.15.2 Application in the "Normal–Normal" Case

In this section, we apply the MMLE method to a linear measurement model on normal distribution assumptions. It is an instance of a so-called normal–normal model for which closed-form marginal likelihood equations exist.

To fix ideas, consider the following example. Suppose the number of hours per week that a panel of N families watch a certain educational television station is monitored. Let the number of hours for family i be y_i, and let the number of weeks that this family remains on the panel be n_i. Our task is to investigate the relationship of successive weeks of panel membership and the number of hours watched per week.

Let us assume that the relationship for any family can be described by a polynomial in the time variable, the coefficients of which are particular to each family. If the panel members are selected randomly from some population, this is a so-called *random regressions* model. In most applications, it is of interest to know the population distribution of the random coefficients. In the present example we might use a cubic polynomial to investigate such features as the average linear trend in viewing over the weeks of the trial, whether the average viewing peaks during the trial, or whether it has an inflection point.

2.15.2.1 First-Stage Estimation

Let the observed number of hours for family i for weeks x_{ij} be y_{ij}, $j = 1, 2, \ldots, n_i$. Assume that the hours of viewing over the n_i weeks can be represented by an $(m - 1)$-degree polynomial in x with coefficients, β,

$$y_i = X_i \beta_i + \epsilon_i. \tag{2.177}$$

The $n_i \times m$ matrix X_i contains the powers of x_{ij} from 0 to $m - 1$, and the n_i-vector, ϵ_i, represents residual variation in viewing behavior from week to week.

Assume that the residual variation is independent normal with mean zero and constant variance; that is,

$$\epsilon_i \sim N\left[0, \sigma_\epsilon^2 I_{n_i}\right]. \tag{2.178}$$

On these assumptions, the likelihood of β_i is expressed as the multivariate normal density,

$$p(Y = y_i | \beta_i, \sigma) = \frac{1}{(2\pi)^{n_i/2}\sigma^{2n}} \exp\left\{-\frac{1}{2}\sigma^{-2} tr\left[(y_i - X_i\beta_i)(y_i - X_i\beta_i)'\right]\right\}. \tag{2.179}$$

For the prior distribution of β, let us assume a multivariate normal distribution with mean μ and covariance matrix Σ; the prior density is then

$$g(\beta; \mu, \Sigma) = \frac{|\Sigma|^{-\frac{1}{2}}}{(2\pi)^{m/2}} \exp\left\{-\frac{1}{2}\text{tr}\left[\Sigma^{-1}(\beta-\mu)(\beta-\mu)'\right]\right\}, \quad |\Sigma| \neq 0. \quad (2.180)$$

On these assumptions, the observations and the coefficients have the joint multivariate normal distribution,

$$\begin{bmatrix} y_i \\ \beta \end{bmatrix} \sim N\left(\begin{bmatrix} X_i\mu \\ \mu \end{bmatrix}, \begin{bmatrix} X_i\Sigma X_i' + \sigma_\epsilon^2 I_{n_i} & X_i\Sigma \\ \Sigma X_i' & \Sigma \end{bmatrix}\right). \quad (2.181)$$

For this linear normal–normal case, it is easy to show that the posterior distribution of β, given y_i, is also normal, with mean

$$\bar{\beta}_i = \sigma^{-2}\left[\sigma^{-2}X_i'X_i + \Sigma^{-1}\right]^{-1}X_i'(y_i - X_i\mu) + \mu \quad (2.182)$$

and covariance matrix

$$\Sigma_{\beta|y_i} = \left[\sigma^{-2}X_i'X_i + \Sigma^{-1}\right]^{-1}. \quad (2.183)$$

Thus, we can easily obtain the Bayes estimate of the trend parameters for viewing time for family i if we have values for σ, μ and Σ. In general, these values will not be known, but they can be estimated with good properties in large samples by the marginal maximum likelihood method.

2.15.2.2 Second-Stage Estimation

The likelihood equations for MMLE of the second-stage parameters are obtained by writing the marginal likelihood for an observation y_i as

$$h(y_i) = \int_\beta f(y_i|\beta; \sigma_\epsilon^2) g(\beta; \mu, \Sigma) d\beta, \quad (2.184)$$

or, briefly, $h_i = \int f_i \cdot g \, d\beta$. Then the posterior density of β is $p_i = p(\beta|y_i) = f_i \cdot g/h_i$, and we obtain the likelihood equations for μ, Σ, and σ_ϵ^2 from general forms of the MML equations as follows. For μ, we have

$$\begin{aligned}
\frac{\partial \log L_M}{\partial \mu} &= \sum_i^N \frac{1}{h_i} \int_\beta f_i \frac{\partial g}{\partial \mu} d\beta \\
&= \sum_i^N \int_\beta \frac{f_i \cdot g}{h_i} \frac{\partial \log g}{\partial \mu} d\beta \\
&= \sum_i^N \int_\beta p_i \Sigma^{-1}(\beta - \mu) d\beta \\
&= \Sigma^{-1} \sum_i^N (\bar{\beta}_i - \mu) = 0. \quad (2.185)
\end{aligned}$$

To obtain the likelihood equation for Σ, we may use the "vec" and "vech" notation (Magnus and Neudecker, 1988):

$$\frac{\partial \log|\Sigma|}{\partial \text{vec } \Sigma} = \text{vec } \Sigma^{-1} \tag{2.186}$$

and

$$\frac{\partial \text{tr } \Sigma^{-1} S}{\partial \text{vec } \Sigma} = -\text{vec } \Sigma^{-1} S \Sigma^{-1}. \tag{2.187}$$

In addition, we will need

$$\begin{aligned}\int_\beta & p_i(\beta - \mu)(\beta - \mu)' d\beta \\ &= \int_\beta p_i[(\beta - \bar{\beta}_i) + (\bar{\beta}_i - \mu)][(\beta - \bar{\beta}_i) + (\bar{\beta}_i - \mu)]' d\beta \\ &= \Sigma_{\beta|y_i} + (\bar{\beta}_i - \mu)(\bar{\beta}_i - \mu)'.\end{aligned} \tag{2.188}$$

Letting

$$w_i = \bar{\beta}_i - \mu, \tag{2.189}$$

we can write the derivatives with respect to the $m(m+1)/2$ functionally independent elements of Σ as

$$\begin{aligned}\frac{\partial \log L_M}{\partial \text{vech } \Sigma} &= G' \text{vec} \sum_i^N \int_\beta p_i \left[-\frac{1}{2}\Sigma^{-1} + \frac{1}{2}(\beta - \mu)(\beta - \mu)'\Sigma^{-1} \right] d\beta \\ &= \frac{1}{2} G' \sum_i^N \text{vec } \Sigma^{-1}(-\Sigma + \Sigma_{\beta|y_i} + w_i w_i')\Sigma^{-1} = 0,\end{aligned} \tag{2.190}$$

where G is the $m^2 \times m(m+1)/2$ full-rank transformation that carries vech Σ into vec Σ; i.e. vec $\Sigma = G$ vech Σ (Magnus and Neudecker, 1988).

Finally, using

$$\begin{aligned}\int_\beta & p_i(y_i - X_i\beta)'(y_i - X_i\beta) d\beta \\ &= \int_\beta p_i[(y_i - X_i\bar{\beta}_i) - X_i(\beta - \bar{\beta}_i)]'[(y_i - X_i\bar{\beta}_i) - X_i(\beta - \bar{\beta}_i)] d\beta \\ &= (y_i - X_i\bar{\beta}_i)'(y_i - X_i\bar{\beta}_i) + \text{tr } X_i \Sigma_{\beta|y_i} X_i',\end{aligned} \tag{2.191}$$

and letting

$$u_i = y_i - X_i\bar{\beta}_i, \tag{2.192}$$

we have

$$\frac{\partial \log L_M}{\partial \sigma_\varepsilon^2} = \sum_i^N \int_\beta \frac{f_i \cdot g}{h_i} \frac{\partial \log f_i}{\partial \sigma_\varepsilon^2} d\beta$$

$$= \sum_i^N \int_\beta p_i \left[-\frac{n_i}{2}\sigma_\varepsilon^{-2} + \frac{1}{2}\sigma_\varepsilon^{-4}\text{tr}(y_i - X_i\beta)(y_i - X_i\beta)' \right] d\beta$$

$$= \frac{1}{2}\sigma_\varepsilon^{-4}\sum_i^N \left[-n_i\sigma_\varepsilon^2 + u_i'u_i + \text{tr}\, X_i\Sigma_{\beta|y_i}X_i' \right] = 0. \tag{2.193}$$

2.15.3 The EM Solution

In agreement with results from Dempster et al. (1981), the simultaneous solution of the likelihood equations yields the MML estimators of the population and structural parameters in terms of the first-stage statistics, $\bar{\beta}_i$ and $\Sigma_{\beta|y_i}$:

$$\hat{\mu} = \frac{1}{N}\sum_i^N \bar{\beta}_i, \tag{2.194}$$

$$\hat{\Sigma} = \frac{1}{N}\sum_i^N (\bar{\beta}_i\bar{\beta}_i' + \Sigma_{\beta|y_i}) - \hat{\mu}\hat{\mu}', \tag{2.195}$$

$$\hat{\sigma}_\epsilon^2 = \left(\sum_i^N n_i\right)^{-1} \sum_i^N \left[(y_i - X_i\bar{\beta}_i)'(y_i - X_i\bar{\beta}_i) + trX_i\Sigma_{\beta|y_i}X_i' \right]. \tag{2.196}$$

The transposed likelihood equations are in a form suitable for a so-called EM solution. From any provisional estimates of σ, Σ, and μ, repeated substitutions of improved estimates on the left into the expressions on the right (Dempster et al., 1977) will eventually, but slowly, converge to the MML estimates of the population values. Because the model is within the exponential family, the likelihood surface is convex and any interior solution point of the likelihood equations is the global maximum.

2.15.4 The Fisher-scoring Solution

The efficient Fisher-scoring solution requires the elements of the information matrix for μ, Σ, and σ_ϵ^2. They have been given in scalar form by Longford (1987) and in matrix form by Bock (1989b). The latter results, expressed for blocks of the information matrix, are as follows:

$$I(\sigma_\epsilon^2) = \mathcal{E}\left[\left(\frac{\partial \log L_M}{\partial \sigma_\epsilon^2} \right)^2 \right]$$

$$= \frac{1}{2}\sigma_\epsilon^{-8}\sum_i^N tr(\sigma_\epsilon^2 I_{n_i} - X_i\Sigma_{\beta|y_i}X_i')^2, \tag{2.197}$$

$$
\begin{aligned}
I(\mu) &= \mathcal{E}\left(\frac{\partial \log L_{\mathrm{M}}}{\partial \mu} \cdot \frac{\partial \log L_{\mathrm{M}}}{\partial \mu'}\right) \\
&= \Sigma^{-1}\sum_{i}^{N}\mathcal{E}[(\bar{\beta}_i - \mu)(\bar{\beta}_i - \mu)']\Sigma^{-1} \\
&= \Sigma^{-1}\left(N\Sigma - \sum_{i}^{N}\Sigma_{\beta|y_i}\right)\Sigma^{-1}, \qquad (2.198)
\end{aligned}
$$

$$
\begin{aligned}
I(\Sigma) &= \mathcal{E}\left(\frac{\partial \log L_{\mathrm{M}}}{\partial \mathrm{vech}\ \Sigma} \cdot \frac{\partial \log L_{\mathrm{M}}}{\partial \mathrm{vech}'\ \Sigma}\right) \\
&= \frac{1}{2}G'(\Sigma^{-1} \otimes \Sigma^{-1})\left[\sum_{i}^{N}(\Sigma - \Sigma_{\beta|y_i}) \otimes (\Sigma - \Sigma_{\beta|y_i})\right](\Sigma^{-1} \otimes \Sigma^{-1})G. \qquad (2.199)
\end{aligned}
$$

Turning to the off-diagonal blocks of the information matrix, we have the following:

$$
I(\Sigma, \mu) = \mathcal{E}\left(\frac{\partial \log L_{\mathrm{M}}}{\partial \mathrm{vech}\ \Sigma} \cdot \frac{\partial \log L_{\mathrm{M}}}{\partial \mu'}\right) = 0, \qquad (2.200)
$$

$$
\begin{aligned}
I(\sigma_\varepsilon^2, \mu) &= \mathcal{E}\left(\frac{\partial \log L_{\mathrm{M}}}{\partial \sigma_\varepsilon^2} \cdot \frac{\partial \log L_{\mathrm{M}}}{\partial \mu'}\right) \\
&= \sum_{i}^{N} n_i \mathcal{E}\left\{\frac{1}{2}\sigma_\varepsilon^{-4}\left[-\sigma_\varepsilon^2 + s_i^2 + n_i^{-1}\mathrm{tr}(X_i\Sigma_{\beta|y_i}X_i')\right](\bar{\beta}_i - \mu)'\Sigma^{-1}\right\}. \\
&= \frac{1}{2}\sigma_\varepsilon^{-4}\sum_{i}^{N}\mathcal{E}\left[(y_{ij} - X_{ij}\bar{\beta}_i)'(y_{ij} - X_{ij}\bar{\beta}_i)(\bar{\beta}_i - \mu)'\right]\Sigma^{-1} = 0, \qquad (2.201)
\end{aligned}
$$

$$
\begin{aligned}
I(\sigma_\varepsilon^2, \Sigma) &= \mathcal{E}\left(\frac{\partial \log L_{\mathrm{M}}}{\partial \sigma_\varepsilon^2} \cdot \frac{\partial \log L_{\mathrm{M}}}{\partial \mathrm{vech}\ \Sigma}\right) \\
&= \frac{1}{2}\sigma_\varepsilon^{-4}G'\mathrm{vec}\ \Sigma^{-1}\left(\sum_{i}^{N}\Sigma_{\beta|y_i}X_i'X_i\Sigma_{\beta|y_i}\right)\Sigma^{-1}. \qquad (2.202)
\end{aligned}
$$

After a number of EM cycles, a few Fisher-scoring cycles are usually sufficient to attain a specified accuracy in the solution of the marginal likelihood equations. The correction to the provisional estimates of μ, Σ, and σ^2 is the product of the inverse information matrix and the gradient vector, just as in maximum likelihood estimation. Similarly, the standard errors of these population parameters are the square-roots of the corresponding diagonal elements of the inverse information matrix at the solution point. Likelihood ratio tests of nested sets of the population parameters may be formed from the respective maxima of the marginal likelihood, just as in maximum likelihood estimation.

2.16 Probit and Logit Analysis

The statistical methods applied in the sequel to item response data are for the most part random-effects extensions of standard fixed-effects procedures for the analysis of binomial and multinomial data. In particular, they are generalizations of well-known methods of probit and logit analysis, which we sketch in this section. Unlike their IRT extensions, the methods here apply to one-stage sampling schemes only. The respondents are the sampling units, and each is assigned to one of two or more mutually exclusive categories. The random-effects versions of these methods, which apply to the classification of responses in a two-stage sampling scheme, appear throughout this text.

2.16.1 Probit Analysis

When data take the form of enumerations of discrete events, least-squares methods do not strictly apply. The counts are not adequately described by the "treatment-effect plus random-error" model upon which these methods depend.

But in many situations, it proves nevertheless reasonable to assume that the *latent* effect of the treatment can be so described, and that the observed qualitative outcome is the manifest expression of a latent random variable exceeding some fixed threshold. We discussed in Chapter 1 some of the background of this assumption in the fields of bioassay and psychophysics. In these applications, the outcome of an experimental trial under condition k is classified as a success or failure, represented respectively by the values $u_k = 1$ or 0 of a Bernoulli variable. According to the threshold concept, probability of success is then

$$P_k = P(u_k = 1) = \int_{\gamma}^{\infty} f_k(y)dy, \tag{2.203}$$

and the probability of failure is $Q = 1 - P_k$. In the integrand, $f(y)$ is the density function of the latent variable under condition k, and the limit of integration γ is the threshold for the response.

Probit analysis is based on the assumption that the latent variable is normally distributed with mean μ_k and common variance σ^2. In that case, the probability of success is given by

$$P_k = \int_{\gamma}^{\infty} \phi_k \, dt = \frac{1}{\sqrt{2\pi}\sigma} \int_{\gamma}^{\infty} e^{-\frac{1}{2}\left(\frac{t-\mu_k}{\sigma}\right)^2} dt, \tag{2.204}$$

which with the change of variable, $y_k = (t - \mu_k)/\sigma$, becomes the standard normal distribution function $P_k = \Phi(y_k)$.

If N_k subjects randomly selected from the population respond independently in condition k, the observed numbers of successes, r_k, for $k = 1, 2, \dots, g$, are distributed as binomial variables. Now suppose these g conditions correspond to levels of an independent variable x_k. Let the deviate y_k for condition k depend on x through the linear relationship

$$y_k = a(x_k - b), \tag{2.205}$$

or, equivalently,

$$y_k = c + ax_k, \tag{2.206}$$

where $c = -ab$.

The parameter b is the point on the x continuum corresponding to a 50% response (called the “median effective dose” in the bioassay literature). The parameter a is a measure of the sensitivity of the respondents to variation in dose. Because the intercept and slope parameters c and a are better conditioned for estimation than b and a, we will assume the latter form of y_k in estimation. The estimate of b, if required, can be calculated from the estimated values of c and a.

Proceeding with the maximum likelihood estimation of these parameters, we express the likelihood, given the data, in terms of the product-binomial frequency function:

$$L(c, a) = \prod_{k=1}^{g} \frac{N_k!}{r_k!(N_k - r_k)!} P_k^{r_k} (Q_k)^{N_k - r_k}; \tag{2.207}$$

whence the log likelihood is

$$\log L(c, a) = C + \sum_{k=1}^{g} [r_k \log P_k + (N_k - r_k) \log(Q_k)]. \tag{2.208}$$

Differentiating with respect to the parameters and equating to zero, we have as the likelihood equations the gradient vector,

$$G(c, a) = \sum_{k=1}^{g} \frac{r_k - N_k P_k}{P_k Q_k} \cdot \frac{\partial P_k}{\partial y_k} \begin{bmatrix} 1 \\ x_k \end{bmatrix} = \begin{bmatrix} 0 \\ 0 \end{bmatrix}, \tag{2.209}$$

where $\partial P_k / \partial y_k = \phi_k$ is the normal ordinate corresponding to the area P_k.

The solution of these nonlinear equations in c and a cannot be expressed explicitly, but it can in general be obtained efficiently by the Fisher scoring method. This requires the information matrix,

$$I(c, a) = -\mathcal{E} \frac{\partial^2 \log L}{\partial c \partial a}$$

$$= -\mathcal{E}\sum_{k=1}^{g}\left(\frac{-P_kQ_k - (r_k - N_kP_k)(1 - 2P_k)}{(P_kQ_k)^2}\left(\frac{\partial P_k}{\partial yk}\right)^2\right)$$

$$+ \left[\frac{r_k - N_kP_k}{P_kQ_k} \cdot \frac{\partial^2 P_k}{\partial y_k^2}\right]\begin{bmatrix} 1 & x_k \\ x_k & x_k^2 \end{bmatrix}$$

$$= \sum_{k=1}^{g} N_k W_k \begin{bmatrix} 1 & x_k \\ x_k & x_k^2 \end{bmatrix}, \tag{2.210}$$

where $W_k = \phi^2/P_kQ_k$ is referred to in the psychometrics literature as the *Müller–Urban* weight. (See (Bock and Jones, 1968), Chapter 3, for details.)

The likelihood equations can be solved by the iterations

$$\begin{bmatrix} \hat{c} \\ \hat{a} \end{bmatrix}_{i+1} = \begin{bmatrix} \hat{c} \\ \hat{a} \end{bmatrix}_i + I_i^{-1}(\hat{c}, \hat{a})G_i(\hat{c}, \hat{a}). \tag{2.211}$$

Satisfactory starting values are $(\hat{c}, \hat{a})_0 = (0, 1)$. The iterations should be continued until the corrections effectively vanish, which requires only four or five such steps.

When the maximum has been obtained, the goodness-of-fit of the model may be tested by the likelihood ratio chi-square,

$$X_{\mathrm{LR}}^2 = -2\sum_{k=1}^{g} \log L(\hat{c}, \hat{a}). \tag{2.212}$$

On the null hypothesis, this test statistic is distributed in large samples as a central chi-square on $m - 2$ degrees of freedom.

Large-sample variances and covariances of the estimator are the elements of the inverse information matrix, namely:

$$\mathcal{V}\begin{bmatrix} \hat{c} \\ \hat{a} \end{bmatrix} = \frac{1}{\sum N_k W_k \sum N_k W_k x_k^2 - (\sum N_k W_k x_k)^2} \times \begin{bmatrix} \sum N_k W_k x_k^2 & -\sum N_k W_k x_k \\ -\sum N_k W_k x_k & \sum N_k W_k \end{bmatrix}. \tag{2.213}$$

2.16.2 Logit Analysis

R.A. Fisher and others (Berkson, 1956; Anscombe, 1956) called attention to the useful properties of the logistic response function,

$$P(u = 1) = \Psi(z) = \frac{1}{1 + e^{-z}}, \tag{2.214}$$

in maximum likelihood bioassay. With the change of scale, Dz, where $D = 1.7$, the logistic function nowhere differs from the normal distribution function by more than 0.01. Thus it can be freely substituted for the normal response function in most applications. Its advantage lies in the property,

$$\Psi = \int_{-\infty}^{z} \Psi(1 - \Psi)dz, \tag{2.215}$$

which implies that the density function can be expressed in terms of the distribution function:

$$\frac{d\Psi}{dz} = \Psi(1 - \Psi). \tag{2.216}$$

The effect of this term in the likelihood equations is to cancel the denominator, thus leading to the extremely simple gradients,

$$G(c) = \sum_{k=1}^{g}(r_k - N_k\Psi_k) = 0, \tag{2.217}$$

$$G(a) = \sum_{k=1}^{g}(r_k - N_k\Psi_k)x_k = 0, \tag{2.218}$$

where $\Psi_k = \Psi(z_k)$, and $z_k = a(x_k - b) = c + ax_k$ is a logistic deviate or *logit*.

Moreover, the second derivatives of the log likelihood do not depend on the data, so their negatives comprise the information matrix,

$$I(c, a) = \sum_{k=1}^{g} N_k W_k \begin{bmatrix} 1 & x_k \\ x_k & x_k^2 \end{bmatrix}, \tag{2.219}$$

where $W_k = \Psi_k(1 - \Psi_k)$ is now the *logistic* weight.

This means that the Fisher scoring solution of the equations is identical to the Newton–Raphson solution, and so enjoys full second-order convergence. It also means that the likelihood surface is convex and any interior maximum is unique: a finite solution of the equations is therefore the global maximum.

The test of fit is the same as in probit analysis substituting Ψ_k for P_k; standard errors and confidence intervals are the same substituting logistic weights for the Müller–Urban weights.

The inverse function

$$z = \log_e \frac{P}{1 - P} \tag{2.220}$$

shows that the logit is a "log-odds ratio."

2.16.3 Logit-Linear Analysis

Logit analysis can be expanded into a versatile system for the analysis of binomial data by introducing more complex linear models into the exponent of the response function. The investigation of qualitative response relations by this method is referred to as *logit-linear* analysis. (It differs from log-linear analysis only in modeling population proportions rather than expected frequencies. See (Goodman, 1968).)

For example, if the groups described above represented successive yearly cohorts of children, we could study trends in the probability of some occurrence, such as contracting a certain childhood disease, by substituting for the logit the degree $q-1$ polynomial model ($q < g$),

$$z_k = b_0 + b_1 x_k + b_2 x_k^2 + \cdots + b_{q-1} x_k^{(q-1)}. \tag{2.221}$$

Stepwise likelihood ratio tests of fit will aid in choosing the degree of the polynomial.

Alternatively, the g groups might represent experimental treatments and a control. Logit linear analysis can test the hypothesis of overall differences in response probabilities of the several groups and also provide estimates of contrasts between the treatments and the control. Their significance can be examined by Dunnettś (1964) procedure. For a control and two treatments, the logits are represented by a design model such as,

$$\begin{bmatrix} z_0 \\ z_1 \\ z_2 \end{bmatrix} = \begin{bmatrix} 1 & 1 & 0 & 0 \\ 1 & 0 & 1 & 0 \\ 1 & 0 & 0 & 1 \end{bmatrix} \cdot \begin{bmatrix} \mu \\ \alpha_0 \\ \alpha_1 \\ \alpha_2 \end{bmatrix}, \tag{2.222}$$

where μ is the population mean, α_0 is the effect of the control, and α_1, α_2 are the effects of the treatments.

Because only three of these four parameters are identified, it is convenient to reparameterize in terms of contrasts between the treatments and the control. The result is a full-rank model such as the following:

$$\begin{bmatrix} z_0 \\ z_1 \\ z_2 \end{bmatrix} = \begin{bmatrix} 1 & -\frac{1}{3} & -\frac{1}{3} \\ 1 & \frac{2}{3} & -\frac{1}{3} \\ 1 & -\frac{1}{3} & \frac{2}{3} \end{bmatrix} \cdot \begin{bmatrix} \mu \\ \gamma_1 \\ \gamma_2 \end{bmatrix}. \tag{2.223}$$

The same reparameterizations that put experimental design models into full-rank form for ANOVA also serve in log-linear analysis (see (Bock, 1975), Chapter 5).

For any full-rank model, the logit of group k can be expressed as product of the $q \times 1$ matrix of parameters, say β, premultiplied by the kth row of the model matrix, say X_k':

$$z_k = X_k' \beta. \tag{2.224}$$

Then the q likelihood equations for estimating β are

$$G(\beta) = \sum_{k=1}^{g} (r_k - N\Psi_k) X_k, \tag{2.225}$$

and the information matrix is

$$I(\beta) = \sum_{k=1}^{g} N_k W_k X_k X_k', \tag{2.226}$$

where W_k is the logistic weight defined above. (See Bock (1975, p. 517.)

As in logit analysis, the information matrix is in general positive-definite independent of the data. Newton–Raphson iterations, starting from zero values for all parameters, will therefore converge quickly to any interior solution point. Likelihood ratio tests of goodness-of-fit are available as above, and the inverse of the information matrix, computed numerically, provides large-sample variances and covariances of the estimator.

2.16.4 Extension of Logit-Linear Analysis to Multinomial Data

When randomly sampled respondents are assigned, according to their attributes, to one of *three or more* mutually exclusive categories, the numbers in the respective categories constitute multinomial data.[4] For m categories, the probability of the observed frequencies is given by the multinomial law discussed in Section 2.5:

$$P(r_1, r_2, \dots, r_m) = \frac{N}{r_1!, r_2!, \dots, r_m!} P_1^{r_1} P_2^{r_2} \cdots P_m^{r_m}. \tag{2.227}$$

In this formula, the frequencies are $r_1, r_2, \dots, r_m$ and their sum is N.

To relate the population proportions, $P_1, P_2, \dots, P_m$ to some independent variable or variables, we seek a model similar to those of the previous section. There are two possibilities, depending on our interpretation of the categories. On the one hand, we may consider the categories to be graded; that is, to correspond to successive intervals on some implicit continuum. A typical example is the traditional A, B, C, D, F categories for marking school work. The grader has in mind some measure of quality or merit by which students can be coarsely grouped into one of the five groups.

On the other hand, the categories may be purely nominal. They may represent qualitative attributes that have no immediately obvious ordering. There may be a sense in which the categories are ordered, but it is not known prior to the analysis of the data. An example might be flavors of ice-cream, which have no *a priori* ordering but might later be ordered according to their popularity as revealed by a sample survey.

Models for multinomial proportions have been proposed for each of these cases. Both of these models are the source of multicategory item-response functions introduced in Chapter 3.

---

4 This section is adapted from Bock (1975, Chapter 8).

2.16.4.1 Graded Categories

In 1952, Thurstone proposed an extension of the threshold concept discussed in Chapter 1 to account for frequencies of response in graded categories (see (Edwards & Thurstone, 1952)). He assumed the observed response to depend upon a latent variable falling in one of m successive intervals defined by $m-1$ thresholds on the latent continuum. He further assumed that the latent variable corresponding to an experimental condition or demographic group h is distributed normally with mean μ_h and variance σ_h^2. Then the probability of observing a response in category k is

$$\begin{aligned} P_{hk} &= \int_{\gamma_k}^{\infty} \phi_h(t)dt - \int_{\gamma_{k-1}}^{\infty} \phi_h(t)dt \\ &= \Phi\left(\frac{\mu_h - \gamma_k}{\sigma_h}\right) - \Phi\left(\frac{\mu_h - \gamma_{k-1}}{\sigma_h}\right), \end{aligned} \tag{2.228}$$

where γ_k is the upper threshold and γ_{k-1} the lower threshold of category k. The model assumes that the same categories and category thresholds apply in all conditions or groups.

Edwards and Thurstone suggested a graphical method for estimating μ_h, σ_h, and γ_k, $h = 1, 2, \dots, g$, and $k = 1, 2, \dots, m-1$. (The lower threshold of category 1 is taken as $-\infty$, and the upper threshold of category m as $+\infty$.) Bock and Jones (1968) gave a least-squares version of this method, and Bock (1975) gave the following maximum likelihood solution for the special case where σ_h is constant and the response function logistic.

In that case, setting $\sigma_h = 1$ by choice of scale on the latent dimension, we have

$$\begin{aligned} P_{hk} &= \Psi_{hk} - \Psi_{h,k-1} \\ &= \left[1 + \exp(\mu_h - \gamma_k)\right]^{-1} - \left[1 - \exp(\mu_h - \gamma_{k-1})\right]^{-1} \end{aligned} \tag{2.229}$$

with $\Psi_{ho} = 0$ and $\Psi_{hm} = 1$.

The vector of μ values, $\mu = [\mu_1, \mu_2, \dots, \mu_g]$ reflects the response propensity of members of the corresponding group. It may depend upon independent variables through the linear model

$$\mu = K\theta, \tag{2.230}$$

where K is a $g \times r$ full rank model matrix consisting of contrasts among the μś. The model may *not* contain an intercept term: the intercept is absorbed in the γ parameters.

In that case, the log likelihood is

$$\ln L = C + \sum_{h=1}^{g} \sum_{k}^{m} r_{jk} \ln P_{hk}, \tag{2.231}$$

and the derivative with respect to γ_k is

$$G(\gamma_k) = \sum_{h-1}^{g} \left(\frac{r_{hk}}{P_{hk}} - \frac{r_{h,k+1}}{P_{h,k+1}} \right) \frac{\partial P_{hk}}{\partial \gamma_k}, \tag{2.232}$$

where

$$\frac{\partial P_{hk}}{\partial \gamma_k} = \Psi_{hk}(1 - \Psi)_{hk},$$

$$\frac{\partial P_{h,k+1}}{\partial \gamma_k} = -\Psi_{hk}(1 - \Psi_{hk}),$$

and

$$\frac{\partial P_{hk}}{\partial \gamma_{k-1}} = -\Psi_{h,k-1}(1 - \Psi_{h,k-1}).$$

Similarly,

$$G(\theta) = \sum_{h=1}^{g} \sum_{k=1}^{m} \frac{r_{hk}}{P_{jk}} \frac{\partial P_{hk}}{\partial \beta}, \tag{2.233}$$

where

$$\frac{\partial P_{jk}}{\partial \theta} = - \left[\Psi_{hk}(1 - \Psi_{hk}) - \Psi_{h,k-1}(1 - \Psi_{h,k-1}) \right] K_h.$$

K_h is the hth row of K written as a column vector.

The elements of the corresponding information matrix are

$$I(\gamma_k) = \sum_{h=1}^{g} N_h \frac{\Psi_{h,k+1} - \Psi_{h,k-1}}{P_{hk} P_{h,k+1}} \Psi_{hk}^2 (1 - \Psi_{hk})^2,$$

$$I(\gamma_{k+1}, \gamma_k) = - \sum_{h=1}^{g} N_h \frac{\Psi_{h,k+1}(1 - \Psi_{h,k+1})}{P_{h,k+1}} \Psi_{hk}(1 - \Psi_{hk}),$$

$$I(\gamma_k, \gamma_\ell) = 0 \quad (|k - \ell| > 1),$$

$$I(\gamma_k, \theta) = - \sum_{h=1}^{g} N_h \left(\frac{\Psi_{hk}(1 - \Psi_{hk}) - \Psi_{h,k-1}(1 - \Psi_{h,k-1})}{P_{jh}} \right.$$
$$\left. - \frac{\Psi_{h,k+1}(1 - \Psi_{h,k+1}) - \Psi_{hk}(1 - \Psi_{hk})}{P_{h,k=1}} \right) \Psi_{hk}(1 - \Psi_{hk}) K_h,$$

$$I(\theta) = \sum_{h=1}^{g} N_h \frac{[\Psi_{hk}(1 - \Psi_{hk}) - \Psi_{h,k-1}(1 - \Psi_{h,k-1})]^2}{P_{hk}} K_h K_h'.$$

Provided $\Psi_{h,k+1} > \Psi_{hh}$, the $(k - 1 + r) \times (k - 1 + r)$ information matrix is positive definite and the solution of the likelihood equations can be obtained by the iteration

$$\begin{bmatrix} \hat{\gamma} \\ \hat{\theta} \end{bmatrix}_{i+1} = \begin{bmatrix} \hat{\gamma} \\ \hat{\theta} \end{bmatrix}_i + I_i^{-1} \begin{pmatrix} \hat{\gamma} \\ \hat{\theta} \end{pmatrix} G_i \begin{pmatrix} \hat{\gamma} \\ \theta \end{pmatrix}. \tag{2.234}$$

Satisfactory starting values are

$$\gamma_k = \ln \sum_h^g r_{hk} \Bigg/ \left(\sum_h^g N_h - \sum_h^g r_{hk} \right) \tag{2.235}$$

and $\theta = 0$.

A test of fit of the model is provided by the likelihood-ratio chi-square statistic

$$\chi^2_{\text{LR}} = \sum_h^g \sum_k^m r_{hk} \ln \frac{r_{hk}}{N_h P_{hk}}, \tag{2.236}$$

where $N_h P_{hk}$ is the expected frequency in category k of group h as calculated from the logit model and the maximum likelihood estimates of the parameters. The degrees of freedom are $(m-1)(g-1)-r$.

2.16.4.2 Nominal Categories

Some examples of responses that give rise to nominal categories data are

1. a voter casts a ballot for one candidate in a slate of several
2. an examinee marks one alternative of a multiple-choice item
3. a consumer purchases one of several competing brands of a product.

Again, we limit our consideration here to one-stage sampling: the respondent makes only one choice and is assigned to a category of the response classification accordingly. (In Chapter 3, these results are generalized to the case of multiple nominal category items.)

As a basis for a statistical model of the response process in these situations, we invoke an *extremal* concept. We assume that the stimuli correspond to vector valued random variable with a multivariate distribution in the populations from which the groups of respondents are sampled. A given respondent chooses the stimulus that has the largest component. With the multivariate distribution assumed continuous and unbounded, there is no probability that two or more components will be equal.

Suppose the random variable is $y = [y_1, y_2, \dots, y_m]$ corresponding to stimuli $X_1, X_2, \dots X_m$. Then the probability that a randomly selected subject will choose stimulus X_1, say, is

$$\begin{aligned} &\text{Prob}(y_1 > y_2 \cup y_3 \cup \cdots \cup y_m) \\ &\quad = \text{Prob}[(y_1 - y_2) \cap (y_1 - y_3) \cap \cdots \cap (y_1 - y_m) > 0] \\ &\quad = \int_0^\infty \cdots \int_0^\infty \int_0^\infty G(\delta)\, d\delta, \end{aligned} \tag{2.237}$$

where G is the distribution of the differences $\delta_{1k} = y_1 - y_k, k = 2, 3, \dots, m$. Note that the dimensionality of this distribution is one less than the number of objects. Bock and Jones (1968), Chapter 9, apply this model to consumer choices of most preferred food in sets of three. In that case, G is bivariate normal and the required orthant probabilities can be easily evaluated. For sets of five or more stimuli, however, the calculations by present methods become very heavy if a general Σ matrix is assumed.

A more tractable case is that of equal variances and equal covariance of the latent variables (and a generalization of Thurstoneś Case V assumptions for comparative judgment). With an arbitrary choice of unit, the covariance matrix of the differences becomes a correlation matrix in which all correlations are equal to $\frac{1}{2}$. Various closed-form expressions for the orthant probabilities exist in this case (see (Gupta, 1963)). In particular, a generalization of one of the bivariate extremal distributions studied by Gumbel (1961) is

$$P(y_1 > y_2 \cup y_2 \cup \cdots \cup y_m) \tag{2.238}$$

$$= \left[1 + e^{-(\mu_1-\mu_2)} + e^{-(\mu_1-\mu_3)} + \cdots + e^{-(\mu_1-\mu_m)}\right]^{-1}$$

$$= \frac{e^{\mu_1}}{e^{\mu_1} + e^{\mu_2} + \cdots + e^{\mu_m}}. \tag{2.239}$$

(This is also a generalization of the so-called Bradley-Terry-Laue model. See (Bock and Jones, 1968).)

To apply this model to the estimation of nominal category probabilities, we define the *multinomial* logit for group h as the vector,

$$z_h = [z_{h1}, z_{h2}, \ldots, z_{hm}]'. \tag{2.240}$$

Then the probability of a response in category k from a respondent randomly selected from group h is

$$P_{hk} = e^{z_{hk}}/D_h; \tag{2.241}$$

$$P_h = e^{z_{h1}} + e^{z_{h2}} + \cdots + e^{z_{hm}}. \tag{2.242}$$

To connect the model to the independent variables, we express the vector logits for the g groups as the multivariate linear model,[5]

$$Z = K\Gamma T, \tag{2.243}$$

where $Z = [z_{hk}]$ is the $g \times m$ matrix of logits,

K a is a $g \times r$ full rank basis matrix for the group effects,
Γ a is an $r \times s$ matrix of parameters to be estimated, and
T a is an $s \times m$ basis matrix for the category effects.

Notice that, because the category probabilities are invariant with respect to translation of the logits, the rows of T must take the form of contrasts. Thus, for

5 This type of model, which is due to Roy (1957), is discussed in greater detail in (Bock, 1975), Chapter 7.

$m = 3$, T might take the form,

$$T = \begin{bmatrix} 1 & -1 & 0 \\ 1 & 0 & -1 \end{bmatrix}, \tag{2.244}$$

contrasting the second and third categories with the first. Then, if K represented quadratic regression on an independent variable with levels 1, 2, 3, and 4 associated with as many groups, it might be coded as

$$K = \begin{bmatrix} 1 & -3 & 9 \\ 1 & -1 & 1 \\ 1 & 1 & 1 \\ 1 & 3 & 9 \end{bmatrix}. \tag{2.245}$$

In this case, Γ would contain the constant, linear, and quadratic coefficients describing the relationship of the independent variable to the two contrasts:

$$\Gamma = \begin{bmatrix} \gamma_{01} & \gamma_{02} \\ \gamma_{11} & \gamma_{12} \\ \gamma_{21} & \gamma_{22} \end{bmatrix}. \tag{2.246}$$

Unlike the model for graded categories, the nominal model belongs to the exponential family and maximum likelihood estimation of its parameters is entirely straightforward. The information matrix is identical to the matrix of second derivatives of the log likelihood, and any interior solution point of the likelihood equations is the global maximum, which can be found efficiently by Newton–Raphson iterations from any finite starting values. Starting the iterations with all γ values equal to zero is quite satisfactory.

The likelihood equations are most conveniently expressed by setting the observed frequencies for group h equal to the vector

$$\boldsymbol{r}_h = [r_{h1}, r_{h2}, \dots, r_{hm}]' \tag{2.247}$$

and the corresponding population proportions equal to

$$\boldsymbol{P}_h = [P_{h1}, P_{h2}, \dots, P_{hm}]'. \tag{2.248}$$

In these terms, the rs likelihood equations are (see Section 2.10)

$$\frac{\partial \log L}{\partial \operatorname{vec} \Gamma} = G(\Gamma) = \sum_h^g T(\boldsymbol{r}_h - N_h \boldsymbol{P}_h) \otimes K_h, \tag{2.249}$$

where $\otimes$ is the Kronecker product and K_h is the transpose of the hth row of K.

To express the information matrix, define

$$W_h = \begin{bmatrix} P_{h1}(1-P_{h1}) & -P_{h1}P_{h2} & \cdots & -P_{h1}P_{hm} \\ -P_{h1}P_{h1} & P_{h2}(1-P_{h2}) & \cdots & P_{h2}P_{hm} \\ \cdots & \cdots & \cdots & \\ -P_{hm}P_{h1} & -P_{hm}P_{h2} & \cdots & P_{hm}(1-P_{hm}) \end{bmatrix}, \tag{2.250}$$

so that

$$I(\Gamma) = \sum_{h}^{m} N_h T W_h T' \otimes K_h K_h'. \tag{2.251}$$

The Newton iterations are then

$$\text{vec } \hat{\Gamma}_{h+1} = \text{vec } \hat{\Gamma}_i + I_i^{-1}(\hat{\Gamma}) G_i(\hat{\Gamma}). \tag{2.252}$$

To test the fit of the model, compute the estimated cell probabilities from the logits $\hat{Z} = K\hat{\Gamma}T$ and compute the likelihood ratio chi-square statistic. The degrees of freedom are $g(m-1) - rs$.

These results provide a general methodology for the analysis of multinomial response relations and contingency tables (see (Bock, 1975; Haberman, 1979)).

2.17 Some Results from Classical Test Theory

Classical test theory develops certain implications of a simple measurement model with additive errors. Applied to a test that consists of multiple subunits, each of which can be scored separately, the model represents the score of ith respondent on the jth test unit as,

$$x_{ij} = \tau_i + \epsilon_{ij}. \tag{2.253}$$

The component τ_i, which is constant in all responses of respondent i to units $j = 1, 2, \ldots, n$, is called a *true score*; the component ϵ_{ij}, specific to respondent i and unit j, is called the *error* or the *respondent-by-test-unit interaction*. In many cases, the units are test items, but they may also be sets of items or more extended exercises.

In the statistical treatment of the model, we suppose there exists a possibly large population of test units from which those of a particular test are drawn. Similarly, we suppose an indefinitely large population of respondents from which the particular sample is drawn. To avoid confusion between the two populations, we usually refer to the former as an *universe*, or *domain*, and latter sample as a "population." The object of classical test theory is to account for the statistical properties of test scores under sampling from either or both of these populations. The minimal assumptions of the theory are as follows:

1. τ is distributed with finite mean μ and variance σ_τ^2 in the population of respondents.
2. ϵ is distributed with mean zero and finite variance σ_ϵ^2 both in the population of respondents and in the population of items.
3. τ and ϵ are uncorrelated in the population of respondents.

These assumptions imply:

$$\begin{aligned} \mathcal{E}_j(x_{ij}) &= \tau_i \quad & \mathcal{E}_i\mathcal{E}_j(x_{ij}) &= \mu, \\ \mathcal{V}_j(x_{ij}) &= \sigma_\epsilon^2 \quad & \mathcal{V}_i\mathcal{V}_j(x_{ij}) &= \sigma_\tau^2 + \sigma_\epsilon^2, \end{aligned} \tag{2.254}$$

$$\mathcal{V}(x_{ij}, x_{ik}) = \sigma(x_{ij}, x_{jk}) = \sigma_\tau^2, \quad \text{for } j \neq k. \tag{2.255}$$

Stated in words, these implications are (i) the true score of respondent i is the expected item score with respect to the sampling of items, (ii) for the same sampling, the variance of the item score is the error variance, (iii) with respect to sampling of respondents, the expected value of the item score is the mean of the population of respondents below for the slightly more general case where the expected scores for the items differ, (iv) similarly, the variance of the scores is the sum of the true score and error variance, and (v) for the sampling of respondents, the covariance of the scores for two different items is the true-score variance.

For a given respondent, the distribution of scores with respect to the sampling of items is the same as that of the error distribution (apart from the translation of the mean from zero to τ). With respect to the sampling of persons, however, the distribution of scores is the convolution of the true-score and error distributions. Its form depends, of course, upon the particular forms of both. The distribution of τ, which is called the *latent* distribution, is often of substantive interest, but can be difficult to estimate separately. An exception is the special situation where the true-score and error distributions are both normal, in which case their convolution is also normal and is fully specified by its mean of μ and variance $\sigma^2 + \sigma_\epsilon^2$. In this situation, the characteristics of the latent distribution can be inferred on classical assumptions. On IRT assumptions, latent distributions can be inferred more generally. Because the results of classical theory are confined to the first and second moments of the score and component distributions, their utility is largely limited to the special "normal–normal" case. But even without that assumption, some useful results, reviewed in this section, follow from the classical model.

Note that the classical theory, as originally formulated by Spearman, Thurstone, Gulliksen, and others, does not consider the case of repeated presentation of the same test unit to the same respondent. Although such studies are common in studies of motor performance and psychophysical research, the interference of memory and learning effects usually makes them unsuitable for psychological testing. A few authors in the psychometric literature – Guttman (1945), Haggard (1958), and Lord and Novick (1968) – have nevertheless discussed the effect of a *propensity* distribution on repeated trials within respondents. In those cases where repeated independent responses to the same test units are possible, variance attributable to sampling from all three of these populations can be separately estimated along the lines discussed in Section 2.11.2. (See also (Haggard, 1958; Lord and Novick, 1968)).

2.17.1 Test Reliability

Consider a test consisting of n units drawn from a specified domain. Let

$$X_i = x_{i1} + x_{i2} + \cdots + x_{in} \tag{2.256}$$

be the *score* of person i responding to these items. Then the variance of the scores for all persons in the population is

$$\sigma^2(X) = n^2\sigma_\tau^2 + n\sigma_\epsilon^2. \tag{2.257}$$

The proportion of this variance attributable to the true score is given by the intraclass correlation,

$$\rho_{XX} = \frac{\sigma_\tau^2}{\sigma_\tau^2 + \sigma_\epsilon^2/n}. \tag{2.258}$$

This strictly nonnegative quantity is the classical measure of test *reliability*. Its square root, called the reliability *index*,

$$\rho_{X\tau} = \sqrt{\rho_{XX}}, \tag{2.259}$$

is the correlation of the test score and the true score.

The quantity $\sigma_\epsilon/\sqrt{n}$ is called the *standard error of measurement* (SEM), and its reciprocal is called the measurement *precision*.

Dividing the numerator and denominator of the reliability formula by $n(\sigma^2 + \sigma_\epsilon^2)$ gives the so-called Spearman–Brown formula

$$\rho_{XX} = \frac{n\rho_{xx}}{1 + (n-1)\rho_{xx}}. \tag{2.260}$$

On the assumption that the unit scores satisfy the classical assumptions, this formula expresses the reliability of a sum of n test unit scores in terms of the reliabilities of the separate units

$$X = \sum_j^n x_j. \tag{2.261}$$

However, we also have

$$\sigma^2(X) = \sum_j \sigma^2(x_j) + \sum\sum_{j\neq k} \sigma(x_j, x_k), \tag{2.262}$$

and, on classical assumptions,

$$\frac{n}{n-1}\sum\sum_{j\neq k} \sigma(x_j, x_k) = n^2\sigma_\tau^2. \tag{2.263}$$

Thus,

$$\rho_{XX} = \frac{n}{n-1}\left(\sum\sum_{j\neq k} \sigma(x_j, x_k)/\sigma^2(X)\right)$$

$$= \frac{n}{n-1}\left(1 - \frac{\sum_j \sigma^2(x_j)}{\sigma^2(X)}\right). \tag{2.264}$$

With sample variances substituted for population variances, this is the so-called "coefficient α" formula for estimating reliability using the scores of a sample of respondents on each of n test units measuring the same true score with equal precision.

2.17.2 Estimating Reliability

The straightforward approach to estimating reliability is via the estimation of the variance components by which it is defined. The random effects ANOVA provides the necessary sample statistics. We will consider three cases. In the first, all the test units are assumed to have the same expected values, which is to say, the same difficulty. In the second case, these expectations are assumed to vary arbitrarily from one unit to another, and are therefore estimated along with the variance components. (In both these cases, the overall expectation for the population of test units is assumed arbitrary and must be estimated.) In the third case, repeated administration of the test units is assumed to be possible, and reliabilities are computed both with respect to sampling from the domain of test units and sampling from the propensity distribution. According to the classical model, the error variances are assumed homogeneous in all three cases.

Case 1: Equal test-unit expected values.

For the purposes of the ANOVA, we write the model for an observed score on the jth test unit as

$$y_{ij} = \mu + \alpha_i + \epsilon_{ij}. \tag{2.265}$$

The fixed parameter μ is the population mean, α is the true score distributed with mean 0 and variance σ_α^2, and ϵ is an independent error distributed with mean 0 and variance σ_ϵ^2. Table 2.5 shows the ANOVA, expressed in terms of the observations, the respondent means, $y_{i\cdot} = \sum_j^n y_{ij}/n$, and the grand mean, $y_{\cdot\cdot} = \sum_i^N \sum_j^n y_{ij}/nN$.

From minimum variance quadratic unbiased estimators of the variance components,

$$\hat{\sigma}_\epsilon^2 = \text{mse},$$
$$\hat{\sigma}_\alpha^2 = (\text{msa} - \text{mse})/n,$$

we obtain the reliability of one test-unit score,

$$\hat{\rho}_{xx} = \frac{\text{msa} - \text{mse}}{\text{msa} + \left(\frac{n-1}{n}\right)\text{mse}}, \tag{2.266}$$

Table 2.5 Analysis of variance: Case 1.

Source of variation	d.f.	Sum of squares	Mean square	Expected mean square
Mean	1	ssm $=Nny_{..}^2$		
Respondents	$N-1$	ssa $=n\sum_i^N y_{i.}$ - ssm	msa = ssa$/(N-1)$	$\sigma_\epsilon^2 + n\sigma_\alpha^2$
Residual	$N(n-1)$	sse = balance	mse = sse$/N(n-1)$	σ_ϵ^2
Total	Nn	sst $= \sum_i^N \sum_j^n y_{ij}^2$		

and the reliability of sum or mean of n test units as

$$\hat{\rho}_n = \frac{\text{msa} - \text{mse}}{\text{msa}}. \tag{2.267}$$

Both of these reliability estimators are consistent in the number of respondents.

To place a confidence interval on ρ, it is preferable to make use of Fisherś z transform,

$$z = \frac{1}{2}\log\frac{1+\hat{\rho}}{1-\hat{\rho}}, \tag{2.268}$$

which in samples of size 25 or greater can be considered normally distributed with mean $(1/2)\log[(1+\tau)/(1-\tau)]$ and variance $1/(N-3)$. An interval of $z \pm 2/\sqrt{N-3}$ therefore has approximately a 95% chance of including the population value of z. The upper and lower bounds on ρ can then be inferred from the inverse transformation

$$\rho = \frac{e^2z-1}{e^2z+1}. \tag{2.269}$$

Example In the archery data of Table 2.3, we may treat the trials as test units for the purpose of estimating the reliability of the contest as a measure of the contestants' abilities in the sport. Confining our attention to the trials at 40 yards, we obtain the sums of squares and mean squares in Table 2.6. Note that the means of the three trials are highly uniform and meet the assumption of equal test-unit expected values.

The estimated variance component for contestants is $(5700-508)/3 = 1731$ and that for error is 508. The square root of the latter, 22.5, is the SEM. The estimated reliability of one trial is

$$\frac{5700-508}{5700+2\times 508/3} = 0.86. \tag{2.270}$$

Table 2.6 Analysis of variance of the scores at 40 yards in the archery data of Table 2.3.

Source of variation	d.f.	Sum of squares	Mean squares
Mean	1	814 045.12	
Respondents		56 998.55	5 700
Residual		11 183.33	508
Total		882 233	

If the sum or mean of the three trials is used as the measure of skill, the reliability is

$$\frac{5700 - 508}{5700} = 0.91. \tag{2.271}$$

Case 2: Unequal test-unit expected values.

The classical estimator of reliability in this case is the product-moment correlation between two test units. But this statistic applies only to scores standardized in the sample, and it does not generalize to more than two units. Estimation by ANOVA does not have these limitations. The model for the observations in this case must include a component, say β_j, attributable to unit j:

$$y_{ij} = \mu + \alpha_i + \beta_j + \epsilon_{ij}. \tag{2.272}$$

For present purposes, β_j may be assumed fixed, although it could also be assumed random as in Section 2.11.2. The ANOVA for fixed β_j is shown in Table 2.7. In addition to the quantities in Table 2.7, the test-unit means, $y_{\cdot j} = \sum_i^N y_{ij}/N$, are also required.

Table 2.7 Analysis of variance: Case 2.

Source of variation	d.f.	Sum of squares	Mean square	Expected mean square
Mean	1	$\text{ssm} = Nny_{\cdot\cdot}^2$		
Respondents	$N-1$	$\text{ssa} = n\sum_i^N y_{i\cdot}^2 - \text{ssm}$	$\text{msa} = \text{ssa}/(N-1)$	$\sigma_\epsilon^2 + n\sigma_\alpha^2$
Test units	$n-1$	$\text{ssb} = N\sum_j^n y_{\cdot j}^2 - \text{ssm}$		
Residual	$(N-1)(n-1)$	$\text{sse} = \text{balance}$	$\text{mse} = \text{sse}/(N-1)(n-1)$	σ_ϵ^2
Total	Nn	$\text{sst} = \sum_i^N \sum_j^n y_{ij}^2$		

Table 2.8 Analysis of variance of the first trial at three distances.

Source of variation	d.f.	Sums of squares	Mean squares
Mean	1	75 818.94	
Contestants	10	56 498.06	5 650
Distances	2	82 066.88	
Residual	20	12 001.06	600
Total	33	908 746	

The calculation of the variance components and reliabilities are the same as in Case 1.

The formula for the reliability of the sum $X = y_1 + y_2 + \cdots + y_n$, namely,

$$\rho_{xx} = \frac{\text{msa} - \text{mse}}{\text{msa}} \tag{2.273}$$

is identically equal to that coefficient α when sample unbiased variance estimates are substituted for population variances.

Example Returning again to the data in Table 2.3, we now consider the first trial at the three distances to be the test units. The difficulty of shooting at these distances obviously differs and must be accounted for by eliminating the distance effect from the estimated error. The sums of squares and mean squares are shown in Table 2.8.

The reliability for one trial at a given distance is therefore

$$\frac{5650 - 600}{5650 + 2 \times 600/3} = 0.84. \tag{2.274}$$

For the mean or sum of scores at the three distances the reliability is

$$\frac{5650 - 600}{5650} = 0.89. \tag{2.275}$$

Case 3: Repeated administrators of the same test units.

If the nature of the test is such that the respondent cannot easily remember earlier responses to particular test units, it may be possible to assess measurement error arising from instability of the attribute in question. In that case, a three-way components of variance analysis discussed in Section 2.11.2 applies, although there may be good reason to regard either or both the test units (methods) or the repeated trials (occasions) fixed rather than random. This will be necessary if the test units are not all equally difficult, or if there are systematic differences between trials as a result of practice or fatigue. The variance components for test-unit and trial main effects will thus be estimated from the expected mean squares, but those for interaction with the random respondent way of classification will remain. Under these conditions, the expected mean squares involved in the assessment of reliability are

$$\begin{aligned}
\mathcal{E}(\text{msa}) &= \sigma_\epsilon^2 + b\sigma_{\alpha\gamma}^2 + c\sigma_{\alpha\beta}^2 + bc\sigma_\alpha^2, \\
\mathcal{E}(\text{msab}) &= \sigma_\epsilon^2 + c\sigma_{\alpha\beta}^2, \\
\mathcal{E}(\text{msac}) &= \sigma_\epsilon^2 + b\sigma_{\alpha\gamma}^2, \\
\mathcal{E}(\text{mse}) &= \sigma_\epsilon^2.
\end{aligned}$$

Four distinct reliability coefficients may be defined from these components of variance depending on whether the response is to a

i. random test unit on a random trial,
ii. random test unit on a fixed trial,
iii. fixed test unit on a random trial, or
iv. fixed test unit on a fixed trial.

The corresponding reliability coefficients are

$$\rho_{\text{i}} = \frac{\sigma_\alpha^2}{\sigma_\alpha^2 + \sigma_{\alpha\beta}^2 + \sigma_{\alpha\gamma}^2 + \sigma_\epsilon^2},$$

$$\rho_{\text{ii}} = \frac{\sigma_\alpha^2}{\sigma_\alpha^2 + \sigma_{\alpha\beta}^2 + \sigma_\epsilon^2},$$

$$\rho_{\text{iii}} = \frac{\sigma_\alpha^2}{\sigma_\alpha^2 + \sigma_{\alpha\gamma}^2 + \sigma_\epsilon^2},$$

$$\rho_{iv} = \frac{\sigma_\alpha^2}{\sigma_\alpha^2 + \sigma_\epsilon^2}.$$

To obtain the coefficients for the *sum* or *mean* of b test units, we divide $\sigma_{\alpha\beta}^2$ and σ_ϵ^2 by b; similarly for a trials, divide $\sigma_{\alpha\gamma}^2$ and σ_ϵ^2 by c; for b test units and c trials, divide $\sigma_{\alpha\beta}^2$ by b, $\sigma_{\alpha\gamma}^2$ by c, and σ_ϵ^2 by bc.

All of the above formulae assume that sample differences between units and between trials have been corrected for in the ANOVA as in Section 2.11.1. If random main effects are not eliminated, the corresponding components, σ_β^2 and σ_γ^2, must appear in the denominators of the reliability coefficients. The component σ_β^2 would be required if we wish to compute confidence intervals on, for example, the difference in true scores of two respondents administered different random forms on the same occasion, or σ_γ^2 for the same form and different occasions, or both for different forms and different occasions.

Example The archery scores in Table 2.3 are an example of physical performance data in which components of variation from several sources can be estimated. In the archery data, the distances correspond to different conditions of performance and the trials to replications on successive occasions. The various reliability coefficients defined above are readily computed from the variance component estimates in the example of Section 2.11.2.

$$\rho_i = \frac{1524}{1524 + 152 + 72 + 506} = 0.68,$$
$$\rho_{ii} = \frac{1524}{1524 + 152 + 506} = 0.70,$$
$$\rho_{iii} = \frac{1524}{1524 + 72 + 506} = 0.73,$$
$$\rho_{iv} = \frac{1524}{1524 + 506} = 0.75.$$

The reliability for the total score of each contestant at all distances and on all trials is

$$\rho = \frac{1524}{1524 + 152/3 + 7213 + 506/9} = 0.92.$$

This is the reliability relevant to the archerś overall standing in the contest.

2.17.2.1 Bayes Estimation of True Scores

Suppose a respondent drawn from an $N(\mu, \sigma_\theta^2)$ distribution has observed score X on a test with a known reliability coefficient ρ. Then, if the classical assumptions apply, the least squares or maximum likelihood estimate of the respondentś true score is X and its standard error is $\sigma_X^2(1-\rho)$. But as we have seen in the example in Section 2.13, the Bayes, or maximum mean-square error estimate, is

$$\bar{\theta}/y = \rho y + (1-\rho)\mu, \tag{2.276}$$

and its PSD is $\sigma_X^2(1-\rho)$ (Kelley, 1947). Evidently, the Bayes estimator incorporates the knowledge that the respondent was drawn from $N(\mu, \sigma_\theta^2)$, and this knowledge improves the precision of estimation by the factor σ_θ/σ_X.

Although the expected value of the sample mean of both estimates is μ, the expected value of the sample variance of the Bayes estimator is smaller than that of the least-squares estimator by the factor ρ^2. Note that neither of these sample variances estimate the true-score variance σ_θ^2.

But under the assumptions of classical test theory, the MML estimator of σ_θ^2 can be obtained using the sample variance of the Bayes estimator as a statistic. Applying the results of Section 2.15.1, we have

$$\hat{\sigma}_\theta^2 = \frac{1}{N}\left[\sum_i^N \bar{\theta}(X)^2\right] + \sigma_\theta^2(1-\rho). \tag{2.277}$$

To show that this estimator is consistent, we substituted expected values and obtain

$$\begin{aligned}\mathcal{E}(\hat{\sigma}_\theta^2) &= \rho^2\sigma_X^2 + \sigma_\theta^2(1-\rho) \\ &= \left(\frac{\sigma_\theta^2}{\sigma_\theta^2+\sigma_\epsilon^2}\right)^2(\sigma_\theta^2+\sigma_\epsilon^2) + \sigma_\theta^2\left(\frac{\sigma_\epsilon^2}{\sigma_\theta^2+\sigma_\epsilon^2}\right) \\ &= \sigma_\theta^2. \end{aligned} \tag{2.278}$$

2.17.3 When are the Assumptions of Classical Test Theory Reasonable?

In applied work, the conditions under which the assumptions of classical test theory are reasonable are rather narrow. The assumption that the error variances of the test units are homogeneous is especially problematic. But in at least one important class of applications, even this strong assumption is satisfied. For if the test units correspond to *randomly parallel* test forms, the expected values and variances of the unit scores will be equal as assumed. By "randomly parallel" is meant that the test forms are constructed by assigning to each equal numbers of items drawn randomly from a defined item universe (or in practice from a given item pool). If these forms are administered to respondents randomly selected from the population, the classical definition of test reliability and the above formulas for estimating it are justified. Because test constructors typically produce parallel forms in this way, the classical reliability estimates computed from randomly parallel forms have genuine relevance to practical testing applications.

Such is not the case, however, when the test units are binary scored items and the reliability is computed from the item scores from one administration of a single test. It is then highly improbable that the assumptions of equal test-unit expectation and variance will obtain. Test items almost always vary in difficulty in the population, and the differences imply different means and variances of the Bernoulli variables that are the item scores (see Section 2.5). Although the differences in the item means (p-values) is allowed for in the Kuder–Richardson formula 20 (KR-20) reliability, the differences in variance are not.

The best that can be said of the KR-20 reliability (or equivalent value computed from estimated variance components) is that the value obtained is a *lower-bound* on the test reliability. This result, which is due to Guttman (1945), makes use of the Cauchy–Schwartz inequality for the inner-product to two vectors, namely

$$x'y \le \sqrt{xx'yy'}. \tag{2.279}$$

Because variances are just inner-products of vectors of measurements deviated about their mean value, the Cauchy–Schwartz inequality implies that the product of the variances of two variables is greater than or equal to the square of the covariance between the variables,

$$\sigma_x^2\sigma_y^2 \geq \left(\sigma_{xy}\right)^2, \tag{2.280}$$

which in turn implies that the product-moment correlation between two variables lies in the closed interval ± 1.

Using this result, we derive Guttmanś lower bound for the case of two test units. (For the general case see Lord and Novick (1968, pp. 88–89).)

We have seen that the reliability coefficient, defined as $\rho_{XX} = \sigma_\tau^2/(\sigma_\tau^2 + \sigma_\epsilon^2)$, is the square of the correlation, $\rho_{X\tau}$, between the observed score and the true score.[6] Also, in the case of two test units with scores x_1 and x_2, the reliability coefficient of the composite score $X = x_1 + x_2$ can be expressed by the formula for coefficient α,

$$\rho_{XX} = (\rho_{X\tau})^2 = 2\left(1 - \frac{\sigma^2(x_1) + \sigma^2(x_2)}{\sigma_X^2}\right). \tag{2.281}$$

Now suppose the true scores corresponding to x_1 and x_2 are τ_1 and τ_2 with variances $\sigma^2(\tau_1)$ and $\sigma^2(\tau_2)$ not necessarily equal (that is not necessarily satisfying the classical assumptions). Then

$$\left[\sigma(\tau_1) - \sigma(\tau_2)\right]^2 \geq 0, \tag{2.282}$$

or

$$\sigma^2(\tau_1) + \sigma^2(\tau_2) \geq 2\sigma(\tau_1)\sigma(\tau_2). \tag{2.283}$$

Applying the Cauchy–Schwartz inequality, we see that the sum of the true-score variances is greater than or equal to twice the covariance of the true scores:

$$\sigma^2(\tau_1) + \sigma^2(\tau_2) \geq 2\sigma(\tau_1, \tau_2). \tag{2.284}$$

Then, since

$$\sigma_{(X)}^2 = \sigma^2(\tau_1) + \sigma^2(\tau_2) + 2\sigma(\tau_1, \tau_2) + \sigma^2(\epsilon_1) + \sigma^2(\epsilon_2) \tag{2.285}$$

and

$$\sigma^2(\tau_1) + \sigma^2(\tau_2) + 2\sigma(\tau_1, \tau_2) \geq 4\sigma(\tau_1, \tau_2), \tag{2.286}$$

the reliability of X, which is the ratio of the true-score variance component to the total variance, satisfies the inequality

$$\rho_{XX} \geq \frac{4\sigma(\tau_1, \tau_2)}{\sigma^2(X)} = 2\left[\frac{\sigma_X^2 - \sigma^2(x_1) - \sigma^2(x_2)}{\sigma^2(X)}\right]. \tag{2.287}$$

More generally, coefficient α is a lower bound of reliability when the classical assumption of equal true-score variances is not met, but it equals reliability when

6 This definition of the reliability coefficient differs from (2.258) which indexes several items.

the assumption is met. When it is computed from the variances of binary-scored items, it should be considered a lower bound. When it is computed from scores of parallel tests, it can generally be considered an equality. In item response theory, test reliability can be defined in an average sense that does not require the assumption of variance homogeneity and applies directly to scale scores based on binary item scores (see Chapter 4).

3

Unidimensional IRT Models

There we measure shadows, and we search among ghostly errors of measurement for landmarks that are scarcely more substantial.

(Source: Edward Powell Hubble)

Item response theory (IRT) deals with the statistical analysis of data in which responses of each of a number of respondents to each of a number of items or trials are assigned to defined mutually exclusive categories. Although its potential applications are much broader, IRT was developed mainly in connection with educational measurement, where the main objective is to measure individual student achievement. Prior to the introduction of IRT, the statistical treatment of achievement data was based entirely on what is now referred to as "classical" test theory. That theory is predicated on the test score (usually the student's number of correct responses to the items presented) as the observation. It assumes that the number of items is sufficiently large to justify treating the test score as if it were a continuous measurement with specified origin and unit. On the further assumption that the items are randomly sampled from a larger item domain, a classical method of estimating measurement error due to item sampling is to score random halves of the test items separately, compute the product-moment correlation of the two scores, and apply the so-called *Spearman–Brown* formula to extend the correlation to that of the full test to obtain its "split-half" reliability. The complement of the reliability coefficient is then proportion of the test variance attributable to measurement error. Various elaborations of the classical theory appear in texts such as Gulliksen (1950), Lindquist (1953), or Lord and Novick (1968).

The results of classical test theory are necessarily limited in application. They are not accurate for short tests, and they ignore sources of error such as variation due to rater effects when responses to the items must be judged subjectively. Although a variance-component extension of classical theory, called generalizability theory, treats multiple sources of error (Brennan 2001), it also assumes that the

Item Response Theory, First Edition. R. Darrell Bock and Robert D. Gibbons.

test scores are continuous measurements. The greatest limitation of the classical theory is, however, its dependence on dichotomous scoring. There is no provision for responses scored in three or more categories, even though this is common practice for performance tests or problem-solving exercises in which differing degrees of student accomplishment are recognized. Classical methods are often applied to this type of item response by arbitrary assignment of numerical values to the categories, a device difficult to justify when items are scored in different numbers of categories. Scaling procedures for assigning values to the categories that maximize the ratio of between to within respondent sums of squares are better motivated, but they collapse the data over items without regard to differences in item characteristics (see (Nishisato and Nishisato 1994), for a review).

IRT overcomes these limitations by assuming a continuous latent variable representing the student's proficiency in responding to the test items. The probability of a response in any one of two-or-more mutually exclusive categories of an item is assumed to be a function of the student's location on the latent continuum and of certain estimable parameters characteristic of the item. This approach leads to statistical methods of scoring tests, whether with large or small numbers of items, without the assumption that the items are sampled from a defined item domain to which the results generalize. Item response models that yield these probabilities are central to these methods. We discuss in Section 3.2 the main models in current use.

A further critical assumption in IRT is that item responses, given the respondent's location in the latent space, are statistically independent. Lazarsfeld made use of this principle of "conditional" independence in an analysis of contingency table data as early as 1950, but first stated it clearly in Lazarsfeld (1958). It is important in IRT because it allows straightforward calculation of the likelihood of the model parameters, given the item responses of a sample of respondents. When the number of respondents is large, it leads to efficient, likelihood-based methods for IRT analysis in a wide variety of applications, including not only measurement of individual differences, such as in education or clinical psychology, but also estimation of group effects or population quantiles without explicit calculation of respondent locations or "scores." Moreover, it makes possible new modes of test-item administration, most importantly computerized adaptive testing, in which items maximally informative for the individual respondent are selected dynamically during computer-implemented testing sessions (see (Weiss 1985)). Extended to multiple latent dimensions, conditional independence allows full-information estimation of factor loadings in the item response models directly from the item response patterns, rather than limited-information factor analysis of interitem tetrachoric or polychoric correlations. If exogenous quantitative measurements of the respondents are available, IRT can be further extended to include estimation of relationships between the latent variables and external variables. All of these applications are amenable to a likelihood-based approach in estimation and hypothesis testing. In Section 3.1, we present a general formulation of IRT that supports these applications.

3.1 The General IRT Framework

In IRT, we distinguish among observed variables (the respondent's categorical choices), explanatory variables (covariates) and latent variables (unobserved proclivities, proficiencies, or factors). IRT models can be seen as having a measurement part where latent variables together with covariates have an effect on some function of the items and a structural part where covariates affect the latent variables. If a linear structure is adopted for the structural part of the model, that implies a shift in the mean of the latent variable for different values of the covariates. An alternative to the structural part is a multiple group model in which respondents are assigned to mutually exclusive groups and parameters of the corresponding latent distributions are estimated. In some applications, group-specific item parameters may also be estimated. It should be noted that multigroup analysis requires a sufficient number of sample members in each group to estimate the model.

Let the groups be indexed in any order by $\upsilon = 1, 2, \ldots, V$; let $\mathbf{u}$ be the n-vector of categorical responses to an n-item instrument, with m_j the number of response categories of item j; let $\mathbf{x}$ the r-vector of covariate values. We write $u_j = k$ to mean that u_j belongs to the category k, $k = 1, \ldots, m_j$. The values assigned to the categories are arbitrary, but in the case of ordinal manifest variables, they must preserve the ordinality property of the variable. Finally, let $\boldsymbol{\theta}$ be the vector of latent variables with elements $\theta_q, q = 1, 2, \ldots, p$. Both the item responses and the latent variables are random variables: the latent variables are random in population of respondents; the responses are random both in the population of respondents and in the population of potential responses of a given respondent.

There are $\prod_{j=1}^{n} m_j$ possible response patterns. Let $\mathbf{u}_i = (u_1 = k_1, u_2 = k_2, \ldots, u_n = k_n)$ represent any one of these. As only $\mathbf{u}$ can be observed any inference must be based on the joint probability of $\mathbf{u}$ for group υ written as:

$$f_\upsilon(\mathbf{u} \mid \mathbf{x}) = \int_{R(\theta_1)} \cdots \int_{R(\theta_p)} f_\upsilon(\mathbf{u} \mid \boldsymbol{\theta}, \mathbf{x}) g_\upsilon(\boldsymbol{\theta}) d\boldsymbol{\theta}, \tag{3.1}$$

where $g_\upsilon(\boldsymbol{\theta})$ is the distribution of the latent variables $\boldsymbol{\theta}$ in group υ, $f_\upsilon(\mathbf{u} \mid \boldsymbol{\theta}, \mathbf{x})$ is the conditional probability of $\mathbf{u}$ given $\boldsymbol{\theta}$ and $\mathbf{x}$ in group υ and $R(\theta_q)$ is the range space of θ_q. Note that the density functions $g_\upsilon(\boldsymbol{\theta})$ and $f_\upsilon(\mathbf{u} \mid \boldsymbol{\theta}, \mathbf{x})$ are not uniquely determined; further restrictions need to be imposed in the selection of those two density functions.

Modeling manifest variables as functions of latent variables and covariates implies that association among the u's can be explained by a set of latent variables ($\boldsymbol{\theta}$) and a set of explanatory variables ($\mathbf{x}$), which when accounted for, the u's will be independent (conditional or local independence). Therefore, the number of latent variables p and the covariates x must be chosen so that

$$f_v(\mathbf{u} \mid \theta, \mathbf{x}) = \prod_{j=1}^{n} f_{vj}(u_j \mid \theta, \mathbf{x}). \tag{3.2}$$

Substituting (3.2) in (3.1):

$$f_v(\mathbf{u} \mid \mathbf{x}) = \int_{R(\theta_1)} \cdots \int_{R(\theta_p)} \prod_{j=1}^{n} f_{vj}(u_j \mid \theta, \mathbf{x}) g_v(\theta) d\theta. \tag{3.3}$$

The latent variables are assumed to be independent with normal distributions, $\theta_q \sim N(\mu_{vq}, \sigma^2_{vq}), q = 1, \ldots, p$. The normality assumption has rotational advantages in the multidimensional case. To reduce the complexity of the presentation, we limit our discussion in the estimation section to the one-dimensional case. We consider the multidimensional case in Chapter 6.

For a random sample of size N that is classified into V groups the likelihood is written as follows:

$$\prod_{i=1}^{N} f(\mathbf{u}_i \mid \mathbf{x}_i) = \prod_{v=1}^{V} \prod_{i=1}^{N_v} f_v(\mathbf{u}_i \mid \mathbf{x}_i), \tag{3.4}$$

where N_v denotes the number of respondents in group v and $f_v(\mathbf{u} \mid \mathbf{x})$ is given in (3.3).

The log-likelihood is

$$L = \sum_{i=1}^{N} \log f(\mathbf{u}_i \mid \mathbf{x}_i) = \sum_{v=1}^{V} \sum_{i=1}^{N_v} \log f_v(\mathbf{u}_i \mid \mathbf{x}_i). \tag{3.5}$$

Note that (3.5) allows each item to have different parameter values in different groups. However, a primary aim of IRT modeling is to construct items that are measurement invariant. Since measurement invariance (see e.g. (Lord 1980, Meredith 1993, Holland and Wainer 1993, Camilli and Shepard 1994)) requires that item parameters are equal across groups, we therefore constrain all item parameters (e.g. thresholds and slopes) to be equal across groups and allow only the distributions of the latent variables to differ among groups. An exception is when items are suspected of performing differently in the groups – so-called *differential item functioning* (DIF) (Holland and Wainer 1993). To investigate such effects, the constraints are selectively relaxed for certain items or parameters during parameter estimation (see Chapter 9).

3.2 Item Response Models

The following are commonly employed modes of response modeled in IRT, with typical examples of each.

Dichotomous. (i) Responses to educational achievement tests are marked "right" or "wrong."(ii) Respondents to a public opinion survey are asked to answer "yes" if they endorse a presented statement of opinion and "no" otherwise.

Ordered polytomous. (i) Responses to clinically administered items of an intelligence scale are rated by a clinician in specified categories representing increasing levels of aptness. (ii) Respondents to a physical fitness self-report inventory are asked whether they typically engage in some type of physical exercise "once a day," "once a week," "once a month," "almost never." Symptom severity measurements of mental health constructs are either rated by a clinician or the patient to obtain a dimensional severity measurement.

Nominal polytomous. (i) Students taking a multiple-choice vocabulary test are required to mark one, and only one, of a number of alternative definitions and the best describes the meaning of a given word. (ii) Respondents in a consumer preference survey are asked which of several flavors of ice cream they most prefer.

Rankings. (i) Members of an association are asked to rank nominated candidates for office in order of preference. (ii) Judges of an essay-writing contest are asked to read a certain number of essays and rank them in order of merit.

In this section, we present the best-known item response models specifying the probability of response in each category as a function of parameters attributed to items and of a latent value attributed to the respondent. The extension of these models to more than one latent variable is deferred to Chapter 6. Section 3.2 describes the models and some of their properties; their roles in item parameter estimation and respondent scoring are discussed in subsequent sections.

3.2.1 Dichotomous Categories

The most widely used models for dichotomous categories are the normal-ogive model and three forms of the logistic model referred to as one-parameter (1PL), two-parameter (2PL), and three-parameter (3PL) logistic.

3.2.1.1 Normal-Ogive Model

The normal-ogive model first appeared in connection with IRT in papers by Lawley (1943) and Lord (1952). Prior to that, the model had a long history in psychophysics going back to Fechner (1860) (who credits his colleague Mobius for the idea), also in psychological scaling by Thurstone (1927) and others, and in bioassay by Bliss (1935), Fisher and Yates (1938), and others (see (Finney 1952)). In these fields only one item at a time is studied, and the data are frequencies of response aggregated to the group level; the response probabilities are assumed to depend upon a single variable, observable in psychophysics and bioassay and latent in psychological scaling. In IRT, on the other hand, multiple items are studied jointly, the data are the responses of individuals, and the probabilities depend upon one or more latent variables. In the one-dimensional normal-ogive model, respondent i and item j are located, respectively, at points θ_i and b_j on the latent continuum. The observed response is assumed to be controlled by an unobservable random response-process variable, say,

$$Y_{ij} = a_j(\theta_i - b_j) + \epsilon_{ij}. \tag{3.6}$$

Either in the population of respondents or in replicate responses of a given respondent, ϵ_{ij} is independent normal with mean 0 and variance 1; a_j is the regression of the response process on θ.

If Y_{ij} is greater than or equal to a threshold value, γ_j, the response falls in category 2 and is assigned a score $u_{ij} = 1$; otherwise, it falls in category 1 and is assigned $u_{ij} = 0$. These so-called *item scores* may be treated as Bernoulli variables, the distribution of which may be expressed in terms of the normal distribution function, $\Phi(y_j)$ parameterized as the so-called *normit*, $y_j = a_j(\theta - b_j)$. Then the item response model is the function of θ,

$$P_j(\theta) = P(u_j = 1 \mid \theta) = \Phi(y_j), -\infty < y_j < +\infty, \tag{3.7}$$

and

$$Q_j(\theta) = P(u_j = 0 \mid \theta) = 1 - \Phi(y_j). \tag{3.8}$$

For computational purposes, and also to allow a straightforward generalization to multiple latent variables, it is convenient to set $c_j = -a_j b_j$ and work with

$$y_j = a_j\theta + c_j. \tag{3.9}$$

The distribution function of the item score can be expressed concisely in the formula

$$f(u_j \mid \theta) = P_j(\theta)^{u_j} Q_j(\theta)^{1-u_j}. \tag{3.10}$$

The plot of the response function, $P_j(\theta)$, with respect to θ is an ascending normal ogive with inflection point at $\theta = b_j$. There is a corresponding descending curve for $Q_j(\theta)$ usually plotted along with $P_j(\theta)$. The plot of the functions called *item characteristic curves* (ICCs) or *item response functions* (IRFs), are mirror images crossing at 0.5 probability, where their slopes are $a_j/\sqrt{2\pi}$ and $-a_j/\sqrt{2\pi}$, respectively (see Figure 3.1). In this context, the slope parameter a_j in the normit is called the item *discriminating power*, b_j the item *location*, and c_j the item *intercept*.

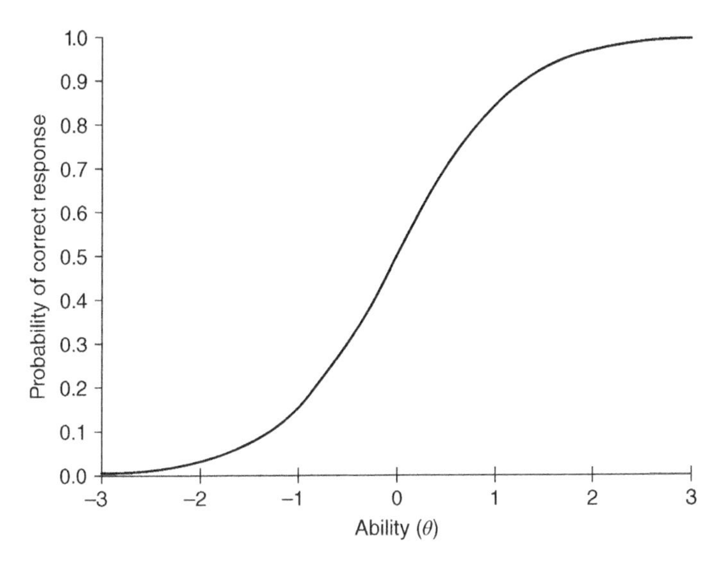

Figure 3.1 Normal ogive model.

The first derivatives of the response function with respect to the item parameters and latent variable are

$$\partial P_j(\theta)/\partial a_j = \theta\phi(y_j),$$
$$\partial P_j(\theta)/\partial c_j = \phi(y_j),$$
$$\partial P_j(\theta)/\partial \theta = a_j\phi(y_j),$$

where $\phi(y_j)$ is the normal ordinate at y_j.

In connection with maximum likelihood estimation of the respondent's θ, the Fisher information with respect to θ conveyed by the item score is the negative expected value of the second derivative of the log likelihood of θ, given u_{ij}. Assuming the responses to a test consisting of n items are conditionally independent, given θ, then for any dichotomous item response model, the first derivative of the log likelihood with respect to θ is

$$\sum_{j=1}^{n} \frac{u_j - P_j(\theta)}{P_j(\theta)Q_j(\theta)} \cdot \frac{\partial P_j(\theta)}{\partial \theta}. \tag{3.11}$$

The second derivative can be expressed as

$$\sum_{j=1}^{n} \frac{(u_j - P_j(\theta)}{P_j(\theta)Q_j(\theta)} \left\{ \frac{\partial^2 P_j(\theta)}{\partial \theta^2} - \frac{(Q_j(\theta) - P_j(\theta))}{P_j(\theta)Q_j(\theta)} \left[\frac{\partial P_j(\theta)}{\partial \theta}\right]^2 \right\}$$
$$- \sum_{j=1}^{n} \frac{1}{P_j(\theta)Q_j(\theta)} \left[\frac{\partial P_j(\theta)}{\partial \theta}\right]^2 \tag{3.12}$$

(see (Bock and Jones 1968), p. 55). Since expectation of the first term vanishes with increasing n, the *information function* of the maximum likelihood estimator of θ is $\sum_j^n I_j(\theta)$ where

$$I_j(\theta) = I(u_j \mid \theta) = \frac{a_j^2 \phi^2(y_j)}{\Phi(y_j)[1 - \Phi(y_j)]} \tag{3.13}$$

is called the *item* information function. See Lord and Novick (1968, p. 449.)

If θ is assumed to be distributed with finite mean and variance in the respondent population, the indeterminacy of location and scale in the model may be resolved by arbitrary choice of the mean and standard deviation of θ; the default choices in IRT are 0 and 1, respectively. With further assumption that θ is normally distributed, the biserial correlation of Y_j and θ in the population is

$$\rho_j = a_j \Big/ \sqrt{1 + a_j^2}. \tag{3.14}$$

This justifies referring to the a-parameter as "discriminating power." (See (Lord and Novick 1968), chapter 16, for this and following results for the normal-ogive model.)

In classical test theory, the proportion of correct responses, π_j, from respondents in a specified population serves as a measure of item *difficulty*. Although this measure increases with increasing ability and might better be called "facility," the conventional term "difficulty" is firmly established in the testing field. Assuming the normal-ogive response model and an $N(0, 1)$ distribution of θ, the difficulty of item j corresponds to the threshold point, γ_j, on the scale of Y_j, and $\pi_j = \Phi(-\gamma_j)$. Then,

$$\gamma_j = a_j b_j \Big/ \sqrt{1 + a_j^2}, \tag{3.15}$$

and $b_j = \Phi^{-1}(\pi_j)/\rho_j$ based on a sample value of π_j is a consistent estimator of the item location. It can be used as a starting value in the iterative estimation of item parameters. For descriptive purposes b_j is preferable to γ_j for characterizing item difficulty because it allows for variation in item discriminating power. In the context of IRT b_j is called "difficulty" and π_j is referred to as the *marginal difficulty* of the item.

A consistent estimator of item discriminating power is available from a one-dimensional factor analysis of the interitem tetrachoric correlation matrix. From the sample factor loading α_j, the slope parameter is obtained from

$$a_j = \alpha_j \Big/ \sqrt{1 - \alpha_j^2}. \tag{3.16}$$

Computationally, however, this method of obtaining a starting value for iterative estimation of a_j is awkward: a value based on the biserial correlation between the item score and the number-right test score is almost always used in its place.

For numerical work with the normal-ogive model, values of the distribution function can be calculated with accuracy 10^{-6} by Hastings (1955) five-term formula or accuracy 10^{-11} by Cooper (1968) algorithm.

The inverse function, $y = \Phi^{-1}(\pi)$, is difficult to approximate accurately with a computing formula; it is better obtained by Newton–Raphson iterations

$$y_{t+1} = y_t - [\Phi(y_t) - \pi]/\phi(y_t), \tag{3.17}$$

with values from the $\Phi(y)$ approximation. See also Kennedy and Gentle (1980).

3.2.1.2 2PL Model

In three US Air Force technical reports, Birnbaum (1957, 1958a,b) obtained numerous results for respondent scoring and classification in which the logistic distribution replaced the normal.[1] Birnbaum's reports were not widely known until (Lord and Novick 1968) included updated versions in their chapters 17–19. We describe some of the results here.

Birnbaum's two-parameter model is

$$P_j(\theta) = \Psi(z_j) = \frac{e^{z_j}}{1 + e^{z_j}}, -\infty < z_j < +\infty, \tag{3.18}$$

and

$$Q_j(\theta) = 1 - \Psi(z_j) = \frac{1}{1 + e^{z_j}}. \tag{3.19}$$

The logit, $z_j = a_j(\theta - b_j) = a_j\theta + c_j$, has the same structure as the normit.

The ICC of the 2PL and normal ogive models can be made almost coincident by simple rescaling of the logit (see Figures 3.2 and 3.3). Although the standard deviation of the logistic distribution, $\pi/\sqrt{3}$ (see (Gumbel 1961)) would suggest 1.81 as the scale factor, 1.7 gives slightly better fit overall and is standard in IRT. The absolute value of $\Phi(y) - \Psi(1.7y)$ is less than 0.01 anywhere in the range (Haley 1952). Most of the above results for the normal-ogive model apply closely to the 2PL model: for example, $\Psi^{-1}(\pi_j)/1.7\rho_j$ is nearly a consistent estimator of b_j.

Unlike the normal distribution function, the logistic function has a simple inverse,

$$z_j = \log\{\Psi(z_j)/[1 - \Psi(z_j)]\}, \tag{3.20}$$

the "log odds ratio." But more important for IRT is the fact that the derivative of the logistic response function can be expressed in terms of the function values:

$$\partial\Psi(z_j)/\partial z = \psi(z_j) = P_j(\theta)Q_j(\theta). \tag{3.21}$$

1 The logistic distribution was first introduced into bioassay by Fisher and Yates (1938), who showed its relationship to his z-transformation of the correlation coefficient.

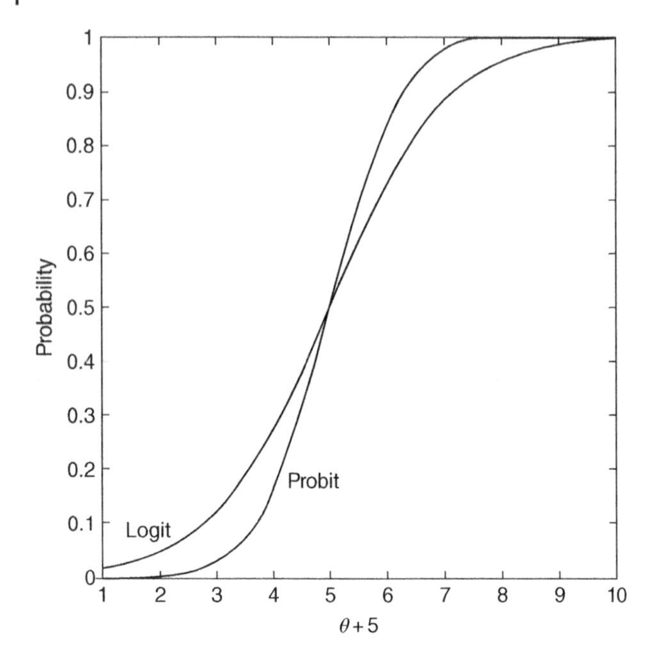

Figure 3.2 Normal (probit) versus logistic (logit) item characteristic curves.

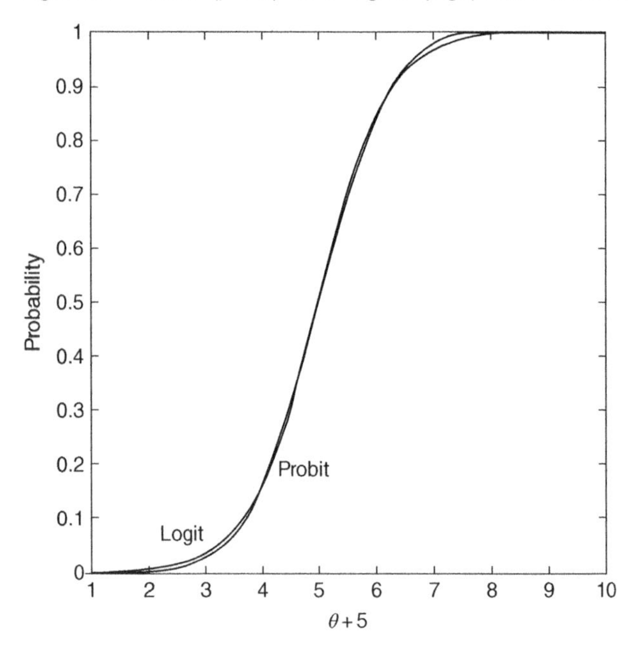

Figure 3.3 Normal (probit) versus logistic (logit) item characteristic curves – rescaled.

Thus, the slope at $\theta - b_j$ is $a_j/4$, and

$$\partial P_j(\theta)/\partial a_j = \theta\psi(z_j),$$
$$\partial P_j(\theta)/\partial c_j = \psi(z_j),$$
$$\partial P_j(\theta)/\partial \theta = a_j\psi(z_j).$$

For a test consisting of n items the likelihood equation for θ is

$$\sum_{j=1}^{n} a_j \left[\frac{u_{ij} - P_j(\theta)}{P_j(\theta)Q_j(\theta)}\right] P_j(\theta)Q_j(\theta) = \sum_{j=1}^{n} a_j[u_{ij} - \Psi_j(\theta)] = 0, \tag{3.22}$$

the solution of which is the scale score θ_i for respondent i. This shows that $t_i = \sum_{j=1}^{n} a_j u_j$ is the sufficient statistic for θ_i.

In applications such as personnel selection and qualification, reporting educational achievement test results in population percentiles, setting scholastic proficiency standards, etc., where only the ranking or classification of respondents is at issue, t_i is a straightforward alternative to θ_i. It has the same efficiency as θ_i, is easy to compute, and is simple to explain to respondents by displaying the item weights.

The item information function is

$$I_j(\theta) = \frac{[\partial P_j(\theta)/\partial\theta]^2}{P_j(\theta)Q_j(\theta)} = a_j^2\psi(z_j). \tag{3.23}$$

That these results depend upon the canceling of the quantity $P_j(\theta)Q_j(\theta)$ in the denominator by $\psi(z_j)$ in the numerator, shows this sufficiency property to be unique to logistic dichotomous item response models. A corresponding result holds for polytomous logistic models (see Section 3.2.2).

3.2.1.3 3PL Model

To allow straightforward machine scoring of item responses, large-scale testing services make use of multiple-choice questions, most commonly with four alternatives. Usually, they are administered with instructions to the examinees not to omit items. If in spite of these instructions some examinees fail to mark all items, the omitted items may be implicitly assigned a random response during test scoring. Examinees who do not know the answer to an item and do not omit will either respond blindly or choose indifferently among remaining alternatives after eliminating one or more that they know to be incorrect. Under these conditions of administration, the IRF cannot approach zero at the lowest levels of ability. Interestingly, in toxicological bioassay, a similar problem arises due to "natural mortality of the controls." The correction of the response function used in that field can also be applied to multiple-choice items except that, absent a control group, the probability of correct response on the part of examinees at the lowest extreme of ability must be estimated from the test data. This correction leads to the three-parameter logistic model of Birnbaum (see (Lord and Novick 1968), p. 404):

$$P_j(\theta) = \Psi(z_j) + g_j[1 - \Psi(z_j)]$$
$$= g_j + (1 - g_j)\Psi(z_j), \tag{3.24}$$
$$Q_j(\theta) = (1 - g_j)[1 - \Psi(z_j)]. \tag{3.25}$$

Examinees of ability θ respond correctly with probability $\Psi_j(\theta)$; those remaining have probability g_j of responding correctly by marking blindly or with partial information. The parameter g_j is the lower asymptote of $P_j(\theta)$ and $1 - g_j$, that of $Q_j(\theta)$. Raising the lower asymptote to g_j moves the inflection point of the ICC to half-way between g_j and 1, where probability is $(g_j + 1)/2$ and the slope is $a_j(1 - g_j)/4$, see Figure 3.4.

The derivatives of the category 2 response function are

$$\partial P_j(\theta)/\partial a_j = (1 - g_j)\theta\psi(z_j),$$
$$\partial P_j(\theta)/\partial c_j = (1 - g_j)\psi(z_j),$$
$$\partial P_j(\theta)/\partial\theta = a_j(1 - g_j)\psi(z_j).$$

The likelihood equation for θ is

$$\sum_{j=1}^{n} a_j\psi(z_j)\frac{u_{ij} - g_j - (1 - g_j)\Psi(z_j)}{g_j[1 - \Psi(z_j)] + (1 - g_j)\Psi(z_j)[1 - \Psi(z_j)]}. \tag{3.26}$$

The item information function is

$$I_j(\theta) = \frac{a_j^2(1 - g_j)\psi^2(z_j)}{g_j[1 - \Psi(z_j)] + (1 - g_j)\Psi(z_j)[1 - \Psi(z_j)]}. \tag{3.27}$$

Note that the derivative with respect to θ in the numerator does not cancel $P_j(\theta)Q_j(\theta)$ in the denominator as it did in the 2PL model: the 3PL model is not logistic and does not have a sufficient statistic for θ_i.

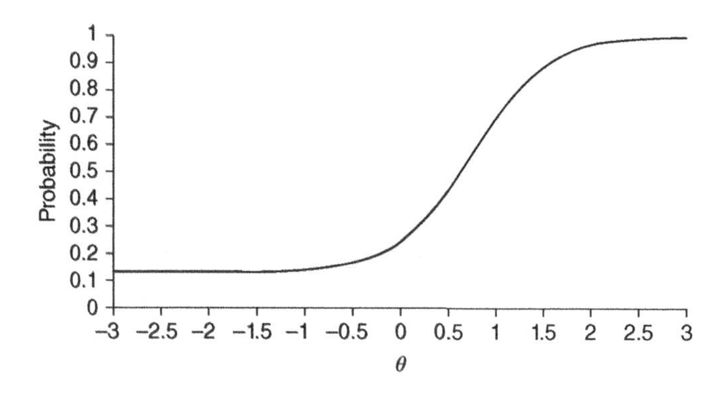

Figure 3.4 Three-parameter logistic model item characteristic curve.

The normal-ogive and the following 1PL model can be modified similarly to account for the effects of guessing.

The g-parameters are difficult to estimate for easy items that produce very few wrong responses and little if any guessing. Values for parameters of such items must be obtained by conditioned estimation. Zimowski et al. (1996) employ a stochastic constraint based on a beta distribution with an assigned weight of observations to determine the influence of the constraint. Thissen (1991) estimates the logit of the guessing parameter and uses a normal stochastic constraint with an assigned variance to determine the influence. (See (DuToit 2003), in this connection.)

3.2.1.4 1PL Model

Rasch (1960) introduced a version of the logistic model in which the logit is defined as $z_j = \theta - b_j$ and $b_j, j = 1, 2, \ldots, n$, is restricted to sum to zero. This way of scaling θ means that different tests will have different scales, but in practice, the scores are adjusted during test development to have the same means and standard deviations in a norming sample from the population of respondents. In applying the model, the test developer must choose from the item pool those items that have the same or nearly the same slope parameters (see Figure 3.5). This requires a

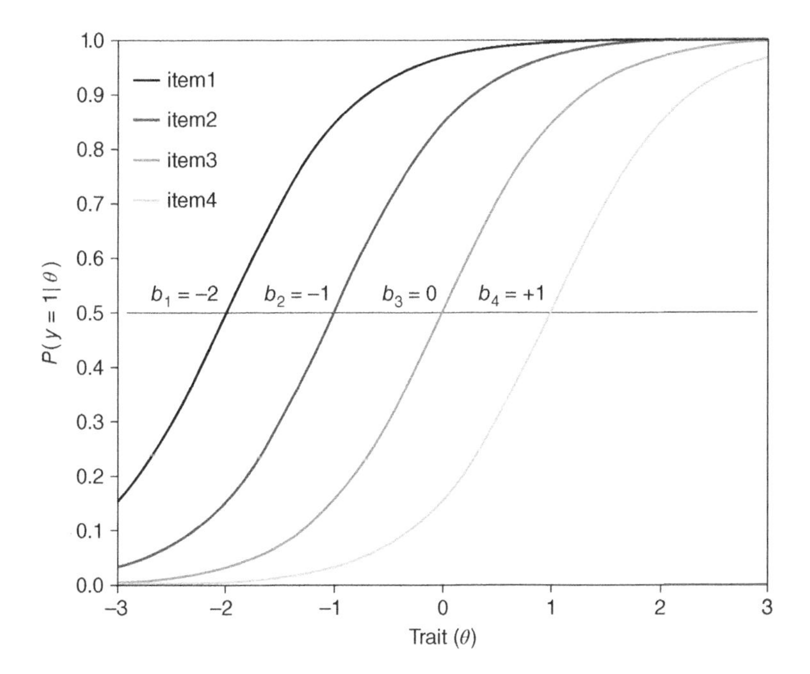

Figure 3.5 Item characteristic curves for the Rasch model for four items.

preliminary analysis of the item pool using the normal ogive or the 2PL model, or inspection of classical item statistics such as biserial correlation of the item-by-test number-right score.

When $a = 1$, the sufficient statistic for θ is the number-right score $t_i = \sum_{j=1}^{n} u_{ij}$, and the likelihood equation becomes

$$t_i - \sum_{j=1}^{n} \Psi(z_j) = 0. \tag{3.28}$$

The item information is equal for all items, and the maximum likelihood estimate of θ is a one-to-one transform of t. This greatly facilitates computation of respondent scale scores, which can be obtained by table lookup. Unlike the scale-scores based on the other models, those of 1PL cannot produce discrepancies between the rankings of respondents' scale scores and their number-right scores. This feature is sometimes required in achievement-testing programs. Note that in applications that require only ranking of respondents, the model adds no information to the number-right scores.

3.2.1.5 Illustration

To illustrate the 1PL, 2PL, and 3PL models, we reanalyze the LSAT-6 example originally described by Bock and Lieberman (1970). We present the item intercept and slope parameters for the parameterization $z(\theta_i) = c_j + a(j)\theta_i$, where $c_j = -a_j b_j$. Note that these estimates are not transformed so that the intercepts sum to zero and the product of the slopes is equal to 1 as in Section 11.5.4 and in Table 3 of Bock and Aitkin (1981) (Table 3.1).

For these data, the 1PL minimizes Bayesian Information Criterion (BIC, (Schwarz, Gideon 1978)) and is therefore selected as the most parsimonious fitting model for these data. The 1PL and 2PL models produce quite similar parameter estimates, with only small deviations from the common slope parameter for the 2PL model. Adding the guessing parameters for the 3PL model produces changes in the intercept and slope parameters. The differences in the chi-square statistics are distributed χ^2 with degrees of freedom equal to the difference in degrees of freedom between the two models being compared. None of these are statistically significant, with all models having similar fit to the data, albeit with increasing complexity in terms of number of parameters. The third item is the most difficult item, and the first and fifth items are the easiest.

Figure 3.6 displays the ICC and item information function for LSAT-6 item 3 for the 1PL model. The threshold for this item is $b_j = -0.32$ which corresponds to the peak of the item information function and the point on the ICC at which there is a 50% chance of a positive response. Figure 3.7 displays the ICC and item information function for LSAT-6 item 3 for the 3PL model (the 2PL and 1PL models are virtually identical for item 3). Due to the inclusion of the guessing parameter,

Table 3.1 LSAT-6 data 1PL, 2PL, and 3PL model parameter estimates intercepts (c_j), slopes (a_j) and guessing g_j parameters.

Model	Item	c_j	a_j	g_j	BIC	χ^2	df	p
1PL	1	2.73	0.76		4975.32	18.35	25	0.83
	2	1.00	0.76					
	3	0.24	0.76					
	4	1.31	0.76					
	5	2.10	0.76					
2PL	1	2.77	0.83		5002.38	18.14	21	0.64
	2	0.99	0.72					
	3	0.25	0.89					
	4	1.28	0.69					
	5	2.05	0.66					
3PL	1	2.50	0.79	0.20	5037.37	18.27	16	0.31
	2	0.66	0.82	0.19				
	3	−0.21	1.10	0.18				
	4	0.98	0.75	0.20				
	5	1.79	0.68	0.20				

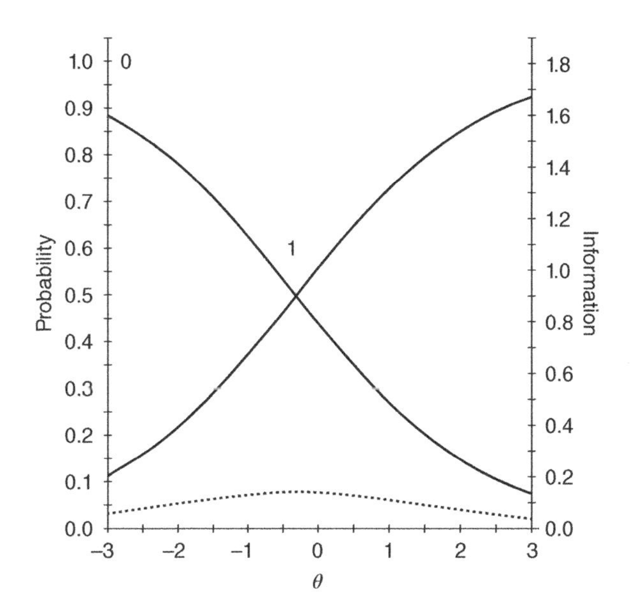

Figure 3.6 Item characteristic curve and information for 1PL model LSAT6 item 3.

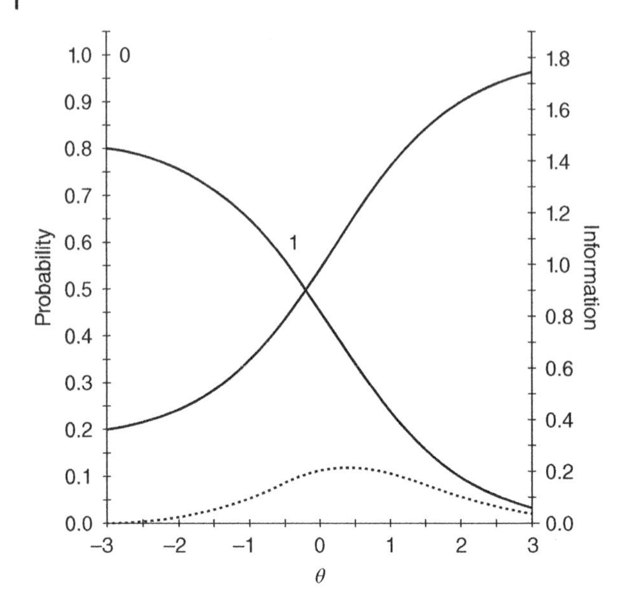

Figure 3.7 Item characteristic curve and information for 3PL model LSAT6 item 3.

the intercept has moved up from a probability of 0.1 to a probability of 0.2. The threshold is also shifted to 0.19 and the item information function has its mode at the threshold and has negative (left) skewness.

3.2.2 Polytomous Categories

The most widely applied models for polytomous item responses are (i) the *graded* models generalized from either the normal-ogive model or the 2PL model, (ii) the *partial credit* models generalized from the 1PL or 2PL models, (iii) the *nominal* models based on a logistic model for first choices, and (iv) the *rating scale* models as special cases of (iii).

Models for polytomous items can be expressed in the same manner as those for dichotomous items by the following formal device.

For an item with m_j categories, let $\mathbf{u}_j$ be an m_j-variate vector such that, if the respondent chooses category k of item j, the element $u_{jhk} = 1$ when $h = k$ and 0 when $h \neq k$. The vector is then a (multinomial) Bernoulli variate serving as the item score of the respondent, and a generic polytomous response model can be expressed as

$$P_{jk}(\theta) = P(\mathbf{u}_j \mid \theta) = \prod_{h=1}^{m_j} P_{jh}^{u_{jhk}}(\theta), \tag{3.29}$$

on condition that

$$\sum_{h=1}^{m_j} P_{jh}(\theta) = 1. \quad (3.30)$$

This device is merely for consistency of notation; in recording data and in computation, only the item scores $k = 1, 2, \ldots, m_j$ are used.

3.2.2.1 Graded Categories Model

Samejima (1969) formulated an item response model for graded item scores as a generalization of the normal-ogive model or the 2PL model. In place of a single threshold of the response process, Y_j, has $m - 1$ thresholds, γ_{jk}, dividing the range of the distribution into successive intervals corresponding to ordered categories. In early work of Thurstone (1929), the same concept appeared in a graphical procedure for locating stimulus medians and category boundaries on a latent variable measuring attitude. Bock and Jones (1968) gave a minimum normit chi-square method for estimating these quantities.

In Samejima's model assuming a normally distributed response process, the normit for category k of item j is

$$y_{jk} = a_j(\theta - b_{jk}), \quad k = 1, 2, \ldots, m - 1, \quad (3.31)$$

where b_{jk} is the location parameter for the upper boundary of category k. Letting $\gamma_{j0} = -\infty$ and $\gamma_{jm} = +\infty$, the response probabilities given θ are

$$P_{jk} = P(\gamma_{j,k-1} \le Y_j < \gamma_{jk} \mid \theta) = \begin{cases} \Phi(y_{jk}) - \Phi(y_{j,k-1}) & k = 1, 2, \ldots, m_j - 1 \\ 1 - \Phi_{j,m_j-1}, & k = m_j, \end{cases} \quad (3.32)$$

where $\Phi(y_{j,0}) = 0$. The b-parameters divide the area under the unit normal curve into proportions $\pi_1, \pi_2, \ldots, \pi_{m_j}$. This means that neighboring categories can be collapsed in the model merely by ignoring the boundary between them.

In practical work, it proves helpful to express the normit in the form

$$y_{jk} = a_j(\theta - b_j + d_{jk}), \quad \sum_{k=1}^{m_j-1} d_{jk} = 0. \quad (3.33)$$

This makes b_j an indicator of the overall location of the item on the latent dimension and expresses the category locations as deviations d_{jk}. For computational purposes, however, the parameterization

$$y_{jk} = a_j\theta + c_{jk} \quad (3.34)$$

with $c_{jk} = -a_j(b_j - d_{jk})$ is more convenient.

In terms of the above Bernoulli variate, the distribution of $\mathbf{u}_j$ is

$$f(\mathbf{u}_j \mid \theta) = \prod_{h=1}^{m_j} P_{jh}^{u_{jhk}}(\theta). \tag{3.35}$$

The graded model has m_j ICCs consisting of:

a descending ogive centered at b_{j1} for $k = 1$;
a single-mode symmetric curve going to zero at $-\infty$ and $+\infty$, with median midway between $b_{j,k-1}$ and b_{jk} for $k = 2$ through $m - 2$; and
an ascending ogive centered at $b_{j,m-1}$ for $k = m$, see Figure 3.8.

The derivatives of the corresponding curves are

$$\begin{aligned}
\partial P_{jk}(\theta)/\partial a_j &= -\theta[\phi(y_{jk}) - \phi(y_{j,k-1})], \\
\partial P_{jk}(\theta)/\partial c_{jk} &= -\phi(y_{jk}) - \phi(y_{j,k-1}), \\
\partial P_{jk}(\theta)/\partial \theta &= -a_j[\phi(y_{jk}) - \phi(y_{j,k-1})],
\end{aligned}$$

where $\phi(y_{j,0}) = 0$ and $\phi(y_{j,m_j}) = 0$.

All category response functions have the same discriminating power and the intervals marked out by the category boundaries determine the partition of the response probability among the categories at a given θ.

The item information function is

$$I_j(\theta) = a_j^2 \sum_{k=1}^{m_j} \frac{[\phi(y_{jk}) - \phi(y_{j,k-1})]^2}{\Phi(y_{jk}) - \Phi(y_{j,k-1})}. \tag{3.36}$$

However, the information conveyed by response in a given category depends upon the expected frequency of use of the category by respondents of ability θ.

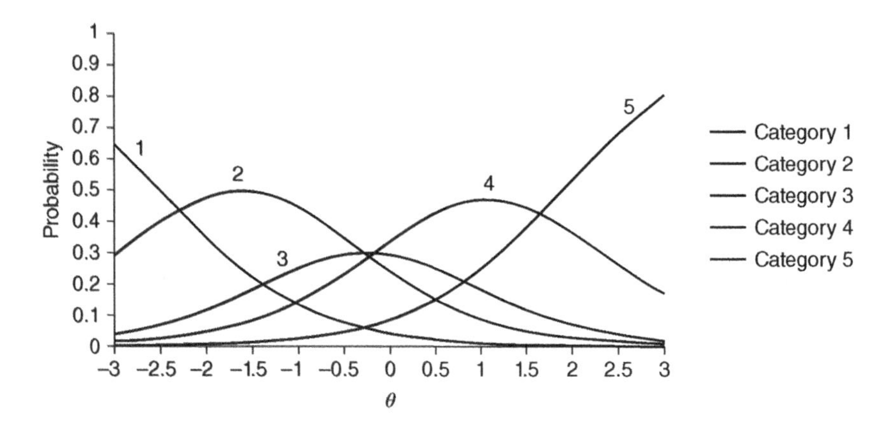

Figure 3.8 Item characteristic curves for the logistic graded category model.

The overall information can therefore be partitioned as

$$I_{jk}(\theta) = I_j(\theta)P_{jk}(\theta). \tag{3.37}$$

These results apply *mutatis mutandis* to the logistic version of the graded model. But the derivative of the category response function with respect to θ does not cancel the denominator in the likelihood equation, and there is not a sufficient statistic for θ.

3.2.2.2 Illustration

We illustrate the graded model using 6 Likert-type response items from the 20 item State-Trait Anxiety Inventory (STAI, Spielberger 1983). The items are the following:

1. I feel calm.
2. I am tense.
3. I am regretful.
4. I feel at ease.
5. I feel anxious.
6. I feel nervous.

The response categories are the following:

1. not at all
2. very little
3. somewhat
4. moderately
5. very much

Items 1 and 4 are reverse scored so that all items are scored in the same directions of increasing psychopathology. Table 3.2 displays the parameter estimates (intercepts and slopes) for the graded response model. Table 3.3 displays factor loadings and thresholds for the graded response model.

Factor loadings reveal generally strong loadings on the latent anxiety dimension, with the anxiousness and being regretful being the most severe items and tension and not being at ease the mildest symptoms. Figures 3.9 and 3.10 display ICC and item information functions for the least severe item, tension, and the most severe item, anxious. There is a clear shift of each of the category curves to the right (more severe) between the tension and anxious items as reflected in the item parameters. The points of intersection between the category curves correspond to the b_{jk} in Table 3.3. BIC for this model is 7710.14.

3.2.2.3 The Nominal Categories Model

When there is no *a priori* order of the response categories, an obvious approach to modeling the choice of one alternative out m_j would be to generalize the

Table 3.2 Six item STAI anxiety scale data graded model parameter estimates intercepts (c_{jk}) and slopes (a_j).

Item	a_j	c_{j1}	c_{j2}	c_{j3}	c_{j4}
Not calm	2.29	2.17	−0.80	−3.57	−6.31
Tense	2.26	2.34	−0.46	−2.60	−5.37
Regretful	1.33	1.03	−0.85	−2.22	−4.08
Not at ease	2.42	2.89	−0.32	−3.19	−6.63
Anxious	1.80	1.31	−1.13	−3.07	−6.60
Nervous	1.71	1.46	−0.61	−2.20	−4.28

Table 3.3 Six item STAI anxiety scale data graded model parameter estimates thresholds (b_{jk}) and factor loadings (λ_j).

Item	λ_j	b_{j1}	b_{j2}	b_{j3}	b_{j4}
Not calm	0.80	−0.95	0.35	1.56	2.76
Tense	0.80	−1.04	0.20	1.15	2.38
Regretful	0.62	−0.77	0.64	1.67	3.08
Not at ease	0.82	−1.20	0.13	1.32	2.75
Anxious	0.73	−0.73	0.63	1.70	3.67
Nervous	0.71	−0.85	0.36	1.28	2.50

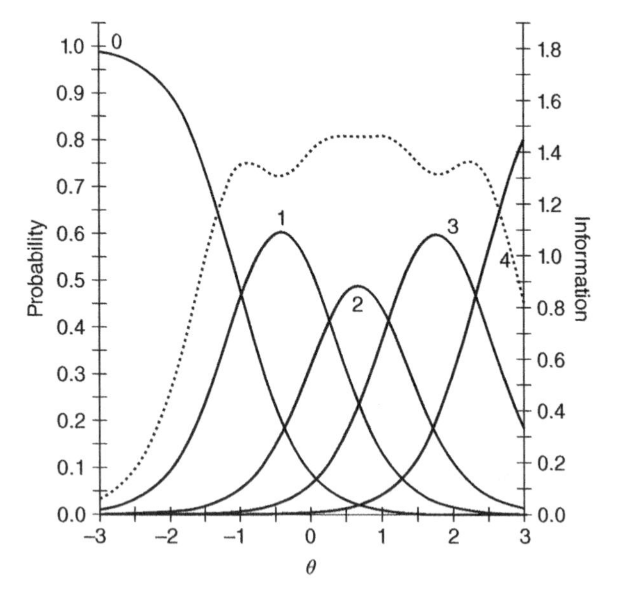

Figure 3.9 Item characteristic curve and information for the graded model – tension item.

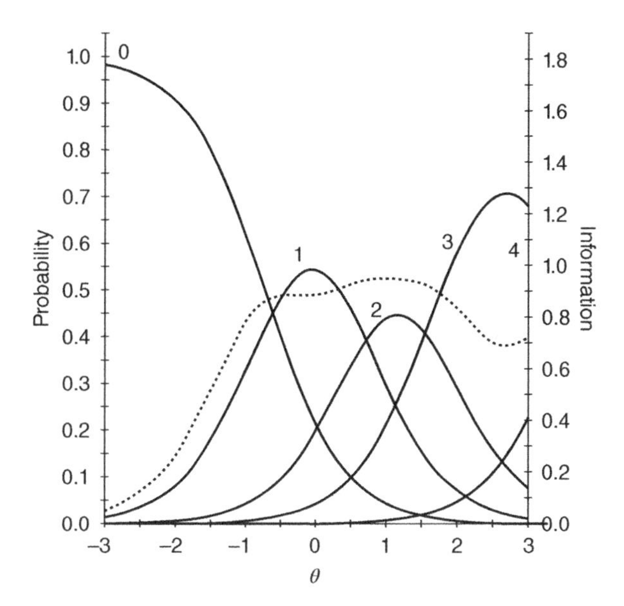

Figure 3.10 Item characteristic curve and information for the graded model – anxious item.

response process distribution of the normal-ogive model to a multivariate normal distribution. Suppose that associated with the response categories there are processes $Y_h, h = 1, 2, \ldots, m_j$ such that the respondent will choose alternative k if $Y_k > Y_h, h \neq k$ (or equivalently $Y_k - Y_h > 0$). Then if the distribution of Y_k and Y_h is, say, $N(\mu_k, \mu_h, \sigma_k, \sigma_h, \rho_{kh})$, the distribution of $Y_k - Y_h$ is normal with, say, mean $\mu_{kh} = \mu_h - \mu_k$ and variance $\sigma_{kh}^2 = \sigma_k^2 + \sigma_h^2 - 2\sigma_k\sigma_h\rho_{kh}$. In terms of the process differences, the problem of evaluating the probability of first choice of one m_j alternatives reduces to evaluating the positive orthant of an $(m_j - 1)$-variate normal distribution.

Consider the case of three alternatives and corresponding Y_h, Y_j, Y_k: the covariance of $Y_k - Y_h$ and $Y_k \geq Y_j$ is then

$$\sigma_{kh,kj} = \sigma_k^2 - \sigma_k\sigma_h\rho_{kh} - \sigma_k\sigma_j\rho_{kj} - \sigma_h\sigma_j\rho_{hj}, \tag{3.38}$$

and the probability of a first choice of alternative k is

$$\begin{aligned} P(Y_k > Y_h \cup Y_k > Y_j) &= \int_0^\infty \int_0^\infty \phi(\mu_{kh}, \mu_{kj}, \sigma_{kh}, \sigma_{kj}, \rho_{kh,kj}) dv\, dw \\ &= \int_{-\mu_{kh}/\sigma_{kh}}^\infty \int_{-\mu_{kj}/\sigma_{kj}}^\infty \phi(0, 0, 1, 1, \rho_{kh,kj}) dx\, dy. \end{aligned} \tag{3.39}$$

If the separate response processes are uncorrelated, the correlation of the differences equals 1/2. In that special case, the above $(m_j - 1)$-dimensional integrals are not difficult to evaluate numerically (Stuart 1958), but there is another approach, based on a generalization of the two-category logistic model, that is more attractive for computation and for other reasons.

The two-category model may be thought of arising from a primitive choice model (Bradley and Terry 1952),

$$\frac{\pi_1}{\pi_1 + \pi_2}$$

for positive π_1 and π_2, giving the probability that alternative 1 will be chosen from alternatives 1 and 2. This suggests as a model for the choice of alternative k from among m_j (Bock 1972)

$$\frac{\pi_k}{\pi_1 + \pi_2 + \cdots + \pi_{m_j}}.$$

To avoid inequality constraints in estimation, let $\pi_h = e^{z_h},\ h = 1, 2, \ldots, m_j$ so that

$$\Psi(z_k) = \frac{e^{z_k}}{\sum_{\ell=1}^{m_j} e^{z_\ell}}. \tag{3.40}$$

McFadden (1973) derived the multinomial logit model by assuming independent error terms.

Gumbel (1961) investigated in some detail the properties of the bivariate logistic distribution expressed as,

$$\Psi(z_1, z_2) = (1 + e^{-z_1} + e^{-z_2})^{-1}.$$

In particular, he showed that the marginal distributions are logistic with mean 0 and variance $\pi/3$, and the joint density function is

$$\psi(z_1, z_2) = 2\Psi^3(z_1, z_2)e^{-z_1-z_2}. \tag{3.41}$$

The expectation of $z_1 z_2$ is $\pi^2/6$, and the correlation is

$$\rho = \frac{\pi^2}{6} \cdot \frac{3}{\pi^2} = 1/2.$$

The density is asymmetric and has curved regression lines that intersect at (0.5413…, 0.5413…) rather than at (0,0).

Nevertheless, the volume of the positive quadrant is in good approximation to that of the bivariate normal distribution with correlation 1/2. Bock and Jones (1968) applied both models to prediction of first choices in a food preference study conducted earlier by Thurstone (1959). In choice of three dishes in four different menus, the average absolute difference between the normal and logistic probabilities was 0.025.

In terms of the logit $z_{jk} = a_{jk}\theta + c_{jk}$, the probability of response in category k of item j is (Bock 1972, 1997)

$$P_{jk}(\theta) = \Psi(z_{jk}) = \frac{e^{z_{jk}}}{\sum_{\ell=1}^{m_j} e^{z_{j\ell}}}. \tag{3.42}$$

Because the probabilities are invariant with respect to translation of the logit, the category parameters are subject to the constraints,

$$\sum_{k=1}^{m_j} a_{jk} = 0, \quad \sum_{k=1}^{m_j} c_{jk} = 0. \tag{3.43}$$

Alternatively, the parameters may be expressed as functions of $m_j - 1$ linearly independent contrasts, or a value of a_{jk} and c_{jk} may be fixed at zero.

The distribution function of the vector-Bernoulli variate can be expressed as

$$f(\mathbf{u}_j \mid \theta) = \prod_{k=1}^{m_j} \Psi^{u_{jk}}. \tag{3.44}$$

To obtain the derivatives for category response function h, let D equal the denominator of (3.42):

$$\begin{aligned}
\frac{\partial P_{jh}(\theta)}{\partial a_{jk}} &= \theta \frac{e^{z_{jh}} D - e^{z_{jh}}}{D^2} = \theta \Psi(z_{jh})[1 - \Psi(z_{jh})], \quad h = k \\
&= \frac{(e^{z_{jh}})^2}{D^2} = -\theta \Psi^2(z_{jh}), \quad h \neq k \\
\frac{\partial P_{jh}(\theta)}{\partial c_{jk}} &= \Psi(z_{jh})[1 - \Psi(z_{jh})], \quad h = k \\
&= -\Psi^2(z_{jh}), \quad h \neq k \\
\frac{\partial P_{jh}(\theta)}{\partial \theta} &= \frac{a_{jh} e^{z_{jh}} D - e^{z_{jh}} \sum_{\ell=1}^{m_j} a_{j\ell} e^{z_{j\ell}}}{D^2} \\
&= \Psi(z_{jh})[a_{jh} - \sum_{\ell=1}^{m_j} a_{j\ell} \Psi(z_{j\ell})].
\end{aligned}$$

To examine the ICC for category k, it is useful to number in the categories in the order of increasing size of the a-parameters and express the response function as

$$P_{jk}(\theta) = \left[1 + \sum_{\ell=1}^{m_j} e^{z_{j\ell} - z_{jk}} \right]^{-1}. \tag{3.45}$$

The derivative with respect to θ reveals the shapes of the category characteristic curves. Consider the case of three categories. In category 1,

$$\partial P_{j1}(\theta)/\partial \theta = -\Psi^2(z_{j1}) \left[(a_{j2} - a_{j1}) e^{z_{j2} - z_{j1}} + (a_{j3} - a_{j1}) e^{z_{j3} - z_{j1}} \right], \tag{3.46}$$

both $a_{j2} - a_{j1}$ and $a_{j3} - a_{j1}$ are positive and $P_{j1}(\theta)$ is strictly decreasing from 1 to 0 with increasing θ.

In the corresponding derivative for category 2, $a_{j1} - a_{j2}$ and $a_{j3} - a_{j2}$ are negative and positive, respectively, and the left and right terms are decreasing and increasing in θ. The curve for $P_{j2}(\theta)$ goes to zero at $\pm\infty$ and has a single mode at the point on the θ continuum, where

$$\frac{|a_{j3}-a_{j2}|e^{z_{j3}}}{|a_{j3}-a_{j2}|e^{z_{j1}}}=1; \tag{3.47}$$

that is, where

$$\theta=\frac{\ln(|a_{j3}-a_{j2}|)-\ln(|a_{j1}-a_{j2}|)+c_{j3}-c_{j1}}{a_{j3}-a_{j2}}. \tag{3.48}$$

Note that when the a values are equally spaced, the mode is at

$$\theta=\frac{c_{j3}-c_{j1}}{a_{j3}-a_{j2}}. \tag{3.49}$$

Finally, in the derivative for category 3 both $a_{j1}-a_{j3}$ and $a_{j2}-a_{j3}$ are negative, and $P_{j3}(\theta)$ is strictly increasing from 0 to 1 with increasing θ. A similar pattern is seen with greater numbers of categories: the curves with respect to all intermediate values of the a-parameter are unimodal.

Similar relationships between the category characteristic curves and their item parameters have been noted previously (Samejima 1972, Andrich 1978). Wainer et al. (1991) observed that increasing θ implies greater probability of response in a higher category rather than lower if and only if the a-parameter of the higher category is greater than that of the lower. In other words, when $\theta_1>\theta_2$ and $k>h$, the odds ratio is

$$\frac{P_k(\theta_1)/P_h(\theta_1)}{P_k(\theta_2)/P_h(\theta_2)}>1, \tag{3.50}$$

and the log odds simplify to

$$a_k(\theta_1-\theta_2)-a_h(\theta_1-\theta_2)>0, \tag{3.51}$$

which implies $a_k>a_h$.

Similarly, Andrich (1978) studied the interpretation of the parameters by examining the log odds ratio of successive categories at the points where their characteristic curves cross – that is, where the respondent is indifferent in choosing between those categories:

$$\log\frac{P_{jk}(\theta)}{P_{j,k-1}(\theta)}=(a_{jk}-a_{j,k-1})\theta+c_{jk}-c_{j,k-1}=0 \tag{3.52}$$

$$\theta=b_{jk}=-(c_{jk}-c_{j,k-1})/(a_{jk}-a_{j,k-1}). \tag{3.53}$$

Although it is plausible that respondents' ability to discriminate between successive categories may differ by category, if discriminating power is equal for all categories, then the differences between the a-parameters will be equal and can be set to a_j, say. In that special case, the b-parameters, which are the inner category boundaries, will correspond to the θ-values that determine the partition of response probabilities among the categories, just as they do in the graded model. In addition, the distance between b_{j1} and b_{j,m_j-1} will be inversely proportional to a_j, and the distances are additive; neighboring categories can then be collapsed just as in the graded model.

More generally, when the spacing of the slopes is unequal, the categories with higher slopes compared to preceding category receive a smaller share of the probability than those with lower slopes.

The likelihood equation for θ given the item score u_{ij} of respondent i is

$$\sum_j^n \sum_k^{m_j} \frac{u_{ijk}}{P_{jk}(\theta)} \cdot \frac{\partial P_{jk}(\theta)}{\partial \theta} = \sum_j^n \sum_k^{m_j} u_{ijk} a_{jk}[1 - \Psi_{jk}(\theta)] = 0. \tag{3.54}$$

The weighted sum of the item scores

$$S_i = \sum_j^n \sum_k^{m_j} a_{jk} u_{ijk} \tag{3.55}$$

is the sufficient statistic for θ_i just as it is in the 2PL and can be similarly used in applications where ranking of the respondents is required.

The item information function is

$$I_j(\theta) = \sum_k^{m_j} a_{jk}^2 \Psi_{jk}(\theta) - \sum_k^{m_j} \sum_h^{m_j} a_{jh} a_{jk} \Psi_{jh}(\theta) \Psi_{jk}(\theta). \tag{3.56}$$

Other parameterizations of the nominal model that greatly extend its versatility and can be expressed in general terms are based on results in chapter 8 of Bock (1975) for estimation of multinomial response relations. Let the parameterization of the transposed $m_j \times 1$ logit vector be $\mathbf{z}_j$ with elements z_{jk} be

$$\mathbf{z}_j' = \mathbf{K}\mathbf{B}_j\mathbf{A}_j, \tag{3.57}$$

where

$$\mathbf{K} = \begin{bmatrix} \theta & 1 \end{bmatrix},$$

$$\mathbf{B}_j = \begin{bmatrix} a_{j1} & a_{j2} & \cdots & a_{jm_j} \\ c_{j1} & c_{j2} & \cdots & c_{jm_j} \end{bmatrix},$$

and $\mathbf{A}_j$ is the $m_j \times m_j$ projection operator that implements the restriction on the parameters; the diagonal elements of the operator are $1 - 1/m_j$ and the off-diagonal elements are $-1/m_j$. The operator can be replaced by

$$\mathbf{A}_j = \mathbf{S}_j \boldsymbol{T}_j, \tag{3.58}$$

where $\mathbf{S}_j$ is an $m_j \times s, s \leq m_j - 1$ matrix of the coefficients of the linear parametric functions for item j. Then

$$\boldsymbol{T}_j = (\mathbf{S}_j'\mathbf{S}_j)^{-1}\mathbf{S}_j'\mathbf{A}_j. \tag{3.59}$$

If s is less than $m_j - 1$, there will be some loss of information in estimating θ, but the loss may be negligible if the parameters are structured favorably.

The quantities estimated after reparameterization are say,

$$\begin{bmatrix} \eta_j \\ \beta_j \end{bmatrix} = \mathbf{B}_j \mathbf{S}_j.$$

The following examples of linear parametric functions are of some interest. They all result in models with no more than $2m_j - 1$ free parameters.

Simple contrasts. The default in most applications is

$$\begin{bmatrix} -1 & -1 & -1 \\ 1 & 0 & 0 \\ 0 & 1 & 0 \\ 0 & 0 & 1 \end{bmatrix}.$$

Parameter trend over categories. Three examination questions pertaining to a reading passage are each scored right–wrong; the number right scorers from 0 to 3 are treated as ordered categories. A degree 2 polynomial trend is observed in the category parameters (see (Wainer et al. 1991)). The **S**-matrix contains the (unnormalized) orthogonal polynomials:

$$\begin{bmatrix} -3 & 1 \\ -1 & -1 \\ 1 & -1 \\ 3 & 1 \end{bmatrix}.$$

Crossed categories. Responses to an essay question are marked pass–fail and pass–fail for content. The matrix is the basis of a 2×2 analysis-of-variance model, excluding the constant term:

$$\begin{bmatrix} 1 & 1 & 1 \\ 1 & -1 & -1 \\ -1 & 1 & -1 \\ -1 & -1 & 1 \end{bmatrix}.$$

Ordered categories. The categories have a prior ordering. The parametric functions force the ordering of the category parameters; the parameters for category 1 are set to zero:

$$\begin{bmatrix} 0 & 0 & 0 \\ 1 & 1 & 1 \\ 0 & 1 & 1 \\ 0 & 0 & 1 \end{bmatrix}$$

For further examples, see Thissen and Steinberg (1986). Matrices for parameterizations 1 and 3 are incorporated in the MULTILOG program of Thissen (1991).

The likelihood equation for estimating θ reduces to

$$\sum_{j=1}^{n} \boldsymbol{\beta}_j' \boldsymbol{T}_j [\mathbf{u}_j - \Psi(\boldsymbol{z}_j)]. \tag{3.60}$$

Note that the sufficient statistic for θ is the sum of the vector-Bernoulli variates premultiplied by $\boldsymbol{\beta}_j' \boldsymbol{T}_j$. The item information function is

$$I_j(\theta) = \boldsymbol{\beta}_j' \boldsymbol{T}_j \mathbf{W}_j \boldsymbol{T}_j' \boldsymbol{\beta}_j, \tag{3.61}$$

where the $m_j \times m_j$ matrix $\mathbf{W}_j$, with diagonal elements $\Psi(z_{jk})[1 - \Psi(z_{jk})]$ and off-diagonal elements $-\Psi(z_{jh})\Psi(z_{jk})$, is the covariance matrix of the vector-Bernoulli variate. The second derivative of the log likelihood is exactly the negative of the information function and does not involve the data. Since the logistic distribution belongs to the exponential family, the likelihood surface is everywhere convex and any finite solution of the likelihood equations is unique and can easily be found numerically by Newton–Raphson or EM iterations.

Thissen et al. (2010) introduced a more general parameterization for the nominal model. They separate the a parameterization into a single overall discrimination parameter and a set of $m - 2$ contrasts among the a parameters, $z_k = a^* a^s_{k+1} \theta + c_{k+1}$ originally termed "scoring functions" by Muraki (1992). Here a^* is the overall slope parameter and a^s_{k+1} is the scoring function for response k and c_{k+1} is the original intercept parameter. For identification, $a^s_1 = 0, a^s_m = m - 1$, and $c_1 = 0$. The reparameterized model is estimated in terms of the parameters α and γ as

$$\boldsymbol{a} = \boldsymbol{T\alpha} \quad \text{and} \quad \boldsymbol{c} = \boldsymbol{T\gamma}. \tag{3.62}$$

However, instead of deviation contrasts used by Bock (1972), they use a Fourier basis as the $\boldsymbol{T}$ matrix of the form

$$\begin{bmatrix} 0 & 0 & \cdots & 0 \\ 1 & f_{22} & \cdots & f_{2(m-1)} \\ 2 & f_{32} & \cdots & f_{3(m-1)} \\ \vdots & \vdots & & \vdots \\ m-1 & 0 & \cdots & 0 \end{bmatrix}, \tag{3.63}$$

where

$$f_{kk'} = \sin[\pi(k' - 1)(k - 1)/(m - 1)]. \tag{3.64}$$

Thissen et al. (2010) note that when the parameters a^*, $[\alpha_2, \dots, \alpha_{m-1}]$, and γ are unrestricted, this is the full-rank nominal model. When $[\alpha_2, \dots, \alpha_{m-1}]$ are set to zero, this is the generalized partial credit (GPC) model.

3.2.2.4 Nominal Multiple-Choice Model

An interesting application of the nominal model is recovery of information in wrong responses to multiple-choice items. The item characteristic curves of the model identify the productive wrong alternatives and the range of data values in which they are most effective (see Figure 3.11). The problem arises however, of how to allow for effects of guessing when the test is administered with instructions not to omit items. Samejima (1979) suggested including a *virtual* response category, labeled 0, representing the hypothetical proportion of respondents who do not know the correct answer. She assumed random guessing with probability of correct response $1/m_j$ and $z_{jh} = a_{jh}\theta + c_{jh}$ and wrote the model as, say,

$$P_{jk}(\theta) = \frac{e^{z_{jk}} + m_j^{-1} e^{z_{j0}}}{\sum_{\ell=0}^{m_j} e^{z_{j\ell}}}, \qquad k = 1, 2, \ldots, m_j, \tag{3.65}$$

under the restrictions $\sum_{\ell=0}^{m_j} a_{j\ell} = 0$ and $\sum_{\ell=0}^{m_j} c_{j\ell} = 0$.

But guessing on multiple-choice items is not likely to be entirely random: experience with the 3PL model indicates that many wrong responders usually rule out a least one of the alternatives before answering. Thissen and Steinberg (1984) therefore substituted a category specific parameter, $g_{jk} \geq 0$, $\sum_{k=1}^{m_j} g_{jk} = 1$. To avoid inequality constraints, they made a second use of the multinomial logit to express g_{jk} as a function of, say, δ_{jk}:

$$g_{jk} = \frac{e^{\delta_{jk}}}{\sum_{\ell=1}^{m_j} e^{\delta_{j\ell}}}, \tag{3.66}$$

under the restriction $\sum_{\ell=1}^{m_j} \delta_{j\ell} = 0$. This generalization of the nominal model has $3m_j - 1$ free parameters.

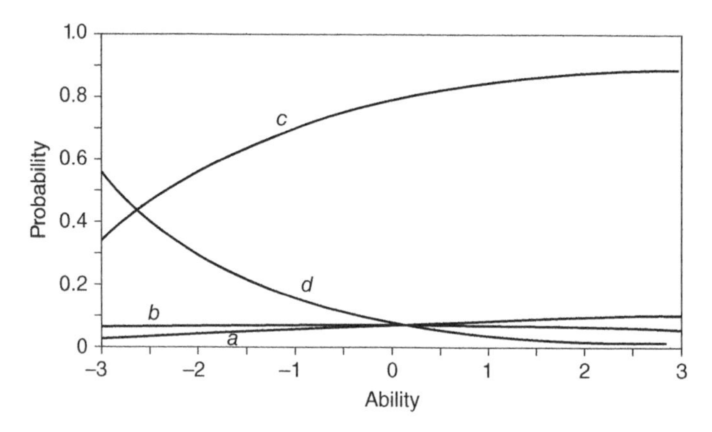

Figure 3.11 Item characteristic curves for the nominal model.

Setting $D_j = e^{z_{j0}} + \sum_{\ell=1}^{m_j} e^{z_{j\ell}}$ and $E_j = e^{z_{jh}} + g_{jh}e^{z_{j0}}$, the derivatives of category h required in estimation of the item parameters and θ can be expressed as follows:

$$\frac{\partial P_{jh}(\theta)}{\partial a_{j0}} = \theta e^{z_{j0}}[g_{jh}D_j - E_j]D_j^{-2}, \tag{3.67}$$

$$\frac{\partial P_{jh}(\theta)}{\partial a_{jk}} = \begin{cases} \theta e^{z_{jk}}[D_j - E_j]D_j^{-2} & h = k \\ -\theta e^{z_{jk}} D_j^{-1} & h \neq k, \end{cases} \tag{3.68}$$

$$\frac{\partial P_{jh}(\theta)}{\partial c_{j0}} = e^{z_{j0}}[D_j - E_j]D_j^{-2}, \tag{3.69}$$

$$\frac{\partial P_{jh}(\theta)}{\partial c_{jk}} = \begin{cases} e^{z_{jk}}[D_j - E_j]D_j^{-2} & h = k \\ -e^{z_{jk}} D^{-1} & h \neq k, \end{cases} \tag{3.70}$$

$$\frac{\partial P_{jh}(\theta)}{\partial \delta} = \begin{cases} g_{jh}(1 - g_{jh})e^{z_{j0}} D_j^{-1} & h = k \\ -g_{jh}g_{jk}e^{z_{j0}} D_j^{-1} & h \neq k, \end{cases} \tag{3.71}$$

$$\frac{\partial P_{jh}(\theta)}{\partial \theta} = \left[(a_{jk}e^{z_{jh}} + g_{jh}a_{j0}e^{z_{j0}})D_j - E_j\left(a_{j0}e^{z_{j0}} + \sum_{l=1}^{m_j} a_{jl}e^{z_{jl}}\right)\right] D_j^{-2}. \tag{3.72}$$

These results do not simplify further for substitution in the general expressions in Section 5.1 for the likelihood equation for θ and corresponding information function. As with the 3PL model, estimation of the g-parameters may require conditioning.

3.2.2.5 Illustration

We illustrate the nominal model using the previously analyzed STAI data (STAI, Spielberger 1983). The traditional nominal model parameters are displayed in Table 3.4. Table 3.5 displays the overall slope a^* and scoring function contrasts α_k for the Thissen et al. (2010) parameterization. Table 3.6 displays the model intercept contrasts γ_k for the Thissen et al. (2010) parameterization. The BIC for this model is 7828.35 which is larger than BIC = 7710.14 for the graded model, suggesting a more parsimonious fit to the simpler graded model under the strong ordinal proportional odds assumption for the slopes. Figures 3.12 and 3.13 provide ICC and information for the nominal model for the tension and anxious test items. Consistent with the BIC, the item characteristic functions and item information functions for these two items are quite similar to the graded model, supporting their ordinal nature.

3.2.2.6 Partial Credit Model

Masters (1982) adapted the rating-scale model proposed by Rasch (1961) and elaborated by Andersen (1977) and Andrich (1978) to apply to ordered categories scored $v_j = 1, 2, \ldots, m_j$ for item j. (In the social science literature, this type of

Table 3.4 Six-item STAI anxiety scale data original (Bock 1972) nominal model parameter estimates intercepts (c_{jk}) and slopes (a_j).

Item	Parameter	Category				
		1	2	3	4	5
Not calm	a	0.00	1.86	3.28	5.42	5.86
	c	0.00	1.65	1.08	−2.05	−4.37
Tense	a	0.00	2.03	3.48	5.00	6.29
	c	0.00	1.91	1.46	−0.14	−3.43
Regretful	a	0.00	1.08	1.51	2.37	3.14
	c	0.00	0.45	−0.20	−1.21	−3.30
Not at ease	a	0.00	2.36	3.97	5.67	6.82
	c	0.00	2.59	2.33	0.09	−3.75
Anxious	a	0.00	1.79	2.87	3.61	5.10
	c	0.00	1.00	0.11	−1.23	−6.46
Nervous	a	0.00	1.45	2.55	3.28	4.14
	c	0.00	0.90	0.33	−0.57	−2.74

Table 3.5 Six-item STAI anxiety scale data Thissen et al. (2010) nominal model parameterization slope (a^*) and scoring function contrasts (α_k).

Item	a^*	Contrasts			
		α_1	α_2	α_3	α_4
Not calm	1.46	1.00	0.46	−0.21	0.22
Tense	1.57	1.00	0.27	0.06	0.06
Regretful	0.78	1.00	0.11	0.18	0.18
Not at ease	1.70	1.00	0.42	0.03	0.09
Anxious	1.28	1.00	0.21	0.29	−0.04
Nervous	1.04	1.00	0.44	0.12	−0.03

category score goes by the name of "Likert scores"; (Likert 1932).) Called the *partial credit* model, Master's version is based on a concept of the respondent processing the item task in successive steps with probability of π_{jk} of success in making the step from k to $k+1$, where $\pi_{j1} = 1$. Step m_j or the first failure terminates the process; the Likert scores count the steps. Generalizing the 1PL

Table 3.6 Six-item STAI anxiety scale data Thissen et al. (2010) nominal model intercept contrast estimates (γ_k).

	Contrasts			
Item	γ_1	γ_2	γ_3	γ_4
Not calm	−1.09	3.04	0.76	−0.23
Tense	−0.86	3.42	0.17	0.25
Regretful	−0.82	1.62	0.01	0.18
Not at ease	−0.94	4.38	0.31	0.17
Anxious	−1.61	3.87	−0.50	0.54
Nervous	−0.68	1.93	0.05	0.24

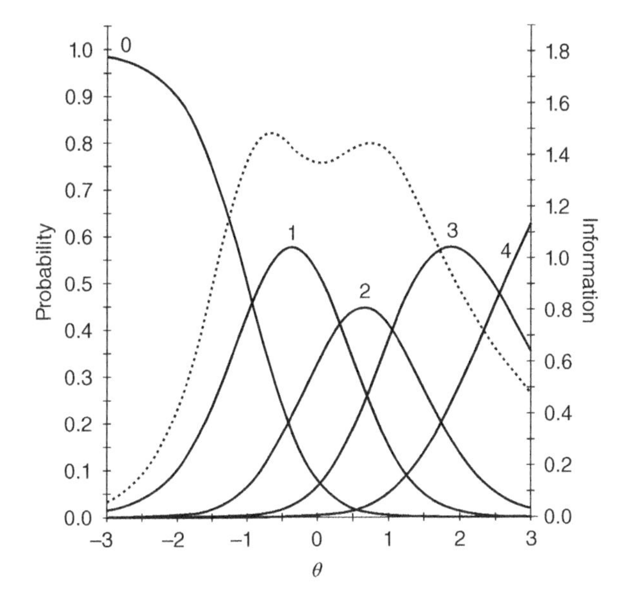

Figure 3.12 Item characteristic curves and information for the nominal model – tension item.

model by defining the logit as $z_{jk} = \theta - b_{jk}$, with $b_{j1} = 0$, and norming the step probabilities to 1 leads to the following model in $(m_j - 1)$ free item parameters:

$$P_{jk}(\theta) = \Psi(z_{jk}) = \frac{\exp \sum_{\ell=1}^{v_j} z_{j\ell}}{\sum_{h=1}^{m_j} \exp[\sum_{\ell=1}^{h} z_{j\ell}]}, \quad k = v_j. \tag{3.73}$$

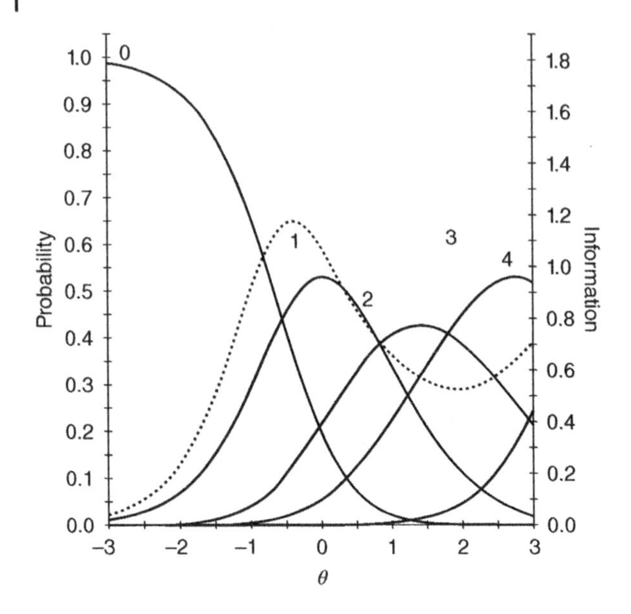

Figure 3.13 Item characteristic curve and information for the nominal model – anxious item.

The derivatives for category h of item j are (see (Masters 1982)):

$$\frac{\partial P_{jh}(\theta)}{\partial b_{jk}} = -\Psi(z_{jk})\left[1 - \sum_{\ell=k}^{m_j} \Psi(z_{j\ell})\right], \tag{3.74}$$

$$\frac{\partial P_{jh}(\theta)}{\partial \theta} = -\Psi(z_{jk})\left[1 - \sum_{\ell=k}^{m_j} \ell\Psi(z_{j\ell})\right]. \tag{3.75}$$

The likelihood equation for estimating θ_i, the location of respondent i on the latent dimension, given the item scores v_{ij}, is

$$\sum_{j=1}^{n}\left[v_{ijk} - \sum_{\ell=k}^{m_j} \ell\Psi(z_{j\ell})\right] = 0, \tag{3.76}$$

which shows that the total step count for the respondent is a sufficient statistic for θ_i. This means that, under the assumptions of the model, the mean Likert score has the same efficiency as the estimated scale score for purposes of ranking respondents.

The corresponding information function is

$$I(\theta) = \sum_{k=1}^{m_j} k^2\Psi(z_{jk}) - \left[\sum_{k=1}^{m_j} k\Psi(z_{jk})\right]^2. \tag{3.77}$$

As parameterized above, the partial credit model assumes that the unit of scale on the latent dimension is equal to 1. In testing applications however, it may be more convenient to fix the unit by assuming that the standard deviation of the latent distribution equals 1. In that case the logit may be expressed as $z_{jk} = a(\theta - b_{jk})$, and the common slope, a, can be estimated along with the category parameters.

3.2.2.7 Generalized Partial Credit Model

Muraki (1992) observed that the partial credit model may be generalized to items with varying discriminating powers simply by defining the logit as $z_{jk} = a_j(\theta - b_{jk})$. He also noted that the exponent of the general model can be expressed for category k as

$$ka_j\theta + c_{jk}, \tag{3.78}$$

where $c_{jk} = \sum_{\ell=1}^{m_j} a_j b_{jk}$, which shows the model to be a special case of the nominal model. The slope parameters in the partial credit model are assigned equally spaced values in the given order of the categories, whereas in the nominal model they are estimated from the sample data. This is also the special case, noted above, of ordered a-parameters with equal differences between successive values that simplifies interpretation of the item parameters and characteristic curves of the nominal model.

The sufficient statistic for θ in the generalized model is the sum of the item step counts weighted by a_j. The model may be formulated in terms of vector-Bernoulli variates rather than Likert scores, such that with the necessary changes to the a-parameters the results for the nominal model apply.

As noted above, when using the Thissen et al. (2010) formulation, the GPC model results when $\alpha_1 = 1$ and $[\alpha_2, \dots, \alpha_{m-1}]$ are set to zero.

There is often confusion regarding the difference between the graded model and the GPC model. The fundamental difference is that the graded model is a proportional odds model, whereas the GPC model is an adjacent odds model, which relaxes the proportionality constraint.

3.2.2.8 Illustration

We illustrate the GPC model using the previously analyzed STAI data (STAI, Spielberger 1983). The model parameters are displayed in Table 3.7. The BIC for this model is 7742.05, which is an improvement over the unrestricted nominal model (BIC = 7828.35), but is still larger than the graded model (BIC = 7710.14), suggesting a more parsimonious fit to the simpler graded model with proportional odds assumption for the slopes. Figures 3.14 and 3.15 provide ICC and information for the nominal model for the tension and anxious test items. Consistent with the BIC, the item characteristic functions and item information functions for these

Table 3.7 Six-item STAI anxiety scale data generalized partial credit model

Item	a	b	d_1	d_2	d_3	d_4	d_5
Not calm	1.69	0.90	0.00	1.82	0.50	−0.65	−1.67
Tense	1.63	0.66	0.00	1.66	0.36	−0.38	−1.64
Regretful	0.76	1.11	0.00	1.59	0.18	−0.19	−1.58
Not at ease	1.87	0.74	0.00	1.92	0.58	−0.54	−1.96
Anxious	1.22	1.37	0.00	1.98	0.64	−0.10	−2.52
Nervous	1.05	0.80	0.00	1.46	0.30	−0.24	−1.52

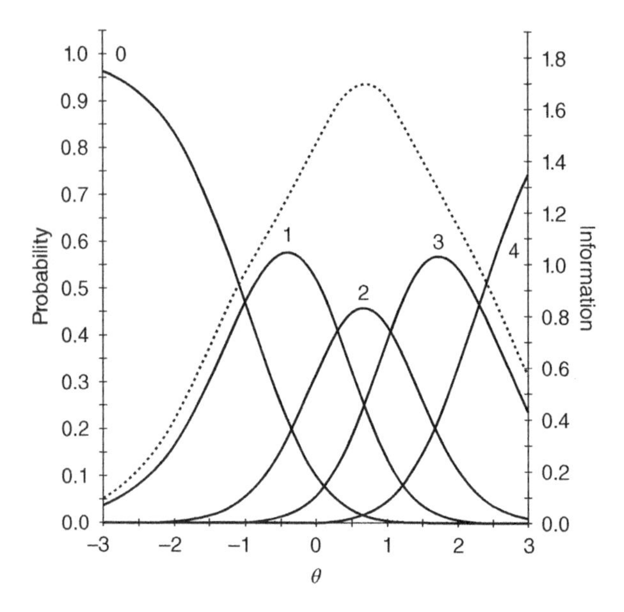

Figure 3.14 Item characteristic curve and information for the GPC model – tension item.

two items are quite similar to the graded model, again supporting their ordinal nature.

3.2.2.9 Rating Scale Models

Multiple item rating scales are widely used in the measurement of respondents' attitudes, opinions, personality traits, health status, and similar personal characteristics. All items have the same format – usually a fixed number of ordered categories labeled in the same way throughout the instrument. In modeling rating scale responses, it is assumed that a respondent's interpretation of the labels and

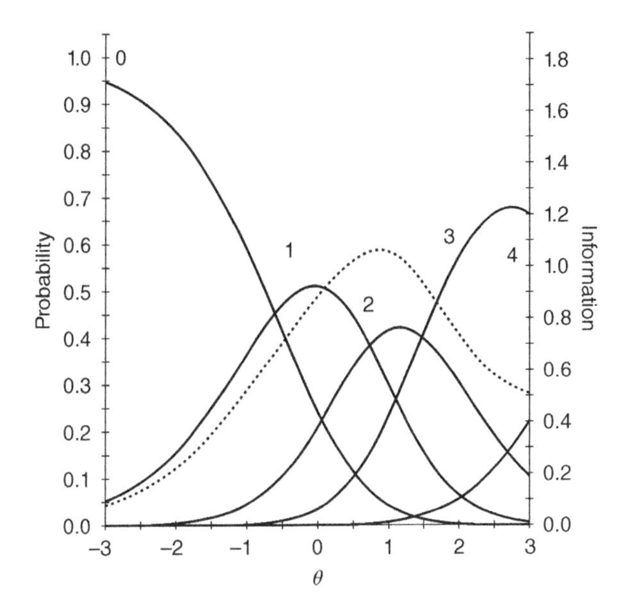

Figure 3.15 Item characteristic curve and information for the GPC model – anxious item.

use of the categories is not influenced by the content of the items. The content itself, however, may vary in the strength of its relationship to the latent variable as reflected in the power of a given item to discriminate among respondents.

These assumptions suggest a response model for, say, m ordered categories with common category parameters and varying discriminating powers. The assumption that the strength of relationship is the same for all items is not plausible in typical applications of rating scales. Among the item response models described above, these assumptions limit the choice to the graded and GPC models, with the added requirement that the category parameters are constrained to equality over items.

For the graded model, the normit is therefore,

$$y_{jk} = a_j(\theta - b_j + d_k), \quad k = 1, 2, \ldots, m-1, \tag{3.79}$$

under the restriction $\sum_{k=1}^{m-1} d_k = 0$.

For the GPC model expressed as a special case of the nominal model, the logit is

$$z_{jk} = a_{jk}\theta - b_j + d_k, \quad k = 1, 2, \ldots, m, \tag{3.80}$$

under the restriction $\sum_{k=1}^{m} d_k = 0$.

Thus, for an n item scale, the rating-scale assumptions reduce the number of free item parameters in these models from nm to $2n + m - 1$. The restriction to common category parameters is easily implemented during estimation of item parameters. The graded and nominal models with varying category parameters may, of course, still be used in preliminary studies to verify the stronger assumptions of the rating-scale model.

3.2.3 Ranking Model

Ranking is of particular interest in measurement of preferences for objects of choice – such as political candidates, proposed public policies, occupational goals, or competing consumer goods. When studying preferences, there is an important advantage to asking the respondents to rank rather than rate the objects: ranking avoids undesirable effects of respondents' idiosyncratic interpretations of category labels such as "agree moderately," "agree strongly," etc., found in rating forms. In an IRT context, the set of objects to be ranked constitute an item, and the objects in the set take the place of categories. For example, in measurement respondents' conservatism, the objects might be proposed alternative solutions to various social problems serving as items. The Ranking model can be used to estimate the probability that a given object will be first, second, or any lower choice in the preference rankings. The nominal model is well-suited for this type of analysis because of its computational simplicity.

Böckenholt (2001) used the Luce (1959) principle of independence of first choices from irrelevant alternatives to decompose a rank-ordering of objects into a succession of first choices among successively smaller sets of objects. The principle implies that the probabilities of the first choices are independent and that the probability of the ranking as a whole is the continued product of those probabilities. To express the conditional probability of a respondent's ranking as a function of θ, let H_j be a set containing integers $h_j = 1, 2, \ldots, m_j$ indexing the objects in item j, $j = 1, 2, \ldots, n$. Let K_{ij} be an ordered subset of elements of H_j corresponding to the q objects ranked by respondent i, $q = 1, 2, \ldots, m_j - 1$. Let $k_{ij}^{(\ell)}$, $\ell = 1, 2, \ldots, q$, be an element in K_{ij} corresponding to the object in position ℓ in the rank ordering, and let $\bar{H}_{ij\ell}$ be the subset of $1 + m_j - \ell$ elements of H_j *excluding* elements corresponding to objects above $k_{ij}^{(\ell)}$ in the rank ordering (note that elements of $\bar{H}_{ij\ell}$ not among the elements of K_{ij} need not be ordered).

Then the probability of K_{ij} is

$$P_j(K_{ij} \mid \theta) = \prod_{\ell=1}^{q} \frac{\exp[z(k_{ij\ell})]}{\sum_{h \in \bar{H}_{ij\ell}} \exp z(h)}, \tag{3.81}$$

where

$$z(h \in \bar{H}_{ij\ell}) = a_{jh}\theta + c_{jh} \tag{3.82}$$

is the logit of the nominal model for the probability of each successive first choice.

Provided there is individual variation in the rankings, the item parameters can be estimated in large samples of respondents for even a single item, the characteristic curves corresponding to objects may then be of interest. The multistage structure of (3.81) imposes constraints on the collection of ranking data. Ideally, rankings should be obtained in the form of successive first choices by requiring respondents to rank the objects from most to least preferred. Deviations from this format that are not consistent with the multistage structure of the model may lead to systematic misfits.

4

Item Parameter Estimation – Binary Data

In God we trust, all others must bring data.

(Source: Edward Deming)

The practical use of scale-score estimation procedures presented in Chapter 3 requires that we know the response functions of the items involved. Since there is no way to determine these functions on theoretical grounds, we have no alternative but to estimate them from a suitable sample of respondents. In the model-based approach to item response theory (IRT), this means estimating the free parameters of the model that represents the response process.

In most applications, the aim of the estimation is to "calibrate" a test or scale – that is, to estimate parameter values that can be used in computing scores for persons who respond to the instrument on future occasions. Used in this way, item calibration plays much the same role as test norming in classical theory. Like the establishing of test norms, it necessarily assumes that conditions remain stable in the population of respondents while the current calibration is in force. This assumption is justified only when the procedure for administering the test or scale is identical at each use. Especially in cognitive testing, even small differences in the format of test booklets, or in the instructions to the respondents or timing of the test session, can appreciably affect the test scores. The effect may not be very apparent for any particular respondent, but if it is in the same direction for all respondents, the bias in the sample mean can invalidate comparisons with other groups tested under different conditions.

It must also be assumed that the operating characteristics of the items remain stable during the period up to the next calibration. Fortunately, only *relative* stability among the items is required; any overall change in item locations or slopes is absorbed into the mean and standard deviation of the population distribution. Relative change in item parameters, which is referred to as "item-parameter drift,"

Item Response Theory, First Edition. R. Darrell Bock and Robert D. Gibbons.

is an aspect of so-called "differential item functioning"or (DIF) and is discussed as such in Chapter 9.

Finally, all of the item calibration procedures in this chapter, like the scoring procedures of the previous chapter, assume conditional independence of item responses among respondents with the same attribute value. Not until in Chapter 6 do we examine the possibility of IRT procedures that do not assume conditional independence. The assumption of conditional independence is not exclusive to IRT: important results in classical theory, such as the Spearman–Brown formula for computing the reliability of an extended test, make the same assumption (see Section 2.17.1).

In Chapter 3, we discuss the calibration of binary-scored items in applications where data are available from a suitable calibration sample and the above assumptions are justified. We limit the discussion here to parametric estimation for the logistic models of Chapter 3, but in later chapters we show how the results generalize to the multicategory and multidimensional models.

4.1 Estimation of Item Parameters Assuming Known Attribute Values of the Respondents

Although not typical of IRT applications, there are some situations in which we know beforehand the attribute values of the respondents. In these cases, parameter estimation is quite straightforward, and the results can be used in the estimation of scores for other members of the population for whom attribute values are *not* known. In the item calibrations, the known values play the role of an *external criterion*, values of which will be predicted when the test or scale is administered to new respondents.

This type of *external* scaling of test items long antedates IRT. The classic example is the scaling of items in the first English-language version of Binet's intelligence test for children. Burt and Thurstone took the age of the child as the criterion and fitted normal ogive response functions to observed percents correct for each item in successive age groups. They then estimated the item location on the chronological age scale in order to arrange the Binet items in a developmental sequence. Thurstone went further and proposed that the observed proportional relationship between the locations and the slopes of the response functions implied a mental growth curve somewhat similar to the physical growth curve of children in the same age range.

A modern, and more important, example of fitting item response functions to an external criterion is the IRT version of the "variant" item technique. Variant items, which do not contribute to the test score, are included in periodically updated tests to obtain data for estimating item statistics for future test construction. If the main test is scored by IRT methods and is sufficiently long to justify treating these scores as known attribute values, the item parameters of the variant items can be estimated with the external criterion. We illustrate this use of variant item data in this section.[1]

Apart from possible practical applications of calibrating items to an external criterion, an exposition of the procedure at this point is also valuable because the results play a role in the more general methods of the later sections, where the attribute values are assumed *unknown*.

4.1.1 Estimation

Fitting the models of Chapter 3 when there is an external criterion is a straightforward application of the logit analysis of Section 2.16.2. The likelihood equations vary according to the number of parameters, but other features of the estimation are similar.

4.1.1.1 The One-Parameter Model

With an external criterion, the attribute values have a nonarbitrary origin and unit depending on the scale on which the criterion is measured. In that case, the one-parameter model must include both a common scale constant, a, as well as a separate location constant, b_j, for each item. The response function for item j and group $k = 1, 2, \dots, m$, with criterion value, x_k, is, therefore,

$$P_{jk} = \Psi(z_{jk}) = \frac{1}{1 + e^{-z_{jk}}}, \tag{4.1}$$

where $z_{jk} = a(x_k - b_j)$, or, equivalently, $z_{jk} = c_j + ax_k$, for $c_j = -ab_j$. Thus, for a test consisting of n items, the number of parameters to be estimated is $n + 1$. For this number of parameters to be identified, $m \geq 2$ is required.

On condition that all item responses of the N_k respondents in group k are independent, the log likelihood is

$$\log L(a, c_j) = C + \sum_{k=1}^{m}\sum_{j=1}^{n}[r_{jk}\log P_{jk} + (N_k - r_{jk})\log(1 - P_{jk})]; \tag{4.2}$$

whence the likelihood equations for $c_j, j = 1, 2, \dots, n$, are the following:

$$G(c_j) = \sum_{k=1}^{m}(r_{jk} - N_k P_{jk}) = 0. \tag{4.3}$$

---

1 Occasionally, in the literature, we also see this method applied to items of a test where the criterion is the *number-right score* for the test obtained from the self-same items. This may have some justification if the test is sufficiently long and the number-right scores fall mostly near the middle of their range. But it cannot be recommended in general because the relationship between the attribute-value continuum and the number-right score is nonlinear and depends on the difficulties of the particular items in the test.

There is an similar equation for a, except that it is also summed over j:

$$G(a) = \sum_{k=1}^{m}\sum_{j=1}^{n}(r_{jk} - N_k P_{jk})x_k = 0. \tag{4.4}$$

The corresponding information matrix will, therefore, be voluminous and sparse if n is large. It has the form,

$$I(c_j, a) = \sum_{k=1}^{m} N_k \begin{bmatrix} W_{1k} & 0 & \cdots & 0 & W_{1k}x_k \\ 0 & W_{2k} & \cdots & 0 & W_{2k}x_k \\ & & \cdots & & \\ 0 & 0 & \cdots & W_{jn} & W_{nk}x_k \\ W_{1k}x_k & W_{2k}x_k & \cdots & W_{nk}x_k & \sum_j W_{jk}x_k^2 \end{bmatrix}, \tag{4.5}$$

where $W_{jk} = P_{jk}(1 - P_{jk})$ is the weight for logit estimation (Section 2.16.2).

Nevertheless, using the result in Section 2.8 for the inverse of a partitioned matrix, we can easily implement the Fisher scoring solution of the likelihood equations. The inverse information matrix has the form,

$$I^{-1}(c_j, a) = \frac{1}{D_1}\begin{bmatrix} \sigma_{x1}^2 + \bar{x}_1^2 & \bar{x}_1\bar{x}_2 & \cdots & \bar{x}_1\bar{x}_n & -\bar{x}_1 \\ \bar{x}_1\bar{x}_2 & \sigma_{x2}^2 + \bar{x}_2^2 & \cdots & \bar{x}_2\bar{x}_n & -\bar{x}_2 \\ & & \cdots & & \\ \bar{x}_1\bar{x}_n & \bar{x}_2\bar{x}_n & \cdots & \sigma_{xn}^2 + \bar{x}_n^2 & -\bar{x}_n \\ -\bar{x}_1 & -\bar{x}_2 & \cdots & -\bar{x}_n & 1 \end{bmatrix}, \tag{4.6}$$

where $D_1 = \sum_j [\sum_k N_k W_{jk} x_k^2 - \bar{x}_j \sum_k N_k W_{jk}]$, $\bar{x}_j = \sum_k N_k W_{jk} x_k / \sum_k N_k W_{jk}$, and $\sigma_{xj}^2 = D_1 / \sum_k N_k W_{jk}$.

Writing in full the premultiplication of the gradient vector by the inverse matrix, we have the following correction terms for the iterations in c_j and a:

$$[\hat{c}_j]_{i+1} = [\hat{c}_j]_i + \frac{1}{D_1}\left[D_1 \sum_k (r_{jk} - N_k P_{jk}) + \bar{x}_j \sum_h \bar{x}_{hk}(r_{jk} - N_k P_{jk}) \right.$$

$$\left. - \sum_j \bar{x}_j \sum_k (r_{jk} - N_k P_{jk}) x_k \right]_i \Big/ D_1^*,$$

$$\hat{a}_{i+1} = \hat{a}_i + \left[\sum_j \sum_k (r_{jk} - N_k P_{jk}) x_k - \sum_j \bar{x}_j (r_{jk} - N_k P_{jk})\right]_i \Big/ D_1.$$

Starting values of 0 for the intercepts and 1 for the slope are quite satisfactory.

The standard errors of the converged estimates are the square roots of the corresponding diagonal elements of the inverse information matrix. The covariances are the corresponding off-diagonal elements.

The maximum likelihood estimate of the location parameter for item j is $b_j = -\hat{c}_j / \hat{a}$; its standard error is obtained by the delta method.

4.1.1.2 The Two-Parameter Model

Estimation for the two-parameter logistic model is considerably simpler because the information matrix is block diagonal. As in Section 2.16.2, the deviate in the logistic function $\Psi(z_{jk})$ is

$$z_{jk} = a_j(x_k - b_j) = c_j + a_j x_k. \tag{4.7}$$

Thus, there are $2n$ likelihood equations,

$$G\begin{bmatrix} c_j \\ a_j \end{bmatrix} = \begin{bmatrix} \sum_k (r_{jk} - N_k P_{jk}) \\ \sum_k (r_{jk} - N_k P_{jk}) x_k \end{bmatrix} = \begin{bmatrix} 0 \\ 0 \end{bmatrix}, \tag{4.8}$$

and the $2n \times 2n$ information matrix is block diagonal consisting of blocks of the following form:

$$I(c_j, a_j) = \begin{bmatrix} \sum_k N_k W_{jk} & \sum_k N_k W_{jk} x_k \\ \sum_k N_k W_{jk} x_k & \sum_k N_k W_{jk} x_k^2 \end{bmatrix}. \tag{4.9}$$

The inverse information matrix is a similar diagonal matrix with the following inverses of these matrices as the blocks:

$$I^{-1}(c_j, a_j) = \frac{1}{D_2}\begin{bmatrix} \sum_k N_k W_{jk} x_k^2 & -\sum_k N_k W_{jk} x_k \\ -\sum_k N_k W_{jk} x_k & \sum_k N_k W_{jk} \end{bmatrix}, \tag{4.10}$$

where $D_2 = \sum_k N_k W_{jk} \sum_k N_k W_{jk} x_k - (\sum_k N_k W_{jk} x_k)^2$.

The Newton–Raphson solution can be carried out separately on each of these blocks from starting values of 0 and 1. At convergence, the elements of each inverse matrix are the large-sample variances and covariances for the estimators of the intercept and slope parameter of each item. The location parameter is then computed as $\hat{b}_j = -\hat{c}_j/\hat{a}_j$, and its standard error is computed as in Section 2.16.2.

4.1.1.3 The Three-Parameter Model

For purposes of estimating item parameters, we use the second of the two expressions for the three-parameter logistic model:

$$P_{jk} = g_j + (1 - g_j)\Psi_{jk}, \tag{4.11}$$

where $\Psi_{jk} = 1/[1 + \exp(z_{jk})]$, with $z_{jk} = c_j + a_j x_k$, and g_j is the probability of a correct response due to guessing.

In this case, $\partial P_k/\partial z_{jk} = (1 - g_j)\Psi_{jk}(1 - \Psi_{jk})$, and with $Q_{jk} = 1 - P_{jk}$ the likelihood equations are the following:

$$G(c_j) = (1 - g_j)\sum_k \frac{r_{jk} - N_k P_{jk}}{P_{jk} Q_{jk}} \Psi_{jk}(1 - \Psi_{jk}) = 0,$$

$$G(a_j) = (1 - g_j)\sum_k \frac{r_{jk} - N_k P_{jk}}{P_{jk} Q_{jk}} \Psi_{jk}(1 - \Psi_{jk}) x_k = 0,$$

$$G(g_j) = \sum_k \frac{r_{jk} - N_k P_{jk}}{P_{jk} Q_{jk}} (1 - \Psi_{jk}) = 0.$$

Unlike the two-parameter model, the guessing model does not lead to second derivatives that are free of the data. The 3×3 blocks of the block diagonal information matrix, therefore, contain the negative of the *expected* second derivatives:

$$I(c_j, a_j, g_j) = \begin{bmatrix} g_j^* \sum_k W_{jk}^* \Psi_{jk} \Psi_{jk}^* & g_j^{*2} \sum_k W_{jk}^* \Psi_{jk}^2 \Psi_{jk}^{*2} x_k & g_j^* \sum_k W_{jk}^* \Psi_{jk} \Psi_{jk}^{*2} \\ g_j^{*2} \sum_k W_{jk}^* \Psi_{jk}^2 \Psi_{jk}^{*2} x_k & g_j^* \sum_k W_{jk}^* \Psi_{jk} \Psi_{jk}^* x_k^2 & g_j^{*2} \sum_k W_{jk}^* \Psi_{jk} \Psi_{jk}^{*2} x_k \\ g_j^* \sum_k W_{jk}^* \Psi_{jk} \Psi_{jk}^{*2} & g_j^{*2} \sum_k W_{jk}^* \Psi_{jk}^2 \Psi_{jk}^{*2} x_k & g_j^* \sum_k N_k Q_{jk}^{-1} \Psi_{jk} \Psi_{jk}^{*2} \end{bmatrix},$$

where $g_j^* = 1 - g_j$, $W_{jk}^* = N_k Q_{jk}^{-1}$, and $\Psi_{jk}^* = 1 - \Psi_{jk}$.

The corresponding inverse matrix can be computed by the standard result for the 3×3 symmetric matrix (see Section 2.8.4). The Fisher-scoring solution for the three-parameter model may be ill-conditioned if the data do not extend into the region of criterion scale where effects of guessing are evident. In that case, the imposition of stochastic constraints on the estimator, such as those discussed in Section 4.2.2 in connection with marginal maximum likelihood (MML) estimation for this model, may be required in order to obtain convergence.

4.2 Estimation of Item Parameters Assuming Unknown Attribute Values of the Respondents

In most IRT applications, we must assume that both the item parameters and the attribute values of the respondents are unknown. Considering how the item parameters might be obtained under this condition, and given the complex nonlinear models involved, we look to the method of maximum likelihood for a solution to the estimation problem. Fortunately, the sample sizes in item calibrations are usually large enough to justify the assumptions of maximum likelihood estimation.

We see immediately, however, that because the number of respondents is effectively unlimited, any direct attempt at joint estimation of the two types of parameters by maximum likelihood would encounter the problem described in Section 2.15 as "estimation of structural parameters in the presence of infinitely many incidental parameters." The structural parameters pertain to the items, which are assumed finite in number, and the incidental parameters pertain to the respondents, who are in principle unlimited in number. MLE of these two sets of parameters jointly is not possible in this situation because the likelihood will increase indefinitely with each additional respondent, and hence an additional parameter, is added.

Three distinct resolutions of this dilemma have been proposed in the IRT literature. The first solution is to assume that the respondents can be sorted into a finite number of homogeneous subgroups within which all respondents have the same attribute value. In that case, the item parameters plus the common value for each group is a finite set, and all parameters can in general be estimated jointly. This method is referred to as *joint* maximum likelihood estimation, or "JML."

The second solution is to look for a simple sufficient statistic for the respondent's attribute value. If it exists, the likelihood function for the sample can be factored into one term involving the attribute, and another, conditional on the statistic, involving only the item parameters. Maximization of the latter with respect to the item parameters provides the required estimates. This is referred to as *conditional maximum likelihood* estimation, or "CML."

The third solution is to assume that the respondents are drawn from a population in which the distribution of attributes is a function of a finite number of parameters. Estimation of the parameters for the items and for the population distribution is then a finite problem and maximum likelihood estimation is possible. This is referred to as *marginal* maximum likelihood estimation, or "MML."

These solutions have in common the assumption of conditional independence introduced in Chapter 3 in order to justify the use of the product rule in computing probabilities of response patterns. We invoke this assumption throughout this chapter as we discuss each solution in turn and illustrate its application to the logistic models of Chapter 3.

4.2.1 Joint Maximum Likelihood Estimation (JML)

Summary data for JML estimation take the convenient form of a groups × items table of binomial frequencies. If there are m groups and n items, the dimensions of the table are $m \times n \times 2$. An entry in the table is the number of respondents in group k who respond "RIGHT" or "WRONG" to item j. Table 4.1 illustrates this form of data. The definition of the respondent groups varies in different situations.

4.2.1.1 The One-Parameter Logistic Model

We have already seen in Chapter 3 that the number-right score of a binary item test is a sufficient statistic for estimating the attribute value of a respondent. This implies that all respondents with the same number-right score, or, briefly, *test score*, will be assigned the same attribute value when the item parameters and attribute values are estimated jointly. As there are at most $n + 1$ distinct test scores of an n item test, there are never more than that number of attribute values assigned to respondents when the one-parameter model is assumed. We show here that, if the respondents are grouped according to their test scores, the frequencies with which items are answered correctly or incorrectly in each group

Table 4.1 LSAT6 data: score-group × item binomial frequencies.

Score-group	Item	RIGHT	WRONG
1	1	10	10
	2	1	19
	3	1	19
	4	2	18
	5	6	14
2	1	62	23
	2	24	61
	3	7	78
	4	28	57
	5	49	36
3	1	212	25
	2	109	128
	3	63	174
	4	139	98
	5	188	49
4	1	342	15
	2	277	80
	3	184	173
	4	296	61
	5	329	28

are sufficient statistics for joint estimation of the groups attribute values and the item location parameters of the model.

Estimation of the item parameters differs from that of Section 4.1, where the attribute values are assumed known, in that the common slope parameter is not required; its value may be set to unity by convention. In that case, the one-parameter logistic model can be expressed in terms of the attribute value, θ_k, for score groups $k = 1, 2, \ldots, n + 1$, and the location parameter, b_j, for items $j = 1, 2, \ldots, n$:

$$P(u_{jk} = 1 \mid \theta) = \frac{1}{1 + e^{-(b_j + \theta_k)}}. \tag{4.12}$$

Apart from the intercept term, μ, we recognize this function as the logit-linear model of Section 2.16.3 representing main-class effects for an $(n + 1) \times n$ table of binomial frequencies. Simply by regarding the group parameters as deviations from μ, however, we can express the exponent in the conventional form for the logit model. Then we can apply standard results in logit-linear analysis to the problem of estimating the θ_k and b_j.

4.2.1.2 Logit-Linear Analysis

The conditions for model identifiability for logit-linear models preclude marginal classes with 0% or 100% response (see Section 2.16.3). This means that the all-correct and the all-incorrect score groups will have to be omitted from the analysis, with the result that the θ values for these groups cannot be estimated. Similarly, any score groups in which there are no respondents, or any items for which all responses are correct or all incorrect, must be omitted for identifiability. If these conditions on the marginal frequencies are met, the maximum likelihood estimates of the item intercepts and score-group attributes will exist and can be obtained efficiently by the Newton–Raphson solution of Section 2.16.3.

In logit-linear analysis of binary item data, the computation of the likelihood depends upon the assumed independence of the observed frequencies over items within score groups. This assumption cannot be strictly valid, even when conditional independence given θ obtains, for some variation in the true θ values within score-groups is inevitable, especially in short tests. This residual association between items within the groups will not necessarily bias the estimates of the item locations, but it may lead to underestimation of the standard errors and overestimation of the goodness-of-fit of the model. These effects become apparent when the data of the following example are reanalyzed by the JML and MML methods later in this section.

Example. The response-pattern data of Bock and Lieberman (1970) for five items from Section 6 of the Law School Aptitude Test (LSAT) can be aggregated into score group frequencies as shown in Table 4.1. Score group 0 and score group 5 have been omitted. The numbers of respondents in retained groups 1 through 4 are 20, 85, 237, and 357, respectively.

To carry out the logit-linear analysis of these data, we require a full-rank model matrix including an intercept term and linearly independent contrasts between score groups and between items. A suitable 20×8 matrix can be constructed in the manner of Section 2.8 from the Kronecker products of selected columns of 4×4 and 5×5 full-rank model matrices for the corresponding one-way designs. The following are the matrices that lead to estimates of the maximum number of independent deviation contrasts among groups and among items:

$$\begin{bmatrix} 1 & 1 & 0 & 0 \\ 1 & 0 & 1 & 0 \\ 1 & 0 & 0 & 1 \\ 1 & -1 & -1 & -1 \end{bmatrix}, \begin{bmatrix} 1 & 1 & 0 & 0 & 0 \\ 1 & 0 & 1 & 0 & 0 \\ 1 & 0 & 0 & 1 & 0 \\ 1 & 0 & 0 & 0 & 0 \\ 1 & -1 & -1 & -1 & -1 \end{bmatrix}.$$

Table 4.2 Basis of the logit-linear model for estimating intercept (μ), score-group ($\theta_k^{(0)}$), and item ($b_j^{(0)}$) effects.

Score-group	Item	Effect							
		μ	θ_1	θ_2	θ_3	b_1	b_2	b_3	b_4
1	1	1	1	0	0	1	0	0	0
	2	1	1	0	0	0	1	0	0
	3	1	1	0	0	0	0	1	0
	4	1	1	0	0	0	0	0	1
	1	1	1	0	0	−1	−1	−1	−1
2	1	1	0	1	0	1	0	0	0
	2	1	0	1	0	0	1	0	0
	3	1	0	1	0	0	0	1	0
	4	1	0	1	0	0	0	0	1
	5	1	0	1	0	−1	−1	−1	−1
3	1	1	0	0	1	1	0	0	0
	2	1	0	0	1	0	1	0	0
	3	1	0	0	1	0	0	1	0
	4	1	0	0	1	0	0	0	1
	5	1	0	0	1	−1	−1	−1	−1
4	1	1	−1	−1	−1	1	0	0	0
	2	1	−1	−1	−1	0	1	0	0
	3	1	−1	−1	−1	0	0	1	0
	4	1	−1	−1	−1	0	0	0	1
	5	1	−1	−1	−1	−1	−1	−1	−1

These matrices correspond to a parameterization of the model in terms of an intercept, deviation contrasts of the θ-values for the first three score groups, and deviation contrasts of the c values of the first four items. The complete basis matrix generated in this way is shown in Table 4.2.

The Newton iterations for JML estimation of the parameters of the model are carried out as described in Section 2.16.3. Because deviation contrasts must sum to zero, the mean deviated estimate for the fourth score group and for the fifth item are equal to the negative of the sum of the preceding estimates. All of these estimates are shown in Table 4.3.

At convergence of the Newton iterations, the square roots of the diagonal elements of the inverse Hessian matrix provide the large-sample standard

Table 4.3 LSAT6 data: JML estimates for the one-parameter logistic model.

Effect		Estimate	Standard error
$\hat{\mu}$		−0.0014	0.0768
	1	−10.7215	0.2084
	2	−0.5192	0.1105
$\hat{\theta}_k$	3	0.5176	0.0910
	4	1.7231	0.0935
	1	1.5488	0.1139
	2	−0.5614	0.0796
$\hat{b}_j$	3	−1.6285	0.0823
	4	−0.1653	0.0816
	5	0.8065	0.0945

errors of the estimated intercept, attribute values of the first three score groups, and locations of first four items. To obtain the standard error for the fourth score group and the fifth item, we need variances and covariances of the estimated deviation contrasts. These appear in the corresponding blocks of the inverse Hessian matrix. For example, the variance–covariance matrix corresponding to the first four items is shown in Table 4.4. The variance for the fifth item is given by result in Section 2.7 for the variance of a linear combination of random variables with coefficients c_j, for $j = 1, 2, \ldots, m$:

$$s_5^2 = \sum_{i=1}^{5} \sum_{j=1}^{5} c_{5i} c_{5j} \sigma_{ij}. \tag{4.13}$$

Table 4.4 Large-sample variances and covariances of the deviation contrast estimators for the first four items.

		Item			
		1	2	3	4
	1	0.01 298			
Item	2	−0.00 309	0.00 634		
	3	−0.00 336	−0.00 061	0.00 678	
	4	−0.00 309	−0.00 089	−0.00 088	0.00 666

Since all the coefficients are −1 in this case, the required variance is just the sum of the elements of the covariance matrix, or 0.0945. This value and similar values for the other estimates are shown in Table 4.4.

Because of the small number of items and the large sample size, the item locations are estimated very accurately, as the standard errors indicate. The same is true of the attribute values of the score groups, except perhaps for score group 1, where the sample size is small.

Notice, however, that the standard errors for the score groups differ from those of the estimated attribute values for individual respondents shown in Chapter 3. This occurs because the standard error for $\hat{\theta}_k$ in Table 4.4 refers to the *group as a whole* and is a function of the number of respondents in the group. Although this estimate of θ_k is the actual attribute value that would be assigned to a respondent in group k if this were a practical application of the one-parameter model, its standard error here does not take into account the errors of classifying the respondent in the true score group on the basis of the observed score. Those errors are accounted for in the Chapter 4 standard errors, which are therefore considerably larger than those in Table 4.4.

When substituted in the one-parameter model, the estimates in Table 4.4 give the expected probabilities for calculating the ratio of likelihoods under the model and under the general product-binomial alternative. The corresponding chi-square value proves to be 4.17. As there are 20 independent frequencies and 8 parameters estimated, the degrees of freedom are 12. That the corresponding p-value is significantly *large* ($p = 0.980$) is an indication that the item responses within score-groups are not entirely independent. Nevertheless, there is nothing to suggest any lack of fit of the one-parameter model to these data.

Joint estimation for the LSAT-7 data does not show the same good fit to the one-parameter model as we found for LSAT-6. The goodness-of-fit chi square is 48.64 on 12 degrees of freedom. The analysis of these data in Section 4.2.2 reveals considerable differences among the estimated slopes for the five items.

4.2.1.3 Proportional Marginal Adjustments

A disadvantage of the logit-linear implementation of JML estimation for the one-parameter model is the excessive size of the information matrix when the number of items and number of score groups is large. If all score groups have at least one member, the order of the information matrix is $2(n-1)$. The number of elements of this matrix that must be stored is roughly one-half the *square* of this number, and the number of multiplications required for its inversion in each iteration of the logit-linear solution is roughly one-half the *cube* of the number. The numbers become prohibitive if the number of items are greater than perhaps 50 or 60, depending on the computing facilities available.

Beyond that range, a good alternative is the method of *proportional marginal adjustments* described in Bishop et al. (1975). In this procedure, the data are treated as a single contingency table – Score-groups by Items by Response Categories. If score-groups 0 and 5 are omitted, the dimensions of the table are $(n-1) \times n \times 2$, and its degrees of freedom are $2n(n-1)-1$.

By means of iterative marginal adjustments, maximum likelihood estimates of the expected cell frequencies are obtained for the no-three-factor association model. This model is equivalent to the logit model above and has the same residual degrees of freedom, viz., $(n-1)(n-2)$. The fit of the model can be checked with the same likelihood ratio test.

If the model is accepted, the estimated expected frequencies for the RIGHT/WRONG categories can be converted to logits by taking the natural logarithm of the ratio of RIGHTs to WRONGS for each item in each retained score group. The estimates of the score group values and the item locations are then just the marginal means of the score-groups by item table of the expected logits after they have been mean-deviated. As the example below shows, by virtue of the invariance of maximum likelihood estimation under nonsingular transformation (see Section 2.12), the expected logits, and the parameter estimates are exactly equal to those obtained by the logit-linear method.

The advantages of marginal adjustments method are that storage is required only for the tables of observed and expected frequencies and their margins, and that the number of multiplications required for each iteration is only about $9n(n-1)$. Somewhat more iterations are necessary than in the Newton method, but not enough to lose the very considerable computational savings of the marginal adjustments method.

Its main disadvantage is that it does not give directly the standard errors of the estimated item parameters. However, a quite satisfactory approximation is available if one considers the group effects to be known values. In that case, the standard error for the intercept parameter in a logit analysis applies:

$$\mathrm{SE}(\hat{b}_j) = \left[\sum_{k}^{m} N_k P_k Q_k \right]^{-\frac{1}{2}}. \tag{4.14}$$

Example. To perform the proportional adjustments procedure on the LSAT6 data of Table 4.1, we rearrange them in the form of a three-way contingency table and margins, as in Table 4.5.

The successive marginal adjustments procedure results in a table of the same configuration containing the maximum likelihood estimates of the cell frequencies under the constraint that the margins of expected

Table 4.5 LSAT6 data: observed frequencies.

	Items					
RIGHTS	**1**	**2**	**3**	**4**	**5**	**Total**
Score 1	10	1	1	2	6	20
Groups 2	62	24	7	28	49	170
3	212	109	63	139	188	711
4	342	277	184	296	329	1428
Total	626	411	255	465	572	2329
WRONGS	1	2	3	4	5	Total
1	10	19	19	18	14	80
2	22	61	78	57	36	255
3	25	128	174	98	49	474
4	15	80	173	61	28	357
Total	73	288	444	234	127	1166
Total	1	2	3	4	5	Total
1	20	20	20	20	20	100
2	85	85	85	85	85	425
3	237	237	237	237	237	1185
4	357	357	357	357	357	1785
Total	699	699	699	699	699	3495

frequencies shall equal the margins of observed frequencies. The calculations are carried out in successive cycles consisting of the three steps involving the expected frequencies, $\hat{e}_{jk\ell}$, the observed frequencies, $r_{jk\ell}$, and marginal frequencies indicated by a dot replacing a subscript summed over. Initially, all expected cell frequencies, $\hat{e}^{(0)}_{jk\ell}$, are set equal to 1. The steps in one cycle of the computations are as follows:

1. Obtain the $A \times B$ margin of the expected cell frequencies in the current step(i). Then compute

$$\hat{e}^{(i+1)}_{jk\ell} = \frac{\hat{e}^{(i)}_{jk\ell} r_{jk\cdot}}{\hat{e}^{(i)}_{jk\cdot}}; \tag{4.15}$$

   i.e. divide the current expected cell frequency times the corresponding observed $A \times B$ margin by the corresponding expected $A \times B$ margin. Example: $\hat{e}^{(1)}_{111} = 20/2 = 10$.

2. Obtain the $B \times C$ margin of the expected cell frequencies from step 1. Then compute

$$\hat{e}_{jk\ell}^{(i+2)} = \frac{\hat{e}_{jk\ell}^{(i+1)} r_{\cdot k\ell}}{\hat{e}_{\cdot k\ell}^{(i+1)}}; \tag{4.16}$$

i.e. divide the current expected cell frequency times the corresponding observed $B \times C$ margin by the corresponding current expected $B \times C$ margin.
Example: $\hat{e}_{111}^{(2)} = 10 \times 626/349.5 = 17.91$.
3. Obtain the $A \times C$ margin of the expected cell frequencies from step 2. Then compute

$$\hat{e}_{jk\ell}^{(i+3)} = \frac{\hat{e}_{jk\ell}^{(i+2)} r_{j\cdot\ell}}{\hat{e}_{j\cdot\ell}^{(i+2)}}; \tag{4.17}$$

i.e. divide the current expected cell frequency times the corresponding observed $A \times B$ margin by the corresponding expected $A \times B$ margin.
Example: $\hat{e}_{111}^{(1)} = 20/2 = 10$.

These cycles are continued until the expected margins agree with the observed to the required number of places. Table 4.6 shows the converged estimates of the expected cell frequencies.

Taking the natural logarithms of the odds ratios of expected RIGHTs to expected WRONGs, we obtain the expected logits in Table 4.7.

Mean deviating the column and row means from the grand mean (−0.0014), gives the following estimates of the item score-group values and item parameters. They agree with the estimates in Table 4.3 within rounding error.

Score groups: −1.7214, −0.5193, 0.5175, 1.7231
Items: 1.5488, −0.5613, −1.6286, −0.1654, 0.8065

The standard errors of the item parameter estimates are approximated, as described above, from the numbers of respondents in the score groups and the expected proportions of RIGHTS and WRONGS (computed from the expected frequencies in Table 4.6). The resulting values are close enough to the more exact standard errors of Table 4.4 for any practical purpose:

Item-parameter SE: 0.1316, 0.0840, 0.0840, 0.0877, 0.1062.

4.2.2 Marginal Maximum Likelihood Estimation (MML)

The MML method of estimating item parameters is easily applied to almost any item response model, including the multicategory models of Chapter 5 and the

Table 4.6 LSAT6 data: Expected frequencies.

	Items					
RIGHTS	**1**	**2**	**3**	**4**	**5**	**Total**
Score 1	9.132	1.849	0.677	2.629	5.714	20.001
Groups 2	62.606	21.515	8.875	28.470	48.532	169.998
3	210.326	115.821	58.639	139.077	187.140	711.003
4	343.936	271.817	186.808	294.824	330.617	1428.002
Total	626.000	411.002	254.999	465.000	572.003	
WRONGS	1	2	3	4	5	Total
1	10.868	18.151	19.323	17.371	14.287	80.000
2	22.392	63.483	76.125	56.532	36.467	254.999
3	26.675	121.182	178.360	97.923	49.863	474.003
4	13.065	85.184	170.193	62.176	26.385	357.003
Total	73.000	288.000	444.001	234.002	127.002	
Total	1	2	3	4	5	Total
1	20.000	20.000	20.000	20.000	20.001	100.001
2	84.998	84.998	85.000	85.002	84.999	424.997
3	237.001	237.003	236.999	237.000	237.003	1185.006
4	357.001	357.001	357.001	357.000	357.002	1785.005
Total	699.000	699.002	699.000	699.002	699.005	

Table 4.7 LSAT6 data: expected logits.

	Items					
	1	**2**	**3**	**4**	**5**	**Mean**
1	−0.1740	−2.2841	−3.3514	−1.8882	−0.9164	−1.7228
Score 2	1.0282	−1.0820	−2.1491	−0.6860	0.2858	−0.5207
Groups 3	2.0649	−0.0452	−1.1124	0.3508	1.3226	0.5161
4	3.2705	1.1603	0.0931	1.5564	2.5282	1.7217
Mean	1.5474	−0.5627	−1.6300	−0.1668	0.8051	

multidimensional models of Chapter 6. But it introduces a new assumption that the form of the distribution of attribute values in the population from which the respondents are drawn is known or can be estimated along with the item parameters. Its main advantages are in wide applicability, consistency of estimation in the limit of the sample size only, absence of any assumption of homogeneous grouping, and ease of implementation. If the need for population assumptions in the measurement of individual differences is accepted (see Chapter 1), the only disadvantage of the method is the lack of a general goodness-of-fit test, although likelihood ratio tests of alternative models are available. We discuss the properties of MML estimation of item parameters at the end of this section.

Up to this point in the text, we have considered only conditional probabilities of item responses. The response function for item j, represented generically as $P_j(\theta)$, gives the probability of a correct or incorrect response conditional on the respondents attribute value, θ. In connection with the estimation of attribute values in Chapter 3, we further defined the likelihood for respondent i,

$$L_i(\theta) = \prod_j^n [P_j(\theta)]^{x_{ij}} [Q_j(\theta)]^{(1-x_{ij})}, \tag{4.18}$$

given the response pattern

$$x_i = (x_{i1}, x_{i2}, \ldots, x_{in}),$$

on the assumption of conditional independence.

Now, we introduce the unconditional, or marginal, probability of observing the response pattern x_i in data from a sample of respondents from a given population. Similar to the marginal density for a measured variable, the marginal probability, $\bar{P}(x_i)$, of the response pattern is expressed as a definite integral of the likelihood weighted by $g(\theta)$, the population density of θ:

$$\bar{P}(x_i) = \int_{-\infty}^{\infty} L_i(\theta) g(\theta) d\theta. \tag{4.19}$$

We will assume for present purposes that the population distribution of θ is normal (we relax this assumption in Chapter 11). The main points of difference between the present application of MML estimation and that under the normal–normal model in Section 2.15 are the following:

1. Because the scale of θ is arbitrary in IRT models, we are at liberty to set the location and scale of the distribution of θ to any convenient values. For distributions that are indexed by location and scale parameters only, this scale-free property makes the item parameters the only free quantities in the marginal likelihood. In the case of a normal population distribution, the convention in IRT is to fix the scale of θ by setting the population mean to 0 and the standard deviation to 1. The scales of both the item parameters and the attribute

values can then easily be set to any other values by the formulas given in Chapter 3.

2. The definite integral above does not have a closed form for a general item response model. It must be evaluated numerically: for $g(\theta)$ normal, the method of choice is the Gauss–Hermite quadrature formula,

$$\bar{P}(x_i) \approx \sum_{k=1}^{q} L_i(X_k)A(X_k), \tag{4.20}$$

where X_k is quadrature point, or "node," and $A(X_k)$ is the corresponding quadrature weight. Values of the points and weights are tabled or can be generated from widely available computer routines. Error bounds of the approximation to the true value of the integral are available.

If the number of respondents in the sample is large relative to the number of items, there is a computational advantage in counting the frequency of response patterns and expressing the marginal likelihood in terms of these counts. They are sufficient statistics for fitting the 2PL or more complex item response models. Although the same result is obtained in unsorted data by setting the frequency of each respondent's vector of item scores to 1.0, it is worthwhile when the number of respondents is very large relative to the number of items to sort or partially sort the patterns and count the duplications. The computer operations required in sorting and counting are considerably less time-consuming than estimating item parameters directly from the item scores. In the remainder of this chapter, we will assume the data are sorted into distinct patterns and counted, but will understand that all of the results apply as well to unsorted data with unit frequencies.

In any event, suppose there are N respondents in the sample, and let the frequency of pattern ℓ be r_ℓ. Then the marginal likelihood for a vector item parameter, ζ, computed from the sample is

$$L_M(\zeta) = \frac{N!}{\prod_{\ell}^{s} r_\ell!} \prod_{\ell=1}^{s} \bar{P}_\ell^{r_\ell}, \tag{4.21}$$

where s is the number of distinct response patterns.

With n binary items, the maximum number of patterns possible in the population is 2^n, but since no more than N distinct patterns can occur in the sample, $s \leq \min(2^n, N)$. In effect, the pattern frequencies comprise a 2^n contingency table in which many entries are zero if 2^n is large relative to N. Since the terms in the likelihood that correspond to zero frequencies equal unity, the index ℓ needs refer only to the patterns that actually occur. The patterns may be enumerated in any order: for example, by their integer values when read as binary numbers.

To obtain the MML estimator of a vector parameter, ζ_j, of item j, we derive the marginal likelihood equations as follows:

$$\begin{aligned}\frac{\partial \log L_M(\zeta)}{\partial \zeta_j} &= \sum_{\ell=1}^{s} \frac{r_\ell}{\bar{P}_\ell} \cdot \frac{\partial \bar{P}_\ell}{\partial \zeta_j} \\ &= \sum_{\ell=1}^{s} \frac{r_\ell}{\bar{P}_\ell} \int_\theta \frac{L_\ell(\theta)}{[P_j(\theta)]^{x_{\ell j}}[1-P_j(\theta)]^{1-x_{\ell j}}} \\ &\quad \cdot \frac{\partial\{[P_j(\theta)]^{x_{\ell j}}[Q_j(\theta)]^{1-x_{\ell j}}\}}{\partial \zeta_j} g(\theta)d\theta \\ &= \sum_{\ell=1}^{s} \frac{r_\ell}{\bar{P}_\ell} \int_\theta \left(\frac{x_{\ell j} - P_j(\theta)}{P_j(\theta)Q_j(\theta)}\right) L_\ell(\theta) \frac{\partial P_j(\theta)}{\partial \zeta_j} g(\theta)d\theta = 0. \end{aligned} \tag{4.22}$$

To put these equations in a form convenient for an EM solution like that of Section 2.15, we reverse the order of summation and integration to obtain say,

$$\frac{\partial \log L_M(\zeta)}{\partial \zeta_j} = \int_\theta \frac{\bar{r}_j - \bar{N}P_j(\theta)}{P_j(\theta)Q_j(\theta)} \cdot \frac{\partial P_j(\theta)}{\partial \zeta_j} d\theta, \tag{4.23}$$

where

$$\bar{r}_j = \sum_{\ell=1}^{s} \frac{r_\ell x_{\ell j} L_\ell(\theta) g(\theta)}{\bar{P}_\ell} \tag{4.24}$$

and

$$\bar{N} = \sum_{\ell=1}^{s} \frac{r_\ell L_\ell(\theta) g(\theta)}{\bar{P}_\ell}. \tag{4.25}$$

Replacing the definite integral with the Gaussian quadrature formula, we have

$$\frac{\partial \log L_M}{\partial \zeta_j} = \sum_{k}^{q} \frac{\bar{r}_{jk} - \bar{N}_k P_j(X_k)}{P_j(X_k)Q_j(X_k)} \cdot \frac{\partial P_j(X_k)}{\partial \zeta_j}. \tag{4.26}$$

We recognize this result as the gradient vector of a probit or logit analysis (depending on the choice of response function) in which

$$\bar{r}_{jk} = \sum_{\ell=1}^{s} \frac{r_\ell x_{\ell j} L_\ell(X_k) A(X_k)}{\bar{P}_\ell} \tag{4.27}$$

plays the role of the number of successes at level k, and

$$\bar{N}_k = \sum_{\ell=1}^{s} \frac{r_\ell L_\ell(X_k) A(X_k)}{\bar{P}_\ell} \tag{4.28}$$

plays the role of the number of trials at level k.

We see also that the latter quantities are, respectively, the posterior expectation (Bayes estimates, Section 2.13) of the number of correct responses to item j at quadrature point k, and the posterior expectation of the number of *attempts*, or number of respondents, at that point. This part of the calculation is the Expectation, or "E," step of the EM algorithm; the probit or logit analysis computed with these expectations is the Maximization, or "M," step.

As in the EM solution of the MML equations for the normal–normal model described in Section 2.15, the E-step depends upon the results of the M-step, and the M-step depends on those of the E-step, and so the computations must be iterative. It is an important strength of the EM algorithm that the starting values of these iterations are not critical. In IRT applications, the classical estimates of item parameters discussed in Chapter 3 are quite satisfactory initial values.

The main weakness of the algorithm is that its convergence is first-order, which means that the improvement at each iteration becomes very small as the solution point is approached; moreover, the changes propagate into higher decimal places so that the accuracy of the result after a finite number of iterations cannot be guaranteed. The other shortcoming is that the EM solution does not provide a direct estimate of the standard errors of the estimates.

The opposite is true of Newton and Fisher-scoring solutions: especially in nonlinear multiparameter estimation where convexity of the likelihood surface cannot be assumed, these methods typically require rather close approximations to the solution as starting points, otherwise, they will diverge. If the starting points are within the circle of convergence, however, the improvement on each iteration is second-order, the changes do not propagate upward, and accuracy of decimal places to the left of the corrections is certain. In addition, these solutions provide in the inverse Hessian matrix or inverse information matrix the large-sample standard errors of the estimates.

For these reasons, the preferred approach to MML estimation of item parameters is a series of EM iterations followed by a few Newton or Fisher-scoring iterations to fix the accuracy of the result and to provide the required standard errors. To obtain the information matrix for the Fisher-scoring solution, we use the general result for multinomial distributions given in Section 2.15. In the present case, the information matrix is

$$I(\zeta) = \sum_{\ell=1}^{2^n} \frac{1}{\bar{P}_\ell} \cdot \frac{\partial \bar{P}_\ell}{\partial \zeta_j} \cdot \frac{\partial \bar{P}_\ell}{\partial \zeta_j'}. \tag{4.29}$$

The problem with this formula is that the summation, which involves only population values, has 2^n terms and is therefore unusable when n is large. For practical computation, we approximate the matrix by summing the cross-products of the first derivatives of the log marginal likelihood over the observed patterns weighted by the pattern frequencies. The result is referred to as the *empirical* information matrix.

Application of the EM and Fisher-scoring methods to MML item parameter estimation for the two-parameter logistic model appears in Section 4.2.2.1.

4.2.2.1 The two-parameter Model

For the two-parameter model, MML estimation employs as sufficient statistics the frequencies of the response *patterns*, not just those of the score-groups. Patterns for three items from the LSAT-6 data appear in this form in Table 4.8, and those of the LSAT-7 data, in Table 4.16.

As previously, the solution of the likelihood equations is better conditioned if we parameterize the model terms of the intercept, c_j, and slope, a_j. Then,

$$P_j(\theta) = 1/(1 + e^{-(c_j + a_j\theta)}),$$
$$Q_j(\theta) = 1 - P_j(\theta) = e^{-(c_j - a_j\theta)}/(1 + e^{-(c_j + a_j\theta)}),$$

and

$$\frac{\partial P_j(\theta)}{\partial \theta} = P_j(\theta)Q_j(\theta). \tag{4.30}$$

The likelihood equations expressed as quadratures, therefore, simplify to

$$G(c_j) = \sum_k^q [\bar{r}_{jk} - \bar{N}_k P_j(X_k)],$$
$$G(a_j) = \sum_k^q [\bar{r}_{jk} - \bar{N}_k P_j(X_k)]X_k,$$

Table 4.8 Response pattern frequencies for items 3, 4, and 5 of the LSAT-6 data.

ℓ	Pattern $[x_{\ell j}]$	Frequency $r_{\ell j}$
1	0 0 0	30
2	0 0 1	99
3	0 1 0	37
4	0 1 1	281
5	1 0 0	15
6	1 0 1	93
7	1 1 0	48
8	1 1 1	397
	Total	1000

where $\bar{r}_{jk}$ and $\bar{N}_k$ are the posterior expectations of number right and number of respondents, as defined above.

EM-Solution Because of the insensitivity of the the EM-method to the choice of starting values, it is quite satisfactory for the 2PL model to set $c_j = 0$ and $a_j = 1$, for all j. For the first E-step, one then proceeds as follows:

1. Substitute the starting values in the 2PL response function to obtain the $P_j(X_k)$ and $Q_j(X_k)$ for $k = 1, 2, \dots, q$, where the X_k are the chosen quadrature points.
2. Pass through the response patterns that occur in the sample and compute the marginal probability of each by the quadrature

$$\bar{P}_\ell = \sum_k^q \prod_j^n [P_j(X_k)]^{x_{\ell j}} [Q_j(X_k)]^{(1-x_{\ell j})} A(X_k), \tag{4.31}$$

   where $A(X_k)$ is the prior weight at the point X_k.
3. At the same time, accumulate for each item and quadrature point, the $n \times q$ table of posterior expected number rights,

$$\bar{r}_{jk} = A(X_k) \sum_\ell^s \frac{r_\ell x_{\ell j}}{\bar{P}_\ell} \prod_j^n [P_j(X_k)]^{x_{\ell j}} [Q_j(X_k)]^{(1-x_{\ell j})}, \tag{4.32}$$

   where r_ℓ is the frequency of pattern ℓ in the sample.
4. At the same time, accumulate the posterior expected number of attempts, $\bar{N}_k$ by the sum in 3, omitting the item score $x_{\ell j}$.

For the first M-step, one performs a logit analysis for the 2-PL model on each item in turn exactly as in Section 4.1 in the case of an external criterion. The quadrature points X_k correspond to the criterion values; the posterior expected number right $\bar{r}_{jk}$, to the group frequencies; and the posterior expected number of attempts $\bar{N}_k$, to the number of respondents per group.

It is not necessary to obtain the maximum likelihood estimates of the item parameters very precisely in any one M-step. Three or four iterations of the Newton solution in each M-step will give increasingly accurate values as the EM cycles continue.

The results of the first M-step provide the new provisional values for the second E-step, computed in the same way as above. The cycles of successive E- and M-steps are then continued until the largest change in any parameter value becomes acceptably small. It is advisable to monitor the quantity

$$\log L_M = \sum_\ell^s r_\ell \log \bar{P}_\ell \tag{4.33}$$

to check on the change of marginal log likelihood in succeeding cycles.

Example. To illustrate the calculations of the EM-solution in a problem small enough to display in detail, we will fit the 2PL model to items 3, 4, and 5 of the LSAT-6 data analyzed above. The sufficient statistics for MML estimation of the item parameters are the frequencies of the eight response patterns given in Table 4.8.

Initiating the first cycle of the EM procedure, we set intercepts for the three items to 0, and the slopes to 1, in the E-step. To keep the calculations simple, we will use only four quadrature points: -1.5, -0.5, 0.5, and -1.5. With this choice of starting values, the logits for the three items at each point have the same value as the point, and the response probabilities shown in section A of Table 4.9 have the same values at each point. In the second EM cycle, this will no longer be the case. (See Table 4.11.)

To compute the expected number of attempts and expected number of right responses at each of these points, we need the likelihoods for each response pattern as shown in Section B. These values are computed from the response probabilities and their complements, depending whether the item score is 1 or 0 in the pattern. For example, the likelihood for pattern 0 1 0 at quadrature point 1 is $0.1824 \times 0.8176 \times 0.1824 = 0.5465$. Because the starting values for the iterations are the same for all items, there are only four distinct values of the likelihoods at each quadrature point in this cycle.

The results of the first E-step are the expected numbers of attempts (i.e. numbers of respondents) at each quadrature point and the expected number of correct responses at each point. These are the sufficient statistics for estimating updated provisional values of the parameters for each item in the first M-step.

The M-step applies the logit analysis of Section 2.16 to these so-called "complete data" statistics. The logit analysis is also iterative but requires only a few iterations to obtain sufficiently accurate estimates. The calculations for the first of these M-step iterations are shown in section A of Table 4.10. The solutions of iterative least-squares equations provide corrections to the initial values of the item parameters. Conveniently, these calculations are performed one item at a time and require minimal storage of intermediate values. The final, updated provisional values and the number of M-step iterations required to produce them are shown in section B of the table.

The updated provisional estimates are then carried into the E-step of cycle 2 as shown in Table 4.11. The calculations repeat those of the previous E-step, but with the changed values of the response probabilities that arise from the new provisional parameter estimates. The results are again the expected attempts and rights that are the statistics for the subsequent M-step, shown here in Table 4.12. That the negative of the

Table 4.9 EM calculations, cycle 1: E-step response patterns, likelihoods, marginal probabilities, expected attempts, and rights. LSAT-6 items, $j = 3, 4, 5$; $\hat{c}_j^{(1)} = (0, 0, 0)$, $\hat{a}_j^{(1)} = (1, 1, 1)$.

A. Quadrature points X_k, weights $A(X_k)$, and 2PL response probabilities $\Psi(z_{jk}), z_{jk} = \hat{c}_j^{(1)} + \hat{a}_j^{(1)} X_k$

X_k	=	−1.5	−0.5	0.5	1.5
$A(X_k)$	=	0.1345	0.3655	0.3655	0.1345
$\Psi(z_{jk})$	=	0.1824	0.3775	0.6225	0.8176
$j = 3, 4, 5$		0.1824	0.3775	0.6225	0.8176
		0.1824	0.3775	0.6225	0.8176

B. Response patterns $x_{\ell j}$, likelihoods $L_\ell(X_k)$, and marginal probabilities $\bar{P}_\ell$

ℓ	$[x_{\ell j}]$	$L_\ell(X_1)$	$L_\ell(X_2)$	$L_\ell(X_3)$	$L_\ell(X_4)$	$\bar{P}_\ell$
1	000	0.5465	0.2412	0.0538	0.0061	0.182 13
2	001	0.1219	0.1463	0.0887	0.0272	0.105 96
3	010	0.1219	0.1463	0.0887	0.0272	0.105 96
4	011	0.0272	0.0887	0.1463	0.1219	0.105 96
5	100	0.1219	0.1463	0.0887	0.0272	0.105 96
6	101	0.0272	0.0887	0.1463	0.1219	0.105 96
7	110	0.0272	0.0887	0.1463	0.1219	0.105 96
8	111	0.0061	0.0538	0.2412	0.5465	0.182 13

C. Expected attempts $\bar{N}_k$, expected rights $\bar{\tau}_{jk}$

$\bar{N}_k$	=	51.82	262.76	454.58	230.84
$\bar{\tau}_{jk}$	=	8.97	93.60	267.90	182.52
$j = 3, 4, 5$		18.87	162.25	369.51	212.38
		30.01	207.31	411.20	221.48

$L_\ell(X_k) = \Pi_{j=3}^{5} [\Psi(z_{jk})]^{x_{\ell j}} [1 - \Psi(z_{jk})]^{1-x_{\ell j}}$
$\bar{P}_\ell = \sum_{k=1}^{4} L_\ell(X_k) A(X_k)$
$\bar{N}_k = \sum_{\ell=1}^{8} \tau_\ell L_\ell(X_k) A(X_k) / \bar{P}_\ell$
$\bar{\tau}_{jk} = \sum_{\ell=1}^{8} \tau_\ell x_{\ell j} L_\ell(X_k) A(X_k) / \bar{P}_\ell$

Table 4.10 EM calculations, cycle 1: M-step first iteration corrections[a] and provisional estimates[b].

A. First M-step iteration corrections

Item 3							
$210.735\delta_c + 62.587\delta_a$	=	-27.340;	$\hat{c}_3$	=	0.0–0.117	=	–0.117
$62.587\delta_c + 137.000\delta_a$	=	-13.306;	$\hat{a}_3$	=	1.0–.044	=	0.956
Item 4							
$210.735\delta_c + 62.587\delta_a$	=	182.660;	$\hat{c}_4$	=	0.0+0.920	=	0.920
$62.587\delta_c + 137.000\delta_a$	=	33.110;	$\hat{a}_4$	=	1.0–0.179	=	–0.179
Item 5							
$210.735\delta_c + 62.587\delta_a$	=	289.660;	$\hat{c}_5$	=	0.0+1.339	=	1.339
$62.587\delta_c + 137.000\delta_a$	=	28.356;	$\hat{a}_5$	=	1.0–0.429	=	0.571

B. Final provisional estimates from cycle 1 M-step

Item	Iterations	$\hat{c}_j^{(2)}$	$\hat{a}_j^{(2)}$
3	2	–0.116	0.961
4	4	0.968	0.995
5	5	1.775	0.945

C. Final marginal log likelihood and largest change

from cycle 1 $-2\log L_M = 3272.5877$

Largest change = 1.775

a) $$\left(\sum_{k=1}^{4} \bar{N}_k \Psi(z_{jk})[1 - \Psi(z_{jk})] \begin{bmatrix} 1 \\ X_k \end{bmatrix} [1, X_k]\right) \begin{bmatrix} \delta_c \\ \delta_a \end{bmatrix} = \sum_{k=1}^{4} \left[r_{kj} - \bar{N}_k \Psi(z_{jk}) \right].$$

b) $$\hat{c}_j^{(2)} = \hat{c}_j^{(1)} + \delta_c$$
$$c_j^{(2)} = a_j^{(1)} + \delta_a.$$

marginal log likelihood has decreased from the previous M-step indicates that the corrections are moving in the desired direction.

The Newton solution The EM cycles can be continued in the manner of the above example until the parameter estimates become stable. The first-order

Table 4.11 EM calculations, cycle 2: E-step response patterns, likelihoods, marginal probabilities, expected attempts and rights. $\hat{c}_j^{(2)} = (-0.116, 0.968, 1.775)$ $\hat{a}_j^{(2)} = (0.961, 0.995, 0.945)$.

A. Quadrature points, weights, and 2PL response probabilities

	1	2	3	4
X_k	−1.5	−0.5	0.5	1.5
$A(X_k)$	0.1345	0.3655	0.3655	0.1345
$\Psi(z_{jk})$	0.1741	0.3553	0.5900	0.7900
$j = 3, 4, 5$	0.3719	0.6156	0.8124	0.9214
	0.5883	0.7863	0.9045	0.9606

B. Response patterns, likelihoods, and marginal probabilities

ℓ	$[x_{\ell j}]$	$L_\ell(X_1)$	$L_\ell(X_2)$	$L_\ell(X_3)$	$L_\ell(X_4)$	$\bar{P}_\ell$
1	000	0.2136	0.0530	0.0073	0.0013	0.050 85
2	001	0.3052	0.1949	0.0696	0.0310	0.139 86
3	010	0.1265	0.0848	0.0318	0.0149	0.060 66
4	011	0.1807	0.3121	0.3013	0.3629	0.273 46
5	100	0.0450	0.0292	0.0106	0.0024	0.020 93
6	101	0.0643	0.1074	0.1001	0.0596	0.092 54
7	110	0.0017	0.0467	0.0458	0.0287	0.041 27
8	111	0.0001	0.1720	0.4335	0.6992	0.320 42

C. Expected attempts, expected rights

$\bar{N}_k$	104.89	342.82	395.17	157.12
r_{jk}	23.55	144.82	255.37	129.26
$j = 3, 4, 5$	45.85	233.86	336.03	147.26
	69.07	284.96	364.25	151.72

Table 4.12 EM calculations, cycle 2: *M*-step first iteration corrections and provisional estimates.

A. First M-step corrections

Item 3

$215.249\delta_c + 24.999\delta_a$	=	55.618;	$\hat{c}_3$	=	$-0.116 + 0.265$	= 0.149
$24.999\delta_c + 136.106\delta_a$	=	-0.685;	$\hat{a}_3$	=	$0.961 - 0.054$	= 0.907

Item 4

$177.237\delta_c - 30.123\delta_a$	=	47.170;	$\hat{c}_4$	=	$0.968 + 0.262$	= 1.230
$-40.914\delta_c + 93.478\delta_a$	=	-10.455;	$\hat{a}_4$	=	$0.955 - 0.022$	= 0.973

Item 5

$123.096\delta_c - 40.914\delta_a$	=	30.370;	$\hat{c}_5$	=	$1.775 + 0.230$	= 2.005
$-40.914\delta_c + 93.478\delta_a$	=	-14.131;	$\hat{a}_5$	=	$0.945 - 0.050$	= 0.895

B. Final provisional estimates from cycle 2 M-step

Item	Iterations	$\hat{c}_j^{(3)}$	$\hat{a}_j^{(3)}$
3	3	0.145	0.922
4	3	1.246	0.986
5	3	2.024	0.897

C. Marginal log likelihood and largest change from cycle 2

$-2 \log L_M = 3226.2669$ Largest change = 0.278

improvement provided by the process at each step soon becomes only minimally productive, however. At that point, we switch to second-order Fisher-scoring (Gauss–Newton) iterations, which are more time-consuming but yield greater improvement in the approximation. They also have the advantage of providing large-sample standard errors of the estimates at convergence. If the largest change in provisional estimates between EM cycles is reduced to 0.01 or less, not more than one or two Newton cycles will be necessary for another order of magnitude in accuracy. The steps in one cycle of the Newton solution are as follows:

1. Using the provisional parameter estimates from the most recent EM cycle, compute the logits and response probabilities of the items at each quadrature point (Table 4.13).

2. Evaluate the likelihoods (Table 4.13) for each response pattern, and from them and the quadrature weights, compute the estimated marginal probability of the each pattern.
3. Compute for each pattern the derivatives of the log marginal probabilities with respect to the item parameters (Table 4.14).
4. Sum the derivatives over the patterns weighted by the frequencies of the patterns to obtain the first derivatives (gradient vector) of the log marginal likelihood (Table 4.15).
5. At the same time, compute the sum of cross-products of the derivatives of the log probabilities weighted by the pattern frequencies to obtain the empirical information matrix. If there are more distinct patterns than free parameters, this matrix will in general be positive-definite. With unfavorable data, it may be necessary to condition the matrix by the addition of small positive quantities to the diagonal elements.
6. Invert the empirical information matrix and premultiply the gradient vector by the inverse matrix to obtain the correction vector.
7. Add the corrections to the corresponding provisional parameter estimates to obtain improved values.
8. Compute the marginal log likelihood from the improved values to verify that it has increased from the previous value.
9. Recompute the information matrix from the improved values and take the square roots of the diagonal elements of its inverse to obtain the large-sample standard errors of the corresponding parameter estimates.

The calculations of a full Newton solution become impractical when n is large because of the large size of the information matrix.

Example. We continue the previous example with a Newton step beginning from the provisional estimates of the second EM cycle. Calculations of the pattern likelihoods and marginal probabilities are shown in Table 4.13.

From these quantities, the derivatives of the log marginal probabilities of each pattern are computed as shown in Section A of Table 4.14. The sum of their cross-products weighted by the pattern frequencies from Table 4.14 is the empirical information matrix shown in Section B.

The weighted sum of the derivatives from Table 4.14 is the gradient vector in Section A of Table 4.15. Its product with the inverse information matrix gives the correction terms shown in Section C. Adding the corrections to the corresponding provisional parameter estimates gives the improved values in Section D. Since these are not fully converged values, we do not compute standard errors at this point, but would do so at convergence.

Table 4.13 Newton calculations, cycle 3: Response patterns, likelihoods, and marginal probabilities.

$\hat{c}_j^{(3)} = (.145, 1.246, 2.024); \hat{a}_j^{(3)} = (0.922, 0.986, 0.897)$

A. Quadrature points, weights, and 2PL response probabilities

	1	2	3	4
X_k	−1.5	−.5	0.5	1.5
$A(X_k)$	0.1345	0.3655	0.3655	0.1345
$\Psi(z_{jk})$	0.2251	0.4219	0.6470	0.8216
$j = 3, 4, 5$	0.4420	0.6798	0.8506	0.9385
	0.6635	0.8286	0.9222	0.9667

B. Response patterns, likelihoods, and marginal probabilities

ℓ	$[x_{\ell j}]$	$L_\ell(X_1)$	$L_\ell(X_2)$	$L_\ell(X_3)$	$L_\ell(X_4)$	$\bar{P}_\ell$
1	000	0.1455	0.0317	0.0041	0.0004	0.03 273
2	001	0.2869	0.1534	0.0486	0.0106	0.11 388
3	010	0.1153	0.0674	0.0234	0.0056	0.04 943
4	011	0.2273	0.3256	0.2769	0.1619	0.27 260
5	100	0.0423	0.0232	0.0075	0.0017	0.01 713
6	101	0.0833	0.1119	0.0891	0.0488	0.09 124
7	110	0.0335	0.0492	0.0428	0.0257	0.04 156
8	111	0.0660	0.2376	0.5075	0.7454	0.03 811

The negative marginal log likelihood has again decreased as required. The largest change is larger than in the previous EM cycle because Newton steps are generally larger than EM steps as the solution point is approached.

Test of Fit The test of fit of the model in comparison with the general multinomial alternative is also available in the two parameter case, but the sufficient statistics are the frequencies of the 2^n score groups rather than of the $n - 1$ score groups. Because this test breaks down when the expected frequencies become small, however, it is not usable in applications where the sample size is limited compared to the large number of patterns when the items are numerous. Fortunately, the likelihood ratio test of alternative models is not similarly affected by small expected pattern frequencies. In large samples, minus twice the difference in the maximum marginal log likelihood under the null and alternative models is distributed

Table 4.14 Newton calculations, cycle 3: terms in first derivatives; information matrix

A. Unweighted derivatives of the log marginal probabilities with respect to parameters[a)]

Pattern	c_3	a_3	c_4	a_4	c_5	a_5
1	−0.3148	0.2598	−0.5458	0.4955	−0.7343	0.7188
2	−0.3951	0.1521	−0.6291	0.3080	0.2111	−0.2066
3	−0.4049	0.1363	0.3614	−0.3281	−0.7950	0.4171
4	−0.5152	−0.0885	0.2628	−0.1287	0.1442	−0.0756
5	0.6022	−0.4968	−0.6318	0.3010	−0.7906	0.4417
6	0.4932	−0.1899	−0.7303	−0.0197	0.1483	−0.0828
7	0.4816	−0.1621	0.2601	−0.1240	−0.8574	−0.0073
8	0.3682	0.0633	0.1747	0.0047	0.0934	0.0008

B. Empirical information matrix[b)]

	c_3	a_3	c_4	a_4	c_5	a_5
c_3	192.075					
a_3	−5.323	17.114				
c_4	−21.348	1.523	143.312			
a_4	6.452	9.234	−43.928	27.537		
c_5	−16.787	2.154	−7.912	1.393	99.986	
a_5	6.132	4.723	2.495	4.247	−41.535	31.337

a) $\partial \log \bar{P}_\ell / \partial[c_j, a_j] = \frac{1}{\bar{P}_\ell} \sum_{k=1}^{4} \left[x_{\ell j} - \Psi_j(c_k)\right] L_\ell(X_k) A(X_k) \cdot [1, X_k]$.

b) $I[c,a] = \sum_{\ell=1}^{7} r_\ell \frac{\partial \log \bar{P}_\ell}{\partial [c,a]'} \cdot \frac{\partial \log \bar{P}_\ell}{\partial [c,a]}$.

as chi-square with degrees of freedom equal to the difference in the number of fitted parameters in the two models. If the 1PL is the null model and the 2PL the alternative, for example, this chi-square statistic tests the hypothesis that the item-discriminating parameters are equal. The degrees of freedom equal the number of items.

Example. Having found the LSAT-7 data to exhibit poor fit in JML and CML estimation for the 1PL model, let us see if the 2PL model is an improvement. If the 2PL model fits satisfactorily, we will be able to assess the significance of the improvement by inspecting the increase in the log likelihood relative to the corresponding change in the degrees of freedom. The LSAT-7

Table 4.15 Newton calculations, cycle 3: corrections and updated estimates.

A. First derivatives (gradient vector)[a]

c_3	a_3	c_4	a_4	c_5	a_5
15.870	−4.755	13.006	−4.333	7.827	−5.825

B. Inverse information matrix

	c_3	a_3	c_4	a_4	c_5	a_5
c_3	0.00 544					
a_3	0.00 239	0.10 266				
c_4	0.00 037	−0.02 075	0.01 875			
a_4	−0.00 152	−0.06 447	0.03 731	0.12 004		
c_5	0.00 087	−0.01 081	0.00 005	−0.00 315	0.02 495	
a_5	−0.00 009	−0.01 988	−0.00 343	−0.01 341	0.03 495	0.08 334

C. Corrections

c_3	a_3	c_4	a_4	c_5	a_5
0.094	−0.409	0.207	0.302	0.071	−0.105

D. Updated estimates

c_3	a_3	c_4	a_4	c_5	a_5
0.239	0.513	1.453	1.287	2.095	0.792

$-\partial \log L_M = 3221.1974$ Largest change = 0.409

a) $G(c_j, a_j) = \sum_{\ell=1}^{s} r_\ell \dfrac{\partial \log \bar{P}_\ell}{\partial [c_j, a_j]}$.

data are shown in Table 4.16 in the form of response pattern frequencies suitable for a 2PL analysis.

After 17 EM cycles and one Newton cycle, the parameters and their standard errors are estimated as shown in Table 4.17. We show both the intercept and slope estimates and the location parameter estimates calculated from them. Wide variation in the slope estimates is apparent,

Table 4.16 Observed and expected response pattern frequencies: LSAT-7.

Pattern		Observed	Expected	Pattern		Observed	Expected
1	00000	12	10.09	17	10000	7	12.79
2	00001	19	18.50	18	10001	39	32.56
3	00010	1	4.50	19	10010	11	8.03
4	00011	7	10.66	20	10011	34	26.25
5	00100	3	4.94	21	10100	14	13.31
6	00101	19	15.88	22	10101	51	59.74
7	00110	3	3.95	23	10110	15	15.07
8	00111	17	16.44	24	10111	90	89.31
9	01000	10	3.94	25	11000	6	8.07
10	01001	5	10.34	26	11001	25	29.29
11	01010	3	2.55	27	11010	7	7.32
12	01011	7	8.60	28	11011	35	34.54
13	01100	7	4.40	29	11100	18	19.46
14	01101	23	20.41	30	11101	136	130.29
15	01110	8	5.16	31	11110	32	33.42
16	01111	28	31.65	32	11111	308	308.54
					Total	1000	

$G^2 = 31.71, df = 21, p = .0625$

so we would expect an improvement over the 1PL solution. It is indicated by the chi-square test of fit relative to the multinomial alternative, where the probability of the statistic on 21 degrees of freedom is just short of the 0.05 level. The fit is formally acceptable, although we shall see in Chapter 6 that further improvement in fit is possible if a two-dimensional model is adopted.

A direct test of the hypothesis of homogeneous discriminating power is provided by the difference of this chi-square and that previously calculated for the 1PL model. The value

$$43.89 - 31.71 = 12.18$$

with degrees of freedom

$$26 - 21 = 5$$

has probability less than 0.03. The evidence is somewhat stronger than that of the goodness-of-fit test because the hypothesis is more restrictive.

Table 4.17 Item parameter estimates for the LSAT-7 data: 2PL model, logistic metric.

Item	Intercept (SE)	Slope (SE)	Location[a] (SE)
1	1.856	0.987	−1.880
	(0.130)	(0.176)	(0.265)
2	0.808	1.081	−0.748
	(0.091)	(0.169)	(0.110)
3	1.804	1.706	−1.058
	(0.204)	(0.321)	(0.116)
4	0.486	0.765	−0.635
	(0.075)	(0.135)	(0.132)
5	1.854	0.736	−2.521
	(0.115)	(0.153)	(0.455)

a) $b_j = -c_j/a_j$.

Essentially, the same chi-square value is obtained by subtracting −2 times the log marginal likelihoods from the 1PL and 2PL solutions:

5329.8030 − 5317.618 = 12.185.

The latter method has the advantage of relative insensitivity to small expected values or zero observed frequencies, whereas the goodness-of-fit chi-squares becomes unstable under these conditions. The change in $-2 \log L_M$, with degrees of freedom equal to the change in the number of independent parameters estimated, is therefore a useful statistic for investigating the improvement of fit of nested hierarchies of response models.

5

Item Parameter Estimation – Polytomous Data

It is rather surprising that systematic studies of human abilities were not undertaken until the second half of the last century An accurate method was available for measuring the circumference of the earth 2,000 years before the first systematic measures of human ability were developed.

(Source: Nunnally (1967))

In the following we present the results essential to marginal maximum likelihood (MML) estimation of parameters for polytomous item response models. They concisely unify, clarify, and extend scattered results now in the item response theory (IRT) literature. The treatment includes the following polytomous models now in wide use: nominal categories, graded categories, generalized partial credit in both its original and rating scale versions. For all models, derivations of likelihood equations are given for both the EM item-by-item solution and the Newton–Gauss solution for all items jointly. Computing formulas for the gradient vectors and information matrices are included. Also considered are boundary problems in the solutions and issues of failure of assumptions.

5.1 General Results

Suppose that the data record of respondent $i = 1, 2, \dots, N$, consists of integers, $x_{ij} = 1, 2, \dots, m_j$, indicating assignment of a response to one of m_j mutually exclusive categories of item $j = 1, 2, \dots, n$. Let the categories be indexed by $k = 1, 2, \dots, m_j$, and by $h = 1, 2, \dots, m_j$. The values of x_{ij} are referred to as item category *scores* and, collectively for all items, as response *pattern i*.

Item Response Theory, First Edition. R. Darrell Bock and Robert D. Gibbons.

It is convenient in derivations to define an incidence variable, u_{ijhk}, such that when $k = x_{ij}$,

$$u_{ijhk} = \begin{cases} 1, & h = k \\ 0, & h \neq k, \end{cases} \tag{5.1}$$

where h is a designated category of item j and k is the category assigned to item j in pattern i, represented here as the n-element vector, $\mathbf{u}_i = [u_{ijhk}]$.

To model the response process, let the probability of a response in category h of item j be a function of θ,

$$P_{jh}(\theta) = P(y_{jh}), \tag{5.2}$$

under the restriction

$$\sum_{h=1}^{m_j} P_{jh}(\theta) = 1. \tag{5.3}$$

The argument of the function has the form,

$$y_{jh} = f(\theta \mid \mathbf{v}_j), \tag{5.4}$$

where $\mathbf{v}_j$ is an n_j-element vector of parameters of item j, and θ is an unobservable continuous variable attributed to the respondent. The response function is assumed twice-differentiable with respect to θ and to $\mathbf{v}_j$. In the present context, θ is assumed unidimensional but in item factor analysis may be multidimensional.

Under the restriction, only $m_j - 1$ of the category response function are independent. This implies that only $m_j - 1$ of category parameters are estimable, and that sum of the m_j derivatives of the functions with respect to the category parameters equals zero, as does the sum of the corresponding second derivatives.

On the assumption that item responses are conditionally independent, given θ, the conditional likelihood of $\mathbf{u}_i$ with respect to the item parameters is

$$\begin{aligned} \mathcal{L}_i(\theta) &= \prod_j^n \prod_h^{m_j} P_{jh}(\theta)^{u_{ijhk}} \\ &= \exp \sum_j^n \sum_h^{m_j} u_{ijhk} \log P_{jh}(\theta). \end{aligned} \tag{5.5}$$

Now suppose that, for a given population, there is a probability sample of N respondents each of whom respond to each of the n items. Let $g(\theta)$ be the probability density of θ in the population and let the marginal probability of $\mathbf{u}_i$ be

$$\mathcal{P}_i = \int \mathcal{L}_i(\theta) g(\theta) d\theta. \tag{5.6}$$

In computation, this and similar integrations can be approximated as closely as necessary by quadrature:

$$\mathcal{P}_i \simeq \sum_q^Q \mathcal{L}_i(\theta)(X_q)A(X_q), \tag{5.7}$$

where $X_q, q = 1, 2, \ldots, Q$ are the quadrature points and $A(X_q)$ the corresponding weights normed so that $\sum_q^Q A(X_q) = 1$. If $g(\theta)$ is a normal density, the Gauss–Hermite points and weights may be used. If the number of points is relatively large, equally spaced points and corresponding normal densities normed to unity will serve as well.

For purposes of MML estimation, let the marginal likelihood of the sample with respect to the model parameters be

$$L_N = \prod_i^N \mathcal{P}_i^{w_i}, \tag{5.8}$$

where w_i is a case weight of pattern i, or a frequency if the data are aggregated by pattern, and $\sum_i^N w_i = N$.

The derivative of the marginal log likelihood with respect to a general item parameter, v_j, is

$$\frac{\partial \log L_N}{\partial v_j} = \sum_i^N \frac{w_i}{\mathcal{P}_i} \int \left(\sum_h^{m_j} \frac{u_{ijhk}}{P_{jh}(\theta)} \frac{\partial P_{jh}(\theta)}{\partial v_j} \right) \mathcal{L}_i(\theta) g(\theta) d\theta. \tag{5.9}$$

In the E-step of EM item parameter estimation, the order of summation and integration is exchanged to obtain the so-called complete data statistics:

$$\overline{r}_{jh}(\theta) = \sum_i^N \frac{w_i}{\mathcal{P}_i} u_{ijhk} \mathcal{L}_i(\theta), \tag{5.10}$$

$$\overline{N}(\theta) = \sum_i^N \frac{w_i}{\mathcal{P}_i} \mathcal{L}_i(\theta). \tag{5.11}$$

Note that $\sum_h^{m_j} \overline{r}_{jh}(\theta) = \overline{N}(\theta)$ and $\int \overline{N}(\theta) g(\theta) d\theta = N$.

With these statistics, the general form of the likelihood equation for v_j becomes

$$\int \left(\sum_h^{m_j} \frac{\overline{r}_{jh}(\theta)}{P_{jh}(\theta)} \frac{\partial P_{jh}(\theta)}{\partial v_j} \right) g(\theta) d\theta = 0, \tag{5.12}$$

the specific form of which depends upon the assumed response function.

Numerical solution of (5.12) in the M-step is a standard procedure in nonlinear maximum likelihood estimation. It consists of Newton–Gauss iterations (also called Fisher scoring) starting from given initial values of $\hat{\mathbf{v}}_j^{(t)}$:

$$\hat{\mathbf{v}}_j^{(t+1)} = \hat{\mathbf{v}}_j^{(t)} + \mathcal{I}\left(\hat{\mathbf{v}}_j^{(t)}\right)^{-1} \mathcal{G}\left(\hat{\mathbf{v}}_j^{(t)}\right), \tag{5.13}$$

where $\mathcal{G}(\hat{\mathbf{v}}_j)$ is the gradient vector and $\mathcal{I}(\hat{\mathbf{v}}_j)$ is the information matrix.

The M-step gradient vector contains the first derivatives of the log likelihood with the provisional values of the parameter estimates at each iteration. Evaluated by quadrature,

$$\mathcal{G}(\hat{\mathbf{v}}_j) = \sum_q^Q \left(\sum_h^{m_j} \frac{\bar{r}_{jh}(\theta)}{P_{jh}(\theta)} \frac{\partial P_{jh}(\theta)}{\partial \mathbf{v}_j} \right) A(X_q). \tag{5.14}$$

The M-step information matrix is the negative of the expected value of the matrix of second derivatives of the log likelihood, the Hessian matrix:

$$\mathcal{H}(\hat{\mathbf{v}}_j, \hat{\mathbf{v}}_j') = \int \left(-\sum_h^{m_j} \frac{\bar{r}_{jh}(\theta)}{P_{jh}(\theta)^2} \frac{\partial P_{jh}(\theta)}{\partial \mathbf{v}_j} \frac{\partial P_{jh}(\theta)}{\partial \mathbf{v}_j'} + \sum_h^{m_j} \frac{\bar{r}_{jh}(\theta)}{P_{jh}(\theta)} \frac{\partial^2 P_{jh}(\theta)}{\partial \mathbf{v}_j \partial \mathbf{v}_j'} \right) g(\theta) d\theta. \tag{5.15}$$

Upon taking expectation, $\bar{r}_{jh}(\theta) = \overline{N}(\theta) P_{jh}(\theta)$, and for all of the common item response models, the second term in the Hessian matrix vanishes, simplifying the computations with little loss in speed of convergence. With change of sign, the general M-step information matrix evaluated by quadrature is then

$$\mathcal{I}\left(\hat{\mathbf{v}}_j, \hat{\mathbf{v}}_j'\right) = \sum_q^Q \overline{N}(X_q) \left(\sum_h^{m_j} \frac{1}{P_{jh}(X_q)} \frac{\partial P_{jh}(X_q)}{\partial \mathbf{v}_j} \frac{\partial P_{jh}(X_q)}{\partial \mathbf{v}_j'} \right) A(X_q). \tag{5.16}$$

In many applications good starting values for the first M-step can be derived from a preliminary classical item analysis of the data. For subsequent steps, the provisional estimates from the preceding M-step are the starting values, and as few as three or four Newton–Gauss iterations are sufficient.

Successive cycles of the EM steps can be shown to converge to a maximum of the likelihood for all item parameters jointly (Dempster et al. 1977), provided there are no boundary problems in the solution. Because the cross-derivatives between items are ignored, however, the inverses of the information matrices for the separate items at convergence do not give the large sample error variances and covariances for all items jointly. Generally, they underestimate the size of the standard errors. Nevertheless, the between-item derivatives are functions of the sample size and parameter values, and they can be obtained once the EM item parameter estimates have been obtained.

To that end, let v_j and $v_{j'}$ be general parameters of items j and j' and differentiate (5.9) with respect to $v_{j'}$, which occurs only in $\mathcal{P}_i$ and $\mathcal{L}_i(\theta)$:

$$\frac{\partial^2 \log L_N}{\partial v_j \partial v_{j'}} = \int \sum_i^N w_i \left(\frac{\mathcal{P}_i}{\mathcal{P}_i^2} \frac{\partial \mathcal{L}_i(\theta)}{\partial v_{j'}} - \frac{\mathcal{L}_i(\theta)}{\mathcal{P}_i^2} \frac{\partial \mathcal{P}_i}{\partial v_{j'}} \right) \times \left(\sum_h^{m_j} \frac{u_{ijhk}}{P_{jh}(\theta)} \frac{\partial P_{jh}(\theta)}{\partial v_j} \right) g(\theta) d\theta.$$

$$= \int \sum_i^N w_i \left[\frac{\mathcal{L}_i(\theta)}{P_i} \left(\sum_{h'}^{m_{j'}} \frac{u_{ij'h'k'}}{P_{j'h'}} \frac{\partial P_{j'h'}}{\partial v_{j'}} \right) \right.$$

$$\left. - \frac{\mathcal{L}_i(\theta)}{P_i} \int \frac{\mathcal{L}_i(\theta)}{P_i P_{j'h'}(\theta)} \frac{\partial P_{j'h'}(\theta)}{\partial v_{j'}} \right] \left(\sum_h^{m_j} \frac{u_{ijhk}}{P_{jh}(\theta)} \frac{\partial P_{jh}(\theta)}{\partial v_j} \right) g(\theta) d\theta. \tag{5.17}$$

Note, however, that the second term within the brackets contains as a factor the first derivative of the log likelihood with respect to $v_{j'}$, which goes to zero at the solution point attained in the EM cycles. Taking expectations therefore yields

$$\mathcal{I}(\mathbf{v}_j, \mathbf{v}_{j'}) = \int \overline{N}(\theta) \left(\sum_{h'}^{m_{j'}} \sum_h^{m_j} \frac{p_{j'h'jh}}{P_{j'h'}(\theta) P_{jh}(\theta)} \frac{\partial P_{j'h'}(\theta)}{\partial \mathbf{v}_{j'}} \frac{\partial P_{jh}(\theta)}{\partial \mathbf{v}_j} \right) g(\theta) d\theta, \tag{5.18}$$

where $p_{j'h'jh}$ is a joint-occurrence probability.

On the assumption that the distribution of latent response processes of items j and j' is bivariate normal, $\Phi(\beta_j, \beta_{j'}, \rho_{jj'})$,

$$p_{jj'} = \int_{\beta_j}^{\infty} \int_{\beta_{j'}}^{\infty} \frac{1}{2\pi\sqrt{1 - \rho_{jj'}^2}} \exp\left[- \left(z_j^2 + z_{j'}^2 - 2 z_j z_{j'} \rho_{jj'} \right) \Big/ 2 \left(1 - \rho_{jj'}^2 \right) \right] dz_j \, dz_{j'} \tag{5.19}$$

and

$$\rho_{jj'} = \frac{\alpha_j}{\sqrt{1 + \alpha_j^2}} \frac{\alpha_{j'}}{\sqrt{1 + \alpha_{j'}^2}} \tag{5.20}$$

is the product of factor loadings in the one-dimensional case. The slopes must be positive and $\rho_{jj'}$ less than unity. Excellent computing approximations are available for bivariate normal probabilities (see (Divgi 1979b)).

In computation, the elements of the between-item information blocks may be evaluated by quadrature and placed in the information matrix, the diagonal blocks of which have been computed by the EM-formula. If the estimates of the slopes are available from a previous EM solution, the information elements for the joint solution can be evaluated by quadrature for an assumed sample size N by setting $\overline{N}(X_q) = Ng(X_q)$. In this way, anticipated standard errors for items of a putative test or scale can be inspected.

The information matrix is a weighted sum of the cross-products of derivatives of the response functions and as such is grammian. It can be positive definite, however, only if the number of quadrature points is equal to or greater than $2n$. It will then be well conditioned for inversion provided the correlation does not approach unity too closely. When the number of items is large, the information matrix can be of very high order, and the space and time required for inverting the matrix may be considerable. Nevertheless, the fast Cholesky method of inverting the information matrix from the non-redundant elements in their own space (see, e.g. (Bock and Jones 1968), pp. 337–338), give greater scope for working with the full information matrix. The inverse can also be used in performing one or more iterations of the joint solution to improve the precision of parameter estimation. In the latter use, the gradient formulas are the same as in the EM solution. When the inverse is available, it provides exact standard errors for benchmark comparison with approximations.

The general results this section specialize in various ways for specific item response models and parameters. In particular, the sum over categories is not required for parameters specific to a single category.

5.2 The Normal Ogive Model

The normal ogive model (Lord and Novick 1968) is an early example of a response model for cognitive test items, traditionally scored $x_{ij} = 1$ for correct response and $1 - x_{ij} = 0$ for incorrect. Assuming the normal ogive model, $P_{jh}(\theta) = \Phi_{jh}(\theta)$, with $P_{j1} = \Phi(y_j)$ and $P_{j0} = 1 - \Phi_j(y_j)$, where $y_j = \alpha_j(\theta - \beta_j)$ is a standard normal deviate, or *normit*.

The parameter α_j is called the item discriminating power, or *slope*, and β_j is the item *threshold*. To simplify computation in maximum likelihood estimation (MLE), the model may be reparameterized equivalently as $y_j = \alpha_j\theta + \gamma_{j\ell}$, with intercept $\gamma_{j\ell} = -\alpha_j\gamma_{j\ell}$.

For the parameter vector $\mathbf{v}_j = [\alpha_j, \gamma_j]'$, the derivative of the response function is $\partial\Phi_j(\theta)/\partial\mathbf{v}_j = \phi(\theta)\partial y_j/\partial\mathbf{v}_j$, where $\phi_j(\theta)$ is the normal ordinate at y_j. The likelihood of response pattern i is $\mathcal{L}_i(\theta) = \prod_j^n \Phi_j(\theta)^{x_{ij}}[1 - \Phi_j(\theta)]^{1-x_{ij}}$.

The derivative of the log marginal likelihood with respect to $\mathbf{v}_j$ is then

$$\begin{aligned}
&\sum_i^N \frac{w_i}{P_i} \int \left(\frac{x_{ij}}{\Phi_j(\theta)} \frac{\partial\Phi_j(\theta)}{\partial\mathbf{v}_j} + \frac{1 - x_{ij}}{1 - \Phi_j(\theta)} \frac{\partial[1 - \Phi_j(\theta)]}{\partial\mathbf{v}_j} \right) \mathcal{L}_i(\theta)g(\theta)d\theta \\
&= \sum_i^N \frac{w_i}{P_i} \int \frac{x_{ij} - \Phi_j(\theta)}{\Phi_j(\theta)[1 - \Phi_j(\theta)]} \frac{\partial\Phi_j(\theta)}{\partial\mathbf{v}_j} \mathcal{L}_i(\theta)g(\theta)d\theta. \qquad (5.21)
\end{aligned}$$

For this model, the E-step complete data statistics are

$$\begin{aligned}
\bar{r}_{j1}(\theta) &= \sum_i^N \frac{w_i}{P_i} x_{ij}\mathcal{L}_i(\theta), \\
\bar{r}_{j0}(\theta) &= \overline{N}(\theta) - \bar{r}_{j1}(\theta), \\
\overline{N}(\theta) &= \sum_i^N \frac{w_i}{P_i} \mathcal{L}_i(\theta).
\end{aligned}$$

The likelihood equation requires only $\bar{r}_{j1}(\theta)$:

$$\int \frac{\bar{r}_{j1}(\theta) - \overline{N}(\theta)}{\Phi_{j1}(\theta)[1 - \Phi_{j1}(\theta)]} \frac{\partial \Phi_j(\theta)}{\partial \mathbf{v}_j} g(\theta) d\theta = 0. \tag{5.22}$$

The Newton–Gauss solution is essentially the same as that of Bliss (1935) probit analysis (see also (Garwood 1941; Finney 1952)). The gradient vector when $y_j = \hat{\alpha}_j X_q + \hat{\gamma}_j$ may be computed by quadrature from the provisional values of item parameters at each iteration:

$$\mathcal{G}(\hat{\mathbf{v}}_j) = \sum_q^Q \frac{\bar{r}_{jh}(X_q) - \overline{N}(X_q)\Phi_j(X_q)}{\Phi_j(X_q)[1 - \Phi_j(X_q)]} \phi_j(X_q) \frac{\partial y_j}{\partial \mathbf{v}_j} A(X_q). \tag{5.23}$$

Because terms of the second derivatives that involve the residuals, $\bar{r}_{j1}(X_q) - \overline{N}(X_q)\Phi_j$, vanish in expectation, only a single term remains in the elements of the information matrix:

$$\mathcal{I}\left(\hat{\mathbf{v}}_j, \hat{\mathbf{v}}_j'\right) = \sum_q^Q \frac{\overline{N}(X_q)}{\Phi_j(X_q)[1 - \Phi_j(X_q)]} \phi_j^2(X_q) \frac{\partial y_j}{\partial \mathbf{v}_j} \frac{\partial y_j}{\partial \mathbf{v}_j'} A(X_q). \tag{5.24}$$

Starting values $\hat{\alpha}_j = 1$ and $\hat{\gamma}_j = 0$ are quite satisfactory in the initial M-step

For the joint solution with the normal ogive model, differentiate (5.22) with respect to $\mathbf{v}_{j'}$ on the same assumptions as in (5.16):

Let any two items be j and j', and differentiate (5.9) with respect to $\mathbf{v}_{j'}$, which appears only in $\mathcal{L}_i(\theta)$:

$$\int \sum_i^N \frac{w_i}{\mathcal{P}_i} \frac{x_{ij'} - \Phi_{j'}(\theta)}{\Phi_{j'(\theta)}[1 - \Phi_{j'}(\theta)]} \frac{x_{ij} - \Phi_j(\theta)}{\Phi_j(\theta)[1 - \Phi_j(\theta)]} \frac{\partial \Phi_j(\theta)}{\partial \mathbf{v}_j} \cdot \frac{\partial \Phi_{j'(\theta)}}{\partial \mathbf{v}_{j'}} \mathcal{L}_i(\theta) g\theta \, d\theta. \tag{5.25}$$

Taking expectation yields the information element,

$$\mathcal{I}(\mathbf{v}_j, \mathbf{v}_{j'}) = \int \frac{\overline{N}(\theta)[p_{jj'} - \Phi_j(\theta)\Phi_{j'}(\theta)]}{\Phi_j(\theta)[1 - \Phi_j(\theta)]\Phi_{j'}(\theta)[1 - \Phi_{j'}(\theta)]} \frac{\partial \Phi_j(\theta)}{\partial \mathbf{v}_j} \frac{\partial \Phi_{j'}(\theta)}{\partial \mathbf{v}_{j'}} \frac{\partial y_j}{\partial \mathbf{v}_j} \frac{\partial y_{j'}}{\partial \mathbf{v}_{j'}} g(\theta) d\theta, \tag{5.26}$$

where $p_{jj'}$ is the correct-response joint-occurrence probability.

Note that when $j' = j$, $p_{jj'} = \Phi_j$, and a $\Phi_j(\theta)[1 - \Phi_j(\theta)]$ in the denominator cancels, reducing the result to the M-stage information.

5.3 The Nominal Categories Model

As an example of the class of models for first choices, consider the nominal categories model (Bock 1972). The response function for category h of item j is defined as

$$\Psi_{jh}(\theta) = \frac{e^{y_{jh}}}{e^{y_{j1}} + e^{y_{j2}} + \cdots + e^{y_{j,m_j}}}, \tag{5.27}$$

where $y_{jh} = \alpha_{jh}(\theta - \beta_{jh}) = \alpha_{jh}\theta + \gamma_{jh}$. The m_j-element vector, $[y_{jh}]$ is called a *multinomial logit* ((Bock 1972)).

The derivative with respect to a category parameter $\alpha_{j\ell}$ is

$$\frac{\partial \Psi_{jh}(\theta)}{\partial \alpha_{j\ell}} = \begin{cases} \theta \Psi_{jh}(\theta)[1 - \Psi_{j\ell}(\theta)] & \ell = h \\ -\theta \Psi_{jh}(\theta)\Psi_{j\ell}(\theta) & \ell \neq h, \end{cases} \tag{5.28}$$

and similarly for $\gamma_{j\ell}$, without the leading θ.

The log likelihood equations for the parameters of this model are remarkably simple. Substituting the response function $\Psi_{j\ell}(\theta)$ and its derivatives in (5.12), the resulting likelihood equation for $\mathbf{v}_{j\ell} = [\alpha_{j\ell}, \gamma_{j\ell}]'$ is

$$\int \left[\frac{\bar{r}_{j\ell}(\theta)}{\Psi_{j\ell}(\theta)} \Psi_{j\ell}(\theta)[1 - \Psi_{j\ell}(\theta)] - \sum_{h \neq \ell}^{m_j} \frac{\bar{r}_{jh}(\theta)}{\Psi_{j\ell}(\theta)} \Psi_{j\ell}(\theta)\Psi_{jh}(\theta) \right] \frac{\partial y_{j\ell}}{\partial \mathbf{v}_{j\ell}} g(\theta)d\theta$$
$$= \int \left[\bar{r}_{j\ell}(\theta) - \bar{N}(\theta)\Psi_{j\ell}(\theta) \right] \frac{\partial y_{j\ell}}{\partial \mathbf{v}_{j\ell}} g(\theta)d\theta = \mathbf{0}. \tag{5.29}$$

The expectation of the second derivatives of the log likelihood is equally simple due to the vanishing of terms containing residuals. The information elements are, say $\mathcal{I}(\mathbf{v}_{j\ell}, \mathbf{v}'_{jg})$:

$$\int \bar{N}(\theta)\Psi_{j\ell}(\theta)[1 - \Psi_{jg}(\theta)] \frac{\partial y_{j\ell}}{\partial \mathbf{v}_{j\ell}} \frac{\partial y_{jg}}{\partial \mathbf{v}'_{jg}} g(\theta)d\theta \quad g = \ell,$$
$$-\int \bar{N}(\theta)\Psi_{j\ell}(\theta)\Psi_{jg}(\theta) \frac{\partial y_{j\ell}}{\partial \mathbf{v}_{j\ell}} \frac{\partial y_{jg}}{\partial \mathbf{v}'_{jg}} g(\theta)d\theta \quad g \neq \ell. \tag{5.30}$$

Under restriction (5.3), the parameter of any one category can be assigned an arbitrary value. Following Masters (1982), it is therefore convenient to adopt the indexing $h = 0, 1, 2, \ldots, m_j - 1$, $\ell = 0, 1, 2, \ldots, m_j - 1$, and $g = 0, 1, 2, \ldots, m_j - 1$, and set $\alpha_{j0} = 0$ and $\gamma_{j0} = 0$. Then,

$$\Psi_{jh}(\theta) = \frac{e^{y_{jh}}}{1 + e^{y_{j1}} + e^{y_{j2}} + \cdots + e^{y_{j,m_j-1}}}, \tag{5.31}$$

the parameters of which are equivalent to simple contrasts between the parameters of the first category and those of the remaining categories (see (Bock 1975), Section 8.1.3b).

For the slope parameters, $\alpha_{j\ell}$, the $2(m_j - 1)$-elements of the M-step gradient vector based on the $m_j - 1$ data statistics, $\bar{r}_{j\ell}(\theta)$, $\ell \neq 0$, may be evaluated as

$$\mathcal{G}(\alpha_{j\ell}) = \sum_{q}^{Q} X_q[\bar{r}_{j\ell}(X_q) - \bar{N}(X_q)\Psi_{j\ell}(X_q)]A(X_q), \tag{5.32}$$

and similarly for the intercepts $\gamma_{j\ell}$ without the leading X_q.

The elements of the $2(m_j - 1) \times 2(m_j - 1)$ corresponding information matrix, $\ell \neq 0,\ g \neq 0$, may be evaluated as

$$\mathcal{I}(\alpha_{j\ell}, \alpha_{jg}) = \begin{cases} \sum_q^Q X_q^2 \overline{N}(X_q)\Psi_{j\ell}(X_q)[1 - \Psi_{jg}(X_q)]A(X_q) & g = \ell \\ -\sum_q^Q X_q^2 \overline{N}(X_q)\Psi_{j\ell}(X_q)\Psi_{jg}(X_q)A(X_q) & g \neq \ell, \end{cases} \tag{5.33}$$

and similarly $\mathcal{I}(\gamma_{j\ell}, \gamma_{jg})$, without the leading X_q^2; similarly, $\mathcal{I}(\alpha_{j\ell}, \gamma_{jg})$ with X_q leading. The likelihood equations are then solved numerically as in (5.16).

For the information matrix of the joint solution, let the item pair of the second derivatives be j and j', and let g', be the index for the parameters of item j'. Then the second derivatives with respect to $\mathbf{v}_{j\ell}$ and $\mathbf{v}_{j'g'}$ consist of the $4(m_j - 1)(m_{j'} - 1)$ elements of the information matrix,

$$\mathcal{I}(\mathbf{v}_{j\ell}, \mathbf{v}'_{j'g'}) = \int \overline{N}(\theta)[p_{j\ell j'g'} - \Psi_{j\ell}(\theta)\Psi_{j'g'}(\theta)] \frac{\partial y_{j\ell}(\theta)}{\partial \mathbf{v}_{j\ell}} \frac{\partial y_{j'g'}}{\partial \mathbf{v}_{j'g'}} g(\theta)d\theta. \tag{5.34}$$

The joint occurrence probability, $p_{j\ell j'g'}$, is the bivariate normal probability bounded by the limits $\beta_{j,\ell}, \beta_{j,\ell+1}$ and $\beta_{j'g'}, \beta_{j'+1,g'+1}$, where $\beta_{jm_j} = \beta_{j'm_{j'}} = +\infty$. These probabilities can be computed from the quadrant probabilities of the dichotomous case:

$$\begin{aligned} p_{j\ell j'g'} = {} & \Phi(\beta_{j\ell}, \beta_{j'g'}, \rho_{jj'}) - \Phi(\beta_{j\ell}, \beta_{j',g'+1}, \rho_{jj'}) - \Phi(\beta_{j,\ell+1}, \beta_{j'g'}, \rho_{jj'}) \\ & + \Phi(\beta_{j,\ell+1}, \beta_{j',g'+1}, \rho_{jj'}). \end{aligned} \tag{5.35}$$

Note that when $j' = j$ and $g' = \ell$, $p_{j\ell j'g'} = \Psi_{j\ell}(\theta)$, but when $g' \neq \ell$, $p_{j\ell, j'g'} = 0$.

5.4 The Graded Categories Model

As an example of the class of graded models, consider the version of Samejima (1969) in which the category response functions are the probabilities of successive intervals of the cumulative normal distribution, $\Phi_{jh}(\theta) = \Phi(y_{jh}),\ h = 1, 2, \ldots, m_j$, and $y_{jh} = \alpha_j(\theta - \beta_{jh})$. Then

$$P_{jh}(\theta) = \Phi_{jh}(\theta) - \Phi_{j,h-1}(\theta), \tag{5.36}$$

with $\Phi_{j0}(\theta) = 0$, $\Phi_{jm_j}(\theta) = 1$, $\phi_{j0} = \phi_{jm_j} = 0$, and $\beta_{j0} = \beta_{jm_j} = 0$.

As in the other models, it is convenient to reparameterize the normit as $y_{jh} = \alpha_j\theta - \gamma_{jh}$ and to recover the estimated threshold from the maximum likelihood estimates of the slope and the intercept, $\gamma_{jh} = -\alpha_j\beta_{jh}$.

The derivatives of the category response function with respect to α_j and γ_{jh} contain the corresponding normal densities:

$$\frac{\partial P_{jh}(\theta)}{\partial \alpha_j} = \theta[\phi_{jh}(\theta) - \phi_{j,h-1}(\theta)], \tag{5.37}$$

$$\frac{\partial P_{jh}(\theta)}{\partial \gamma_{jh}} = \phi_{jh}(\theta) - \phi_{j,h-1}(\theta). \tag{5.38}$$

Substituting the response function, $P_{jh}(\theta)$, and its derivatives in (5.9) gives the likelihood equations for the slope and intercept parameters:

$$\begin{aligned}\frac{\partial \log L_N}{\partial \alpha_j} &= \sum_i^N \frac{w_i}{\mathcal{P}_i} \int \left(\sum_h^{m_j} \frac{u_{ijhk}}{P_{jh}(\theta)} \frac{\partial P_{jh}(\theta)}{\partial \alpha_j} \right) \mathcal{L}_i(\theta) g(\theta) d\theta \\ &= \int \left(\sum_h^{m_j} \frac{\bar{r}_{jh}(\theta)}{P_{jh}(\theta)} \frac{\partial P_{jh}(\theta)}{\partial \alpha_j} \right) g(\theta) d\theta = 0,\end{aligned} \tag{5.39}$$

$$\begin{aligned}\frac{\partial \log L_N}{\partial \gamma_{j\ell}} &= \sum_i^N \frac{w_i}{\mathcal{P}_i} \int \left(\frac{u_{ij\ell k}}{P_{j\ell}(\theta)} - \frac{u_{ij,\ell+1,k}}{P_{j,\ell+1}(\theta)} \right) \frac{\partial P_{j\ell}(\theta)}{\partial \gamma_{j\ell}} \mathcal{L}_i(\theta) g(\theta) d\theta \\ &= \int \left(\frac{\bar{r}_{j\ell}(\theta)}{P_{j\ell}(\theta)} - \frac{\bar{r}_{j,\ell+1}(\theta)}{P_{j,\ell+1}(\theta)} \right) \frac{\partial P_{j\ell}(\theta)}{\partial \gamma_{j\ell}} g(\theta) d\theta = 0.\end{aligned} \tag{5.40}$$

The following are the corresponding second derivatives[1]:

$$\frac{\partial^2 \log L_N}{\partial \alpha_j^2} = \int \sum_h^{m_j} \left[-\frac{\bar{r}_{jh}(\theta)}{P_{jh}^2(\theta)} \left(\frac{\partial P_{jh}(\theta)}{\partial \alpha_j} \right)^2 + \frac{\bar{r}_{jh}(\theta)}{P_{jh}(\theta)} \frac{\partial^2 P_{jh}(\theta)}{\partial \alpha_j^2} \right] g(\theta) d\theta, \tag{5.41}$$

$$\frac{\partial^2 \log L_N}{\partial \gamma_{j\ell} \partial \gamma_{jg}} = \begin{cases} \displaystyle\int \left[-\left(\frac{\bar{r}_{j\ell}(\theta)}{P_{j\ell}^2(\theta)} + \frac{\bar{r}_{j,\ell+1}(\theta)}{P_{j,\ell+1}^2(\theta)} \right) \left(\frac{\partial P_{j\ell}(\theta)}{\partial \gamma_{j\ell}} \right)^2 \right. \\ \displaystyle\quad \left. + \left(\frac{\bar{r}_{j\ell}(\theta)}{P_{j\ell}(\theta)} - \frac{\bar{r}_{j,\ell+1}(\theta)}{P_{j,\ell+1}(\theta)} \right) \frac{\partial^2 P_{j\ell}(\theta)}{\partial \gamma_{j\ell}^2} \right] g(\theta) d\theta, & g = \ell \\ \displaystyle\frac{\bar{r}_{j\ell+1}(\theta)}{P_{j,\ell+1}^2(\theta)} \frac{\partial P_{j,\ell+1}(\theta)}{\partial \gamma_{j\ell+1}} g(\theta) d\theta, & g = \ell + 1 \\ 0, & |\ell - g| > 1, \end{cases} \tag{5.42}$$

$$\begin{aligned}\frac{\partial^2 \log L_N}{\partial \alpha_j \partial \gamma_{j\ell}} = \int \Bigg[&-\left(\frac{\bar{r}_{j\ell}(\theta)}{P_{j\ell}^2(\theta)} \frac{\partial P_{j\ell}(\theta)}{\alpha_j} - \frac{\bar{r}_{j,\ell+1}(\theta)}{P_{j,\ell+1}^2} \frac{P_{j,\ell+1}(\theta)}{\alpha_j} \right) \frac{\partial P_{j\ell}(\theta)}{\partial \gamma_{j\ell}} \\ &+ \left(\frac{\bar{r}_{j\ell}(\theta)}{P_{j\ell}(\theta)} - \frac{\bar{r}_{j,\ell+1}(\theta)}{P_{j,\ell+1}(\theta)} \right) \frac{\partial^2 P_{j\ell}(\theta)}{\partial \alpha_j \partial \gamma_{j\ell}} \Bigg] g(\theta) d\theta.\end{aligned} \tag{5.43}$$

Taking expectations in the second derivatives by setting $\bar{r}_{jh}(\theta) = \overline{N}(\theta) P_{jh}(\theta)$ and $\bar{r}_{j,h+1}(\theta) = \overline{N}(\theta) P_{j,h+1}$ nullifies the terms involving second derivatives of the

1 Adapted from Bock (1975, Section 8.1.6).

category response functions and leads to a sum of the second derivatives, which equals zero.

Evaluating by quadrature results in the elements of the gradient vector:

$$\mathcal{G}(\alpha_j) = \sum_q^Q X_q \left(\sum_h^{m_j} \frac{\bar{r}_{jh}(X_q)[\phi_{jh}(X_q) - \phi_{j,h-1}(X_q)]}{\Phi_{jh}(X_q) - \Phi_{j,h-1}(X_q)} \right) A(X_q), \tag{5.44}$$

$$\begin{aligned}\mathcal{G}(\gamma_{j\ell}) = \sum_q^Q & \frac{\bar{r}_{j\ell}(X_q)}{\Phi_{j\ell}(X_q) - \Phi_{j,\ell-1}(X_q)} A(X_q) \\ & - \frac{\bar{r}_{j,\ell+1}(X_q)}{\Phi_{j,\ell+1}(X_q) - \Phi_{j\ell}(X_q)} [\phi_{j\ell}(X_q) - \phi_{j,\ell-1}(X_q)] A(X_q).\end{aligned} \tag{5.45}$$

Similarly, change of sign and taking expectation result in the elements of the information matrix:

$$\mathcal{I}\left(\alpha_j^2\right) = \sum_q^q X_q^2 \bar{N}(X_q) \sum_h^{m_j} \left(\frac{[\phi_{jh}(X_q) - \phi_{j,h-1}(X_q)]^2}{\Phi_{jh}(X_q) - \Phi_{j,h-1}(X_q)} \right) A(X_q), \tag{5.46}$$

$$\mathcal{I}(\gamma_{j\ell}, \gamma_{jg}) = \begin{cases} \sum_q^Q \bar{N}(X_q) \left(\frac{1}{\Phi_{j\ell}(X_q) - \Phi_{j,\ell-1}(X_q)} - \frac{1}{\Phi_{j,\ell+1}(X_q) - \Phi_{j,\ell}(X_q)} \right) \\ [\phi_{j\ell}(X_q) - \phi_{j,\ell-1}(X_q)]^2 A(X_q), \quad g = \ell \\ \sum_q^Q \bar{N}(X_q) \frac{\phi_{j,\ell+1}(X_q) - \phi_{j\ell}(X_q)}{\Phi_{j\ell+1}(X_q) - \Phi_{j,\ell}(X_q)} A(X_q), \quad g = \ell + 1 \\ 0, \quad |g - \ell| > 1, \end{cases} \tag{5.47}$$

$$\begin{aligned}\mathcal{I}(\alpha_j, \gamma_{j\ell}) = \sum_q^Q \bar{N}(X_q) & \left(\frac{\phi_{j,\ell}(X_q) - \phi_{j,\ell-1}(X_q)}{\Phi_{j,\ell}(X_q) - \Phi_{j,\ell-1}(X_q)} \right. \\ & \left. + \frac{\phi_{j,\ell+1}(X_q) - \phi_{j,\ell}(X_q)}{\Phi_{j,\ell+1}(X_q) - \Phi_{j,\ell}(X_q)} \right) [\phi_{j,\ell}(X_q) - \phi_{j,\ell-1}(X_q)] A(X_q).\end{aligned} \tag{5.48}$$

Like those of the nominal model, the expected between-item derivatives of the joint solution depend upon the joint occurrence probabilities of the categorical responses. Due to the slope parameter, the number of derivatives is $m_j m_{j'}$ for each item pair. The presence of the slope parameter in all category response functions results in a product of sums over categories in the between-item derivative with respect to slope:

$$\mathcal{I}(\alpha_j, \alpha_{j'}) = \int \bar{N}(\theta) \sum_h^{m_j} \sum_{h'}^{m_{j'}} \frac{p_{jh,j'h'}}{P_{jh}(\theta) P_{j'h'}(\theta)} \frac{\partial P_{jh}(\theta)}{\partial \alpha_j} \frac{\partial P_{j'h'}(\theta)}{\partial \alpha_{j'}} g(\theta) d\theta. \tag{5.49}$$

The joint occurrence probabilities appear in a similar way in the remaining between-item information elements:

$$\mathcal{I}(\gamma_{j\ell},\gamma_{j'g'}) = \int \overline{N}(\theta)\left(\frac{p_{j\ell,j'g'}}{P_{j\ell}(\theta)P_{j'g'(\theta)}} - \frac{p_{j\ell,j'g'+1}}{P_{j\ell}(\theta)P_{j',g'+1}(\theta)} - \frac{p_{j,\ell+1,j'g'}}{P_{j,\ell+1}(\theta)P_{j'g'}(\theta)}\right.$$
$$\left. + \frac{p_{j,\ell+1,j',g'+1}}{P_{j,\ell+1}P_{j'g'+1(\theta)}}\right)\frac{\partial P_{j\ell}(\theta)}{\partial\gamma_{j\ell}}\frac{\partial P_{j'g'}(\theta)}{\partial\gamma_{j'g'}}g(\theta)d\theta, \tag{5.50}$$

$$\mathcal{I}(\alpha_j,\gamma_{j'g'}) = \int \overline{N}(\theta)\left(\sum_h^{m_j}\frac{p_{jh,j'g'}}{P_{jh}(\theta)P_{j'g'}(\theta)}\frac{\partial P_{jh}(\theta)}{\partial\alpha_j}\frac{\partial P_{j'g'}(\theta)}{\partial\gamma_{j'g'}}\right)$$
$$+\left(\sum_h^{m_j}\frac{p_{jh,j',g'+1}}{P_{jh}(\theta)P_{j',g'+1}(\theta)}\frac{\partial P_{jh}(\theta)}{\partial\alpha_j}\frac{\partial P_{j',g'+1}(\theta)}{\partial\gamma_{j',g'+1}}\right)g(\theta)d\theta, \tag{5.51}$$

$$\mathcal{I}(\gamma_{j\ell},\alpha_{j'}) = \int \overline{N}(\theta)\left(\sum_{h'}^{m_{j'}}\frac{p_{j\ell,j'h'}}{P_{j\ell}(\theta)P_{j'h'}(\theta)}\frac{\partial P_{j\ell}(\theta)}{\partial\gamma_{j\ell}}\frac{\partial P_{j'h'}(\theta)}{\partial\alpha_{j'}}\right)$$
$$+\left(\sum_{h'}^{m_{j'}}\frac{p_{j,\ell+1,j',h'}}{P_{j,\ell+1}(\theta)P_{j'h'}(\theta)}\frac{\partial P_{j,\ell+1}}{\partial\gamma_{j,\ell+1}}\frac{\partial P_{j',h'}(\theta)}{\partial\alpha_{j'}}\right)g(\theta)d\theta. \tag{5.52}$$

Note that when $j' = j$ and $h' = h$, the double summation in (5.49) becomes a single summation and $p_{jh} = P_{jh}(\theta)$. This cancels one of the two $P_{jh}(\theta)$ in the denominator and results in the information element for α_j^2 of the M-step. In (5.50), the same canceling occurs and the second, third and fourth terms become identical apart from sign, resulting in the M-step information element for $\gamma_{j\ell},\gamma_{j'g'}$. The two cross-derivatives, (5.51) and (5.52) of slope and intercept become identical and the cancelation results in the corresponding M-step information element.

5.5 The Generalized Partial Credit Model

The generalized partial credit model (GPCM; (Muraki 1992)) differs from the partial credit model (PCM; (Masters 1982)) only in allowing for varying discriminating power among items. Both models exist in two versions – the unrestricted version, in which each item has its own category parameters, and the *rating-scale* version (Andrich 1978), in which the item category parameters are common to all items. Both versions are discussed in this section.

5.5.1 The Unrestricted Version

Adapting Masters' notation for the PCM to the GPCM, the response function of category k of item j may be expressed as,

$$P_{jh}(\theta) = \frac{\exp\left[\sum_{\ell=0}^{h-1} \alpha_j(\theta - \delta_{j\ell})\right]}{\sum_{h'=0}^{m_{j-1}} \exp\left[\sum_{\ell=0}^{h-1} \alpha_j(\theta - \delta_{j\ell})\right]}, \tag{5.53}$$

where α_j is the item discriminating-power parameter.

To resolve indeterminacy with respect to location, it is convenient to adopt the restrictions $\theta - \delta_{j0} = 0$ and $\delta_{j0} = 0$, such that

$$P_{jh}(\theta) = \frac{\exp\left[\sum_{\ell=0}^{h-1} \alpha_j(\theta - \delta_{j\ell})\right]}{1 + \sum_{h'=1}^{m_{j-1}} \exp\left[\sum_{\ell=0}^{h-1} \alpha_j(\theta - \delta_{j\ell})\right]}. \tag{5.54}$$

Note that the numerator of $P_{j1}(\theta)$ then equals 1.

For purposes of maximum likelihood estimation, it is convenient to reparameterize in terms of the multinomial logits,

$$z_{jh} = \alpha_j s_{j\ell}\theta + \kappa_{j\ell}, \tag{5.55}$$

where $s_{j\ell} = h - 1$ and

$$\kappa_{j\ell} = -\sum_{\ell=0}^{s_{jh}} \alpha_j \delta_{j\ell}, \quad \kappa_{j0} = 0. \tag{5.56}$$

The quantity s_{jh} is called the *scoring function* (Andrich 1978).

In this form, the GPCM is seen to be a special case of the nominal model in which the slope parameters are replaced by $\alpha_j s_{j\ell}$ and the intercepts by the $\kappa_{j\ell}$:

$$\Psi_{jh}(\theta) = \frac{\exp(z_{j\ell})}{\sum_{h'=1}^{m_j} \exp(z_{jh'})}, \tag{5.57}$$

where $\ell = h - 1$.

The estimated location parameters $\hat{\delta}_{j\ell}$ can be recovered from the maximum likelihood estimates of $\kappa_{j\ell}$ and α_j by

$$\hat{\delta}_{j\ell} = -(\hat{\kappa}_{j\ell} - \hat{\kappa}_{j,\ell-1})/\hat{\alpha}_j. \tag{5.58}$$

For the derivatives of (5.28) with respect to the parameters are

$$\frac{\partial \Psi_{jh}(\theta)}{\partial \alpha_j} = \theta \Psi_{jh}(\theta) \left[s_{jh} - \sum_{h'=1}^{m_j} s_{jh'} \Psi_{jh'}(\theta) \right]$$
$$= \theta \Psi_{jh}(\theta)(s_{jh} - \bar{s}_j), \tag{5.59}$$

where $h = \ell + 1,\ h' = \upsilon + 1,\quad \ell \neq 0, \upsilon \neq 0,$

$$\frac{\partial \Psi_{jh}(\theta)}{\partial \kappa_{j\ell}} = \begin{cases} \Psi_{jh}(\theta)[1 - \Psi_{jh'}(\theta)], & h' = h \\ -\Psi_{jh}(\theta)\Psi_{jh'}(\theta), & h' \neq h. \end{cases} \tag{5.60}$$

5.5.2 The EM Solution

Substituting the above derivatives in (5.14) gives the M-step gradients, which replace integration with quadrature and replace θ with the quadrature points X_q, $Q \geq n_{v_j}$. The population density, $g(\theta)$, becomes the quadrature weight, $A(X_q)$. If Gauss–Hermite quadrature is employed, the weights are normed to one as required. If equally spaced points are employed, the densities, $g(X_q)$, must be normed to one. The mean scoring function, $\bar{s}_j$, is evaluated as $\sum_{h'=1}^{m_j} s_{jh'} \Psi_{jh'}(X_q)$. Note that the elements of the gradient vector contain the complete data statistics evaluated at X_q:

$$\mathcal{G}(\alpha_j) = \sum_{q=1}^{Q} X_q \left[\sum_{h=1}^{m_j} [\bar{r}_{jh}(X_q) - \bar{N}(X_q)\Psi_{jh}(X_q)](s_{jh} - \bar{s}_j) \right] A(X_q), \tag{5.61}$$

$$\mathcal{G}(\kappa_{j\ell}) = \sum_{q=1}^{Q} [\bar{r}_{jh}(X_q) - \bar{N}(X_q)\Psi_{jh}(X_q)]A(X_q), \quad h = \ell + 1, \ell \neq 0. \tag{5.62}$$

Substituting in (5.17) gives the elements of the information matrix:

$$\mathcal{I}(\alpha_j) = \sum_{q=1}^{Q} X_q{}^2 \bar{N}(X_q) \left[\sum_{h=1}^{m_j} \Psi_{jh}(X_q) \frac{(s_{jh} - \bar{s}_j)^2}{1 - \Psi_{jh}(X_q)} \right] A(X_q). \tag{5.63}$$

$$\mathcal{I}(\kappa_{j\ell}, \kappa_{j\ell'}) = \sum_{q=1}^{Q} \bar{N}(X_q)\Psi_{jh}(X_q)[1 - \Psi_{jh'}(X_q)]A(X_q)$$
$$= -\sum_{q=1}^{Q} \bar{N}(X_q)\Psi_{jh}(X_q)\Psi_{jh'}(X_q)A(X_q), \tag{5.64}$$

$$h = \ell + 1,\ h' = \upsilon + 1,\ \ell \neq 0,\ \upsilon \neq 0,$$

$$\mathcal{I}(\alpha_j, \kappa_{j\ell}) = \sum_{q=1}^{Q} X_q \overline{N}(X_q) \left[\sum_{h=1}^{m_j} \frac{(s_{jh} - \overline{s}j)}{1 - \Psi_{jh}(X_q)} \right] \Psi_{jh}(X_q)[1 - \Psi_{jh}(X_q)]A(X_q),$$
$$h = \ell + 1, \ \ell \neq 0. \tag{5.65}$$

New provisional estimates of the item parameters at each EM cycle are computed from the provisional parameters of the previous cycle:

$$\hat{\boldsymbol{v}}_j^{(t+1)} = \hat{\boldsymbol{v}}_j^{(t)} + \mathcal{I}^{-1}\left(\hat{\boldsymbol{v}}_j^{(t)}, \hat{\boldsymbol{v}}_j^{T(t)}\right)\mathcal{G}\left(\hat{\boldsymbol{v}}_j^{(t)}\right). \tag{5.66}$$

The cycles are repeated until the change in any parameter is less than some specified value. Rough starting values are quite satisfactory: initial slopes may be set to 1.0 and intercepts to equally spaced values between ± 2.0.

5.5.2.1 The GPCM Newton–Gauss Joint Solution

The gradient vector for the Newton–Gauss joint solution may be obtained by adjoining the EM gradients in a single vector of n_v elements. The main block diagonal of the information matrix contains the EM information matrices in the same order. This leaves only the $n_v \times n_v$ off-diagonal blocks, $j' \neq j$ to be evaluated. The elements in these blocks are as follows:

$$\mathcal{I}(\alpha_j, \alpha_{j'}) = \sum_{q=1}^{Q} \overline{N}(X_q) X_q^{\,2} \rho_{jj'}$$
$$\times \left[\sum_{h=1}^{m_j} \sum_{h'=1}^{m_{j'}} \frac{(s_{jh} - \overline{s}_j)(s_{j'h'} - \overline{s}_{j'})}{\sqrt{[1 - \Psi_{jh}(X_q)][1 - \Psi_{j'h'}(X_q)]}} \sqrt{\Psi_{jh}(X_q)\Psi_{j'h'}(X_q)} \right] A(X_q), \tag{5.67}$$

$$\mathcal{I}(\kappa_{j\ell}, \kappa_{j'\ell'}) = \sum_{q=1}^{Q} \overline{N}(X_q)\rho_{jj'} \Psi_{jh}(X_q)[1 - \Psi_{jh}(X_q)]\Psi_{j'h'}(X_q)[1 - \Psi_{j'h'}(X_q)]A(X_q),$$

$$h = \ell + 1, \ h' = \ell' + 1, \ \ell \neq 0, \ \ell' \neq 0, \tag{5.68}$$

$$\mathcal{I}(\alpha_j, \kappa_{j'\ell'}) = \sum_{q=1}^{Q} X_q \overline{N}(X_q) \overline{N}(X_q) \rho_{jj'}$$
$$\times \left\{ \left[\sum_{h}^{m_j} \frac{(s_{jh} - \overline{s}_j)\Psi_{jh}(X_q)}{1 - \Psi_{jh}(X_q)} \sqrt{\Psi_{j'h'}(X_q)[1 - \Psi_{j'h'}(X_q)]} \right] \right\} A(X_q),$$

$$h' = \ell' + 1, \ \ell' \neq 0, \tag{5.69}$$

$$\mathcal{I}(\kappa_{j\ell}, \alpha_{j'}) = \sum_{q=1}^{Q} X_q \overline{N}(X_q) \rho_{jj'} \times \left\{ \left[\sum_{h'}^{m_{j'}} \frac{(s_{jh} - \bar{s}_j) \Psi_{j'h'}(X_q)}{1 - \Psi_{j'h'}(X_q)} \sqrt{\Psi_{jh}(X_q)[1 - \Psi_{jh}(X_q)]} \right] \right\} A(X_q)$$

$$h = \ell + 1, \ \ell \neq 0. \tag{5.70}$$

The n_v-order gradient and information matrix are in the form required for Newton–Gauss iterations of the joint solution.

5.5.3 Rating Scale Models

Rating scale models (RSM) apply to multiple category items, or subsets of items, that have the same number of categories, $m > 2$. The category descriptions are the same for all items in the scale. The arguments of the category response functions are expressed as,

$$\alpha_j(\theta - \beta_j + \delta_\ell), \tag{5.71}$$

where β_j is a location parameter common to all categories, and δ_ℓ is a deviation from that location; i.e.

$$\sum_{\ell=1}^{m-1} \delta_\ell = 0. \tag{5.72}$$

Let the reparameterized multinomial logit model be

$$z_h = \alpha_j s_{jh} \theta + \gamma + \kappa_\ell, \tag{5.73}$$

and set $\kappa_0 = 0$ to resolve indeterminacy of location. Setting $\kappa_1 = 0$ is also convenient; the restriction can be imposed after the $m - 2$ maximum likelihood estimates of κ_ℓ have be obtained.

5.5.3.1 The EM Solution for the RSM

For the RSM response functions, the derivative with respect to α_j is the same as (5.28). The derivative with respect to γ_j is similar,

$$\frac{\partial \Psi_{jh}(\theta)}{\partial \gamma_j} = \Psi_{jh}(\theta)(s_{jh} - \bar{s}_j). \tag{5.74}$$

Again letting $h = \ell + 1$ and $h' = \ell' + 1$, the derivative of κ_ℓ is

$$\frac{\partial \Psi_{jh}(\theta)}{\partial \kappa_\ell} = \begin{cases} \Psi_{jh}(\theta)[1 - \Psi_{jh'}(\theta)], & h' = h \\ -\Psi_{jh}(\theta)\Psi_{jh'}(\theta), & h' \neq h. \end{cases} \tag{5.75}$$

The likelihood equations for κ_ℓ apply to items of any specified scale:

$$\frac{\partial \log L_N}{\partial \kappa_\ell} = \int \frac{\bar{r}_{jk}(\theta) - \overline{N}(\theta)P_{j\ell}(\theta)}{P_{j\ell}(\theta)[1 - P_{jh}(\theta)]} \cdot \frac{\partial P_{j\ell}(\theta)}{\partial \kappa_\ell} g(\theta)d\theta = 0. \tag{5.76}$$

Let n_S, $1 \leq n_S \leq n$ be the number of items in a subset S of n items; when $n_S = 1$, the result is identical to that of the unrestricted model.

In addition to the gradient for α_j, which is the same as (5.32), the gradients for the intercept parameter and the common category parameters are

$$\mathcal{G}(\gamma_j) = \sum_{q=1}^{Q} \sum_{h=1}^{m} [\bar{r}_{jh}(X_q) - \overline{N}(X_q)\Psi_{jh}(X_q)](s_{jh} - \bar{s}_j)A(X_q), \tag{5.77}$$

$$\mathcal{G}(\kappa_\ell) = \sum_{q=1}^{Q} \sum_{j}^{n_S} [\bar{r}_{j\ell}(X_q) - \overline{N}(X_q)\Psi_{jh}(X_q)]A(X_q), \quad h = \ell + 1, \ \ell \neq 0. \tag{5.78}$$

The information elements for α_j are the same as (5.33). Those for γ_j and α_j, γ_j are

$$\mathcal{I}(\gamma_j) = \sum_{q=1}^{Q} \overline{N}(X_q) \sum_{h=1}^{m} \left[\Psi_{jh}(X_q) \frac{(s_{jh} - \bar{s}_j)}{1 - \Psi_{jh}(X_q)} \right] A(X_q), \tag{5.79}$$

$$\mathcal{I}(\alpha_j, \gamma_j) = \sum_{q=1}^{Q} X_q \overline{N}(X_q) \sum_{h=1}^{m} \left[\Psi_{jh}(X_q) \frac{(s_{jh} - \bar{s}_j)^2}{1 - \Psi_{jh}(X_q)} \right] A(X_q). \tag{5.80}$$

Those involving κ_ℓ require $h = \ell + 1, \ h' = \ell' + 1, \ \ell \neq 0, \ \ell' \neq 0$:

$$\begin{aligned} \mathcal{I}(\kappa_\ell, \kappa_{\ell'}) &= \sum_{q=1}^{Q} \overline{N}(X_q) \left[\sum_{j=1}^{n_S} \Psi_{jh}(X_q)[1 - \Psi_{j\ell'}(X_q)] \right] A(X_q), \quad \ell' = \ell \\ &\quad - \sum_{q=1}^{Q} \overline{N}(X_q) \left[\sum_{j}^{n_S} \Psi_{jh}(X_q)\Psi_{j\ell'}(X_q) \right] A(X_q), \quad \ell' \neq \ell \end{aligned} \tag{5.81}$$

$$\mathcal{I}(\alpha_j, \kappa_{\ell'}) = \sum_{q=1}^{Q} X_q \overline{N}(X_q) \left\{ \sum_{j}^{n_S} \left[\sum_{h=1}^{m} \frac{(s_{jh} - \bar{s}_j)}{1 - \Psi_{jh}(X_q)} \right] \Psi_{jh}(X_q)[1 - \Psi_{jh}(X_q)] \right\} A(X_q), \tag{5.82}$$

$$\mathcal{I}(\gamma_j, \kappa_\ell) = \sum_{q=1}^{Q} \overline{N}(X_q) \left\{ \sum_{j}^{n_S} \left[\sum_{h=1}^{m} \frac{(s_{jh} - \bar{s}_j)}{1 - \Psi_{jh}(X_q)} \right] \Psi_{jh}(X_q)[1 - \Psi_{jh}(X_q)] \right\} A(X_q). \tag{5.83}$$

5.5.3.2 The Newton–Gauss Solution for the RSM

As with the GPCM, the gradients of the RSM are those of the EM solution adjoined in a single vector; the information elements of the main block diagonal of the information matrix are the same as those of the EM solution. Those of the nonredundant elements in the off-diagonal blocks are as follows:

The information for $\alpha_j, \alpha_{j'}$ is the same as (5.33); that for $\gamma_j, \gamma_{j'}$ is the same as (5.33) with the leading $X_q^{\,2}$ omitted; those for $\alpha_j, \gamma_{j'}$ and $\gamma_j, \alpha_{j'}$ are the same as (5.33) with the leading $X_q^{\,2}$ replaced by X_q. The remaining elements involve the common category parameters and require $h = \ell + 1,\ h' = \ell' + 1,\ \ell \neq 0,\ \ell' \neq 0$:

$$\mathcal{I}(\kappa_\ell, \kappa_{\ell'}) = \sum_{q=1}^{Q} \overline{N}(X_q)\rho_{jj'} \sum_{j}^{n_S} \left[\sum_{j'}^{n_S} \Psi_{jh}(X_q)[1 - \Psi_{jh'}(X_q)] \right] A(X_q), \tag{5.84}$$

$$\mathcal{I}(\alpha_j, \kappa_{\ell'}) = \sum_{q=1}^{Q} X_q \overline{N}(X_q)\rho_{jj'} \times \left\{ \left[\sum_{j'}^{n_S} \sum_{h}^{m} \frac{(s_{jh} - \bar{s}_j)}{[1 - \Psi_{jh}(X_q)]} \sqrt{\Psi_{j'h}(X_q)[1 - \Psi_{j'h}(X_q)]} \right] \right\} A(X_q), \tag{5.85}$$

$$\mathcal{I}(\kappa_\ell, \alpha_{j'}) = \sum_{q=1}^{Q} X_q \overline{N}(X_q)\rho_{jj'} \times \left\{ \sum_{j'}^{n_S} \left[\sum_{h'}^{m} \frac{(s_{jh} - \bar{s}_j)}{[1 - \Psi_{jh'}(X_q)]} \right] \sqrt{\Psi_{j'h}(X_q)[1 - \Psi_{j'h}(X_q)]} \right\} A(X_q), \tag{5.86}$$

$$\mathcal{I}(\gamma_j, \kappa_{\ell'}) = \sum_{q=1}^{Q} \overline{N}(X_q)\rho_{jj'} \times \left\{ \sum_{j'}^{n_S} \left[\sum_{h}^{m} \frac{(s_{jh} - \bar{s}_j)}{[1 - \Psi_{jh}(X_q)]} \right] \sqrt{\Psi_{j'h}(X_q)[1 - \Psi_{j'h}(X_q)]} \right\} A(X_q), \tag{5.87}$$

$$\mathcal{I}(\kappa_\ell, \gamma_{j'}) = \sum_{q=1}^{Q} \overline{N}(X_q)\rho_{jj'} \times \left\{ \sum_{j'}^{n_S} \left[\sum_{h'}^{m_{j'}} \frac{(s_{jh} - \bar{s}_j)}{1 - \Psi_{jh'}(X_q)} \right] \sqrt{\Psi_{j'h}(X_q)[1 - \Psi_{j'h}(X_q)]} \right\} A(X_q). \tag{5.88}$$

5.6 Boundary Problems

EM procedures have been shown to converge geometrically to a maximum of the likelihood, provided no parameter is on the boundary (Dempster et al. 1977). Boundary problems can arise when sample sizes are small or the data unfavorable.

An intercept parameter may approach plus or minus infinity, an item slope may be negative or approach infinity, or a lower asymptote may become negative or greater than one. In these situations constraints must be imposed on the solution to keep item parameter estimates away from these boundaries. Without effective constraints likelihood-based methods of item parameter estimation are not sufficiently robust for general application.

A solution to these problems that has been found to work well in practice is to treat the estimates as stochastic quantities drawn from distributions in which inadmissible values have vanishingly small probability. The constraint is implemented by the addition of a penalty function to the M-step gradient and a corresponding ridge to the information matrix. The penalty is the first derivative of the log of the density of the assumed distribution, with respect to the item parameter; the ridge is the negative expectation of a corresponding second derivative. This approach has a number of good properties in that (i) it avoids arbitrary abrupt cutoff points, (ii) the distribution can be chosen so that estimates more removed from the boundaries are little influenced by the constraint, (iii) the influence of the penalty decreases with increasing sample size, thus allowing closer approach to the boundary, and (iv) the value of the parameter estimate can be set closely to the assigned mean of the distribution by greatly increasing the magnitude of the ridge. The following constraint functions for intercept, slope, and lower asymptote parameters have these properties.

- *Item-category intercepts.* The intercept c of the given item or item category is assumed normally distributed with assigned mean $\tilde{\mu}_c$ and standard deviation $\tilde{\sigma}_c$, and density,

$$\phi(c) = \frac{1}{\tilde{\sigma}_c\sqrt{2\pi}} \exp\left[-\frac{1}{2\tilde{\sigma}_c^2}(c - \tilde{\mu}_c)^2\right], \quad \tilde{\sigma}_c > 0. \tag{5.89}$$

the penalty function is

$$\frac{\partial \log \phi(c)}{\partial c} = \frac{c - \tilde{\mu}_c}{\tilde{\sigma}_c^2}, \tag{5.90}$$

and the ridge,

$$-\mathcal{E}\left(\frac{\partial^2 \log \phi(c)}{\partial c^2}\right) = \frac{1}{\tilde{\sigma}_c^2}. \tag{5.91}$$

- *Item slopes.* If a is an item slope and $y = \log a$ that is normally distributed, then a is log normal with parameters τ and υ, and density,

$$\phi_L(a) = \frac{1}{a\upsilon\sqrt{2\pi}} \exp\left[-\frac{1}{2\upsilon_a^2}(\log a - \log \tau)^2\right], \quad a > 0,\ \tau > 0,\ 0 < \upsilon < 1. \tag{5.92}$$

Assigning $\tau = \tilde{\mu}_a$ and $\upsilon = \log \tilde{\sigma}_a$, $\tilde{\sigma}_a < e = 2.7182\ldots$ gives the penalty,

$$\frac{\partial \log \phi_L(a)}{\partial a} = -\frac{1}{a} - \frac{1}{a\upsilon^2}(\log a - \log \tau), \tag{5.93}$$

and ridge,

$$-\mathcal{E}\left(\frac{\partial^2 \log \phi_L(a)}{\partial a^2}\right) = \frac{1}{\tau^2}\left(\frac{1}{\upsilon^2} - 1\right). \tag{5.94}$$

- *Item lower asymptotes.* The lower asymptote g, $0 < g < 1$ is assumed to have a beta distribution with parameters α and β, and density

$$f(g) = \frac{(\alpha + \beta + 1)!}{\alpha!\beta!} g^{\alpha}(1 - g)^{\beta}. \tag{5.95}$$

Assigning $\alpha = \tilde{g}W_g$ and $\beta = (1 - \tilde{g})W_g$, where $0 < \tilde{g} < 1$, and $W_g > 0$ is the weight of the constraint, gives penalty

$$\frac{\partial \log f(g)}{\partial g} = -\frac{\alpha}{g} - \frac{\beta}{1 - g}, \tag{5.96}$$

and ridge,

$$-\mathcal{E}\left(\frac{\partial^2 \log f(g)}{\partial g^2}\right) = \frac{W_g}{\tilde{g}(1 - \tilde{g})}. \tag{5.97}$$

Assignments of $\tilde{\mu}_c$, $\tilde{\mu}_a$ and $\tilde{g}$ may be updated during the EM cycles.

5.7 Multiple Group Models

In MML item parameter estimation, multiple group models are required whenever a test or scale is intended for use in population groups that have different distributions of θ. The basic model follows from the assumptions that (i) the distribution within each group is normal with unknown mean and variance, and (ii) the item response functions are common across groups and their parameters are independent of those of the group distributions.

Suppose there are p such groups and the mean and variance for the distribution in group t are μ_t and σ_t. The item parameters are common to all groups, but the sample sizes, N_t, are group specific, as are the incidence variables, $\mathbf{u}_{ti}$, and the population densities, $g_t(\theta) = g(\theta | \mu_t, \sigma_t^2)$.

Because of indeterminacy of location and scale, the parameters of only $p - 1$ of the groups need be estimated. Those of the remaining group may be set arbitrarily, usually to 1 and 0, respectively. The group-specific complete data statistics are

$$\bar{r}_{tjh}(\theta) = \sum_{i}^{N_t} \frac{w_i}{P_i} u_{tijhk} \mathcal{L}_i(\theta), \tag{5.98}$$

$$\overline{N}_t(\theta) = \sum_i^{N_t} \frac{w_i}{\mathcal{P}_i} \mathcal{L}_i(\theta). \tag{5.99}$$

The likelihood equations for the item parameters are then the sum of the equations over groups, with $\overline{r}_{tjh}(\theta)$ and $\overline{N}_t\theta$ substituted for their one-group counterparts and $\overline{N}_t(\theta)$ and $g_t(\theta)$ substituted in the computation of the information matrix.

In the M-step computations, the same substitutions are made, and the quadrature weights, $A_s(X_{sq})$ based on $g_t(\theta)$ are substituted for those $g(\theta)$. Alternatively, if standard Gauss–Hermite weights are used, the response functions may be reparameterized as,

$$f\left(\frac{\theta - \mu_t}{\sigma_t} | \mathbf{v}_j\right), \tag{5.100}$$

and the weights left unchanged.

Following the updating of the provisional item parameters, the MML estimates of the group parameters may be updated from the provisional posterior means of θ, given the response patterns, $\mathbf{u}_{ti}$, $i = 1, 2, \dots, N_t$,

$$\tilde{\theta}_{ti} = \int \theta \frac{\mathcal{L}_i(\theta)}{\mathcal{P}_i} g_t(\theta) d\theta, \tag{5.101}$$

and the corresponding posterior variance,

$$\tilde{\sigma}^2_{ti} = \int (\theta - \tilde{\theta}_{ti})^2 \frac{\mathcal{L}_i(\theta)}{\mathcal{P}_i} g_t(\theta) d\theta. \tag{5.102}$$

Then, approximating by quadrature,

$$\hat{\mu}_t = \frac{1}{N_t} \sum_i^{N_t} w_i \tilde{\theta}_{ti}, \tag{5.103}$$

and

$$\hat{\sigma}^2_t = \frac{1}{N_t} \sum_i^{N_t} w_i [(\tilde{\theta}_{ti} - \hat{\mu}_t)^2 + \tilde{\sigma}^2_{ti}]. \tag{5.104}$$

See Bock and Zimowski (1997).

With less computation, however, the group mean may be evaluated by the quadrature,

$$\hat{\mu}_t = \frac{1}{N_t} \sum_q^{Q} \overline{N}(X_q) A_t(X_q). \tag{5.105}$$

Similarly, the group variance is well approximated by

$$\hat{\sigma}^2_t \simeq \frac{1}{N_t} \sum_q^{Q} X_q^2 \overline{N}(X_q) - N_t \hat{\mu}^2_t. \tag{5.106}$$

Note the absence of quadrature weights in (5.106): the prior distribution is treated as rectangular to avoid the reduction in variance due to a normal prior.

Starting values of $\mu_s = 0$ and $\sigma_s^2 = 1$ are quite satisfactory for initializing the EM cycles.

The multiple group model can be elaborated when the groups are structured in crossed and/or nested designs by expressing the threshold parameters, for example, in terms of up to $p - 1$ estimable design-effects parameters. If exactly $p - 1$ effects are estimated, the item parameter estimates will remain unchanged. The same applies to regression on covariates descriptive of individual respondents. See Moustaki (2000) and Moustaki et al. (2004).

5.8 Discussion

Several points in the chapter may benefit from further clarification: (i) notational conventions, (ii) choice of number of quadrature points, (iii) problem of empty categories, (iv) violations of distribution assumptions, (v) failure of conditional independence, and (vi) multiple group analysis in the presence of differential item functioning (DIF). They are discussed in that order.

1) The notation of u_{ijhk} serves to distinguish category h of the response function of item j from the category, k, selected by respondent i. Although it appears in the derivations, it does not occur anywhere in the numerical computations. There, the item scores, x_{ij}, directly designate the item selections. Statistically, the u_{ijhk} are multivariate Bernoulli variables with population means equal to the observed category proportions, π_{jh}, say; their variances and covariances are $\pi_{jh}(1 - \pi_{jh})$ and $-\pi_{jh}\pi_{jh'}$, the same as those of multinomial variables in a sample of size 1.

   Another notational device is the indexing of category parameters, one less than the number of categories, by the subscript, ℓ. It ranges 1 to $m_j - 1$ or 0 to $m_j - 1$. The zero in the latter is a place holder for subsequent representation of m_j category parameters under the restriction of summing to 0.
2) The question of how many quadrature points are required is clarified by the observation that the integrations involved in (5.6) and elsewhere is over the posterior density, $\mathcal{L}_i(\theta)g(\theta)$, which can become extremely concentrated as the number of items increase. Since at least two points with appreciable density in the posterior are required to inform its location and dispersion, the number of points to cover the effective range of the prior, $g(\theta)$, can be very large. This suggests the use of adaptive quadrature, as introduced by Schilling (see (Bock and Schilling 1997; Schilling and Bock 2005)), in context of high dimensional item factor analysis, to confine the points to the space of the posterior distribution

for pattern i. On the assumption that the posterior is approximately normal, this can be done efficiently in the unidimensional case using a provisional estimate of the mode of the assumed normal posterior and approximate standard deviation given by the square root of the inverse information. Two, or preferably more, quadrature points and their corresponding weights are then chosen for the interval ± 2 standard deviations from the mode.

3) When the proportion of responses occurring in one or more categories of any item is zero, boundary problems will prevent the iterative solution for item from converging. The conventional device of collapsing categories may solve the problem during item calibration, but during scoring that requires changing the corresponding category scores when the item parameters are used in scoring respondents of other samples. A commonly used alternative is to insert one response in the empty category. That avoids the boundary and scoring problems but can lead to slow convergence for that item if the corresponding proportion is very small. The stochastic constraint on the intercept parameter for that category of the item will avoid that difficulty.
4) In large sample estimation of item parameters, deviation of the population distribution of data from strict normality usually has very little effect on the outcome. An exception is when there is a marked bimodality or multimodality of the distribution, which should be apparent in the $\overline{N}(X_q)$ expected numbers of respondents if there are a sufficient number of quadrature points to reveal it. MML estimation can be extended, however, to include resolution of the distribution into Gaussian components (see (Mislevy and Verhelst 1990)). If there is sufficient separation of the component modes, there will be a cutting point, depending on the proportions represented, for best assignment of respondents to groups on the basis of their estimated scores. An interpretation of the components may then be apparent in background information about the sample members dominant in each group.
5) Educational tests and clinical self-report scales are likely to be subject to failure of conditional independence of item responses conditional on θ when items are arranged in sections according to readily recognizable content. To the extent that respondents have different proclivities with respect to the sections in general, responses to items within sections will have greater association than those between sections. The effect of this conditional dependence on estimated values of the item parameters may be relatively small, but it can appreciably underestimate the standard errors of the respondent score on the test or scale and overestimate reliability. A solution to this problem is available, however, in a form of factor analysis introduced by Holzinger and Swineford (1937) as an alternative to the one factor model for measures of intelligence due to Spearman (1904) to account for so-called "group" factors. It assumes a "bifactor"

pattern of loadings with one general factor and additional independent factors loading only on the measures within each group.

Applied to item factor analysis, each item in the test or scale is assumed to have a loading on the general factor and a second loading on a factor representing the item's content section. Gibbons and Hedeker (1992) and Gibbons, Bock, Hedeker, Weiss, Segawa, Bhaumik, Kupfer, Frank, Grochocinski and Stover (2007a) have extended the MML method of item factor analysis, in which item discriminating powers correspond to loadings, to provide item "bifactor" analysis. It has the felicitous property that for any given item in the EM solution requires only two-dimensional quadrature. This means that a large number of quadrature points can be used with relatively little computational burden, and adaptive quadrature is unnecessary. Bifactor analysis has the desired effect on the factor scores of the respondents: it removes variance of the general factor score that arises from association of responses within sections and thus provides the correct standard error of the respondents score and correct reliability of the test or scale. These ideas are more fully examined in Chapter 6.

6) In multiple group applications of item parameter estimation, an obvious methodological question is whether the assumption of common item parameter values across groups is justified. When it is not, the items are said to exhibit "DIF" (see (Holland and Wainer 1993)). As discussed in Thissen et al. (1993), in large-sample MML estimation of item parameters, a statistical test of the null hypothesis of the absence of DIF is available in the standard likelihood ratio method; that is, the maximum log likelihood ignoring group minus the sum of the maxima of the individual groups, times −2, is distributed as chi-square on degrees of freedom equal to the absolute difference of numbers of parameters estimated. If the no-DIF hypothesis is rejected, the next question is whether any given item is contributing to the differences between the within group estimates. Item-by-item likelihood ratios could be computed by maximizing the group log likelihoods with parameters of only one item released at a time. But with many items, the computational burden would be heavy. The more expeditious alternative is to test directly the significance of between group differences in the parameter estimates for individual items. This is easily accomplished using the sampling covariance matrices from the Newton–Gauss steps for estimation in the separate groups. Because DIF is test or scale specific, rarely involving more than perhaps 100 items, calculation of the required inverse information matrix should not be prohibitive.

For any given item j, let $\hat{\boldsymbol{v}}_{tj}$ be the n_{v_j}-element parameter estimate vector in group t, and let

$$\bar{\boldsymbol{v}}_j = \frac{1}{N}\sum_{t}^{p} N_t \hat{\boldsymbol{v}}_{tj} \tag{5.107}$$

be the weighted mean of the estimates over groups. Then

$$\chi^2 = \sum_{t}^{p} (\hat{\mathbf{v}}_{tj} - \overline{\mathbf{v}}_{tj})' \mathbf{V}_{tj}^{-1} (\hat{\mathbf{v}}_{tj} - \overline{\mathbf{v}}_{tj}), \tag{5.108}$$

where $\mathbf{V}_{tj}$ is the sampling covariance matrix of $\hat{\mathbf{v}}_{tj}$ extracted from the corresponding block diagonal of the inverse information matrix for group t.
This statistic is distributed on the null hypothesis as chi-square on $n_{v_j}(p-1)$ degrees of freedom. If it rejects the null hypothesis, there is statistical justification for examining the size of the differences for a particular group relative to either the overall estimate or to a designated reference group. When the sample sizes of the groups are very large, however, the size of difference, though significant, may be too small to be of practical interest. Of concern are substantial differences, especially in threshold parameters, for they indicate that the item is appreciably more difficult for some groups than others. We examine the determination of DIF in Chapter 9.

5.9 Conclusions

In this chapter, in uniform notation and method, we cover the essentials of item parameter estimation for unidimensional polytomous response models. A number of results not previously in the literature of IRT are derived and discussed. They include (i) the general form of the MML equations required in the EM solution, (ii) stochastically constrained estimation to avoid inadmissible boundary values in the EM solution, (iii) derivation of the information matrix for the Newton–Gauss solution of the likelihood equations for all items simultaneously, and (iv) computationally efficient estimation procedures for multiple group models. Statistical and computational aspects of practical implementation of the GPCM are documented.

6

Multidimensional IRT Models

Not all of the characteristics which are conversationally described in terms of "more" or "less" can actually be measured. But any characteristic which lends itself to such description has the possibility of being reduced to measurement.

(Source: L.L. Thurstone)

Item factor analysis[1] plays an essential role in the development of tests or scales to measure behavioral tendencies that are considered to be a matter of degree but are observed only as discrete responses. Typical examples are the following:

- Tests of school science achievement based on responses to a number of exercises marked right or wrong.
- Social surveys in which degree of conservatism of the respondent is assessed by agreement or disagreement with positions on a variety of public issues.
- Patient self-reports of satisfaction with the outcome of a medical treatment rated on a seven-point scale.
- Inventory of activities favorable or unfavorable to general health reported in terms of frequency – *never, up to once a month, up to once a week, more than once a week*.
- Nutrition survey of food preference categorized as *dislike very much, dislike moderately, neither like nor dislike, like moderately, like very much*.

The main problem in constructing these kinds of response instruments is the lack of any definite rules for choosing items that best represent the concept to be measured. The content and wording of items that embody the concept are almost always up to the item writer. Once the instrument is administered to respondents of interest, however, data become available for critical item-by-item examination of their suitability as representatives of the concept. The unique contribution of

1 Much of the material presented in this chapter follows from Bock and Gibbons (2010) with permission.

Item Response Theory, First Edition. R. Darrell Bock and Robert D. Gibbons.

item factor analysis lies in its power to reveal whether the patterns of association among the item responses arise from one dimension of measurement or more than one.

If the instrument is designed to measure individual differences among the respondents in only one dimension, association of the item responses should be accounted for by a single underlying variable. If the factor analysis finds association attributable to more than one dimension, the results show which items are associated with each dimension and with what degree of discriminating power. Items that are most discriminating on the intended dimension can then be selected or augmented and those related to the additional dimensions removed or modified. Once a satisfactory one-dimensional instrument is attained, the scoring procedures of item factor analysis can be used to estimate each respondent's position on that dimension with the best precision possible.

Alternatively, if the investigator's intention is to construct an instrument that measures individual differences in the several dimensions simultaneously, the item factor analysis helps characterize the dimensions and shows where to add or delete items to balance the content. Personality inventories are often constructed in this multidimensional form.

6.1 Classical Multiple Factor Analysis of Test Scores

Multiple factor analysis as formulated by Thurstone (1947) assumes that the test scores are continuous measurements standardized to mean zero and standard deviation one in the sample. (Number right scores on tests with 30 or more items are considered close enough to continuous for practical work.) The Pearson product-moment correlations between all pairs of tests are then sufficient statistics for factor analysis when the population distribution of the scores is multivariate normal. Because the variables are assumed to be standardized, the mean of the distribution is the null vector and the covariance matrix is a correlation matrix. If the dimensionality of the factors space is d, the assumed statistical model for the jth observed score y is

$$y_j = \alpha_{j1}\theta_1 + \alpha_{j2}\theta_2 + \cdots + \alpha_{jd}\theta_d + \epsilon_j, \tag{6.1}$$

where the underlying vector of latent variables attributed to the respondent is

$$\theta = (\theta_1, \theta_2, \ldots, \theta_d). \tag{6.2}$$

Like the observed variables, the latent variables are assumed standard multivariate normal but are uncorrelated; that is, their covariance matrix is the $d \times d$ identity matrix. The residual, ϵ_j, which accounts for all remaining variation in y_j

is assumed normal with mean 0 and variance $1 - \omega_j^2$, where

$$\omega_j^2 = \sum_{v=1}^{d} \alpha_{jv}^2, \tag{6.3}$$

which Thurstone called the *communality* of the item. Estimation of the loadings requires the restriction $1 - \omega_j^2 > 0$ to prevent inadmissible so-called *Heywood* cases. Moreover, if the reliability of the test is known to be ρ, ω_j^2 cannot be greater than ρ.

On the above assumptions, efficient statistical estimation of the factor loadings from the sample correlation matrix is possible and available in published computer programs. In fact, only the item communality need be estimated: once the communalities are known, the factor loadings can be calculated directly from the so-called "reduced" correlation matrix, in which the diagonal elements of the sample correlation matrix are replaced by the corresponding communalities (see (Harman 1967)).

6.2 Classical Item Factor Analysis

In the item factor analysis, the observed item responses are assigned to one of two-or-more predefined categories. For example, test items marked right or wrong are assigned to dichotomous categories; responses to essay questions may be assigned to ordered polytomous categories (grades) *A, B, C, D* in order of merit; responses in the form of best choice among multiple alternatives may be assigned to nominal polytomous categories.

To adapt the factor analysis model for test scores to the analysis of categorical item responses, we assume that the *y*-variables are also unobservable. We follow Thurstone in referring to these underlying variables as *response processes*. In the dichotomous case, a process gives rise to an observable correct response when y_j exceeds some threshold γ_j specific to item j. On the assumption that y_j is standard normal, γ_j divides the area under the normal curve in two sections corresponding to the probability that a respondent with given value of θ will respond in the first or second category. Designating the categories 1 and 2, we may express these conditional probabilities given θ as

$$P_{j1}(\theta) = \Phi(z_j - \gamma_j) \quad \text{and} \quad P_{j2}(\theta) = 1 - \Phi(z_j - \gamma_j),$$

where Φ is the cumulative normal distribution function and

$$z_j = \sum_{v=1}^{d} \alpha_{jv}\theta_v. \tag{6.4}$$

The *unconditional* response probabilities, on the other hand, are the areas under the standard normal curve above and below $-\gamma_j$ in the population from which the sample of respondents is drawn. The area above this threshold is the classical *item difficulty*, b_j, and the standard normal deviate at b_j is a large sample estimator of $-\gamma_j$ (see (Lord and Novick 1968) Chapter 16).

These relationships generalize easily to ordered polytomous categories. Suppose item j has m_j ordered categories: we then replace the single threshold of the dichotomous case with m_{j-1} thresholds, say, $\gamma_{j1}, \gamma_{j2}, \dots, \gamma_{j,m_{j-1}}$. The category response probabilities conditional on θ are the m_j areas under the normal curve corresponding to the intervals from minus to plus infinity bounded by the successive thresholds:

$$P_{jh}(\theta) = \Phi(z_j - \gamma_{jh}) - \Phi(z_j - \gamma_{j,h-1}),$$

where $\Phi(z_j - \gamma_{j0}) = 0$ and $\Phi(z_j - \gamma_{jm_j}) = 1 - \Phi(z_j - \gamma_{j,m_j-1})$.

Because product-moment correlations of the response processes cannot be calculated numerically, classical methods of multiple factor analysis do not apply directly to item response data. However, an approximation to the correlation can be inferred from the category joint-occurrence frequencies tallied over the responses in the sample. Assuming in the two-dimensional case that the marginal normal distribution of the processes is standard bivariate normal, the correlation value that best accounts for the observed joint frequencies can be obtained by a computing approximation. If both items are scored dichotomously, the result is the well-known tetrachoric correlation coefficient, an approximation for which was given by Divgi (1979a). If one or both items are scored polytomously, the result is the less common polychoric correlation, which can also be calculated by computing approximation (Jöreskog 1994). The correlations for all distinct pairs of items can then be assembled into a correlation matrix and unities inserted in the diagonal to obtain an approximation to the item correlation matrix. Because the calculation of tetrachoric and polychoric correlations breaks down if there is a vacant cell in the joint occurrence table, a small positive value such as 0.5 (continuity correction) is added to each cell of the joint frequency table.

The accuracy of these approximations to the population correlation matrix can be improved by computing their principal components and reproducing the off-diagonal elements of the matrix from the sums of cross products of all real-valued components. With ones placed in the diagonal, this "smoothed" correlation matrix is in a form suitable for further statistical operations such as multiple regression analysis or multiple factor analysis that require a strictly nonsingular matrix. In particular, it can be subjected to the method of principal components analysis with iteration of communalities ((Harman 1967); page 87). In those computations, some provisional number of factors is chosen and the principal vectors corresponding to the d largest principal values are calculated from the reduced correlation matrix by the Householder–Ortega–Wilkinson method for eigenvectors and eigenvalues (see (Bock 1975), for a summary and references). The correlation matrix is then reconstructed from the sum of cross products of each of the principal vectors weighted by the square roots of their principal values. The diagonal of this matrix will then contain the first approximation to the item communality's. If the process is repeated on that matrix, the communality's will be smaller and in better approximation to their final values. These iterations are continued until all differences and commonalities between steps are less than, say 0.01 – accurate enough for practical work. The principal vectors at that stage multiplied by the square roots of their principal values are then large sample estimates of the factor loadings.

The principal factor solution has merits similar to those of principal components of the original unreduced correlation matrix. Whereas the first d principal components account for the maximum total variance of the variables, the corresponding principal factors account for the maximum total association among the variables in the sense of minimizing the sum of squares of the residual correlations (see (Harman 1967)). Principal factor loadings are also mutually orthogonal; that is, the pairwise sums of their cross-products are zero. This is a valuable property in later estimation of factor scores. Classical principal factor analysis of item responses can be useful in its own right, or as a preliminary to more exact and more computationally intensive IRT procedures such as marginal maximum likelihood item factor analysis. In the latter role, the classical method provides a quick way of giving an upper bound on a plausible number of factors in terms of the total amount of association accounted for. It also gives good starting values for the iterative procedures discussed in the following section.

6.3 Item Factor Analysis Based on Item Response Theory

IRT-based item factor analysis makes use of all information in the original categorical responses and does not depend on pairwise indices of association such as tetrachoric or polychoric correlation coefficients. For that reason, it is referred to as *full information* item factor analysis. It works directly with item response models giving the probability of the observed categorical responses as a function of latent variables descriptive of the respondents and parameters descriptive of the individual items. It differs from the classical formulation in its scaling, however, because it does not assume that the response process has unit standard deviation and zero mean; rather it assumes that the *residual* term has unit standard deviation and zero mean. The latter assumption implies that the response processes have zero mean and standard deviation equal to

$$\sigma_{y_j} = \sqrt{1 + \sum_{v}^{d} \alpha_{jv}^2}. \tag{6.5}$$

Inasmuch as the scale of the model affects the relative size of the factor loadings and thresholds, we rewrite the model for dichotomous responses in a form in which the factor loadings are replaced by factor slopes, a_{jv}, and the threshold is absorbed in the intercept, c_j:

$$y_j = \sum_{v=1}^{d} a_{jv}\theta_v + c_j + \epsilon_j. \tag{6.6}$$

To convert factor slopes into loadings, we divide by the above standard deviation and similarly convert the intercepts to thresholds:

$$\alpha_{jv} = a_{jv}/\sigma_{y_j} \quad \text{and} \quad \gamma_j = -c_j/\sigma_{y_j}.$$

Conversely, to convert to factor analysis units, we change the standard deviation of the residual from 1 to

$$\sigma_{\epsilon_j}^* = \sqrt{1 - \sum_{v}^{d} \alpha_j v^2}, \tag{6.7}$$

and change the scale of the slopes and intercept accordingly:

$$a_{jv} = \alpha_{jv}/\sigma_{\epsilon_j}^* \text{ and } c_j = -\gamma_j/\sigma_{\epsilon_j}^*.$$

For polytomous responses, the model generalizes as follows:

$$z_j = \sum_{v=1}^{d} a_{jv}\theta_v, \tag{6.8}$$

$$P_{jh}(\theta) = \Phi(z_j + c_{jh}) - \Phi(z_j + c_{j,h-1}), \tag{6.9}$$

where $\Phi(z_j + c_{j0}) = 0$ and $\Phi(z_j + c_{jm_j}) = 1 - \Phi(z_j + c_{j,m_j-1})$ as previously.

In the context of item factor analysis, this is the multidimensional generalization of the *graded* model introduced by Samejima (1969). Similarly, the *rating scale* model of Andrich (1978), in which all items have the same number of categories and the thresholds are assumed to have the same spacing but may differ in overall location, can be generalized by setting the above linear form to $z_j + e_j + c_h$, where e_j is the location intercept.

6.4 Maximum Likelihood Estimation of Item Slopes and Intercepts

There is a long history, going back to Fechner (1860), of methods for estimating the slope and intercept parameters of models similar to the above – that is, models in which the response process is normally distributed and the deviate is a linear form. These so-called *normal transform* models differ importantly from the IRT models, however, in assuming that the θ variables are manifest measurements of either observed or experimentally manipulated variables. In Fechner's classic study of the sensory discrimination thresholds for lifted weights, the subjects were required to lift successively each of a series of two small, identical appearing weights differing by fixed amounts and say which feels heavier. Fechner fitted graphically the inverse normal transforms of the proportion of subjects who answered correctly and used the slope of the fitted line to estimate the standard deviation as a measure of sensory discrimination. Much later, R.A. Fisher (Bliss 1935) provided a maximum likelihood method of fitting similar functions used in the field of toxicology to determine the so-called "50% lethal dose of pesticides." This method eventually became known as probit analysis (see (Finney 1952); see (Bock and Jones 1968), for behavioral applications).

To apply Fisher's method of analysis to item factor analysis one must find a way around the difficulty that the variable values (i.e. the thetas) in the linear form are unobservable. The key to solving this problem lies in assuming that the values have a specifiable distribution in the population from which the respondents are drawn (Bock and Lieberman 1970). This allows us to integrate numerically over that distribution to estimate the expected numbers of respondents located at given points in the latent space who respond in each of the categories. These expected values can then be subjected to multidimensional version of probit analysis. The so-called "Expectation-Maximization (EM) method" of solving this type of estimation problem (Bock and Aitkin 1981) is an iterative procedure starting from given initial values. It involves calculating expectations (the E-step) that depend on both the parameters and the observations, followed by likelihood maximization (the M-step) that depends on the expectations. These iterations can be shown to converge on the maximum likelihood estimates under very general conditions (Dempster et al. 1977). As noted earlier, in IRT and similar applications, this approach is called *marginal maximum likelihood* estimation because it works with the marginal probabilities of response rather than the conditional probabilities.

The likelihood-based approach to item factor analysis depends on two main assumptions:

1. That the probability of a response in category h of an m_j-category item may be expressed as an item response function (or *model*) depending on parameter

vector $\mathbf{v}_j$ specific to item j and conditional on a vector latent variable θ attributable to the respondent:

$$P_{jh}(\theta) = P(u_j|\mathbf{v}_j, \theta) = \prod_{h=1}^{m_j} P_{jh}^{u_{jhk}}(\theta), \tag{6.10}$$

where $u_{jhk} = 1$ if the response falls in category k and 0 otherwise, and θ is distributed d-dimensional multivariate in the population of respondents.

2. That the n item responses to a test or scale observed in pattern l

$$\mathbf{u}_l = [u_{ljhk}], \quad j = 1, 2, \ldots, n \tag{6.11}$$

are stochastically independent, conditional on θ; implies that the conditional probability of pattern l is

$$P(\mathbf{u}_l|\theta) = \prod_{j=1}^{n}\prod_{h=1}^{m_j} P_{jh}^{u_{ljhk}}(\theta). \tag{6.12}$$

It follows from these assumptions that the unconditional (marginal) probability of $\mathbf{u}_l$ in the population is the integral,

$$\overline{P}_l = \int_\theta L_l(\theta)\, g(\theta) d\theta \tag{6.13}$$

over the range of θ.

6.4.1 Estimating Parameters of the Item Response Model

Assuming members of a sample of N respondents respond independently to one another, the marginal likelihood of the parameter vector $\mathbf{v}_j$ of item j is,

$$L_N = \frac{N!}{r_1!\, r_2! \cdots r_s!} \prod_{l=1}^{s} \overline{P}_l^{\,r_l}, \tag{6.14}$$

where r_l is the number of occurrence of pattern l and s is the number of distinct patterns observed in the sample. The likelihood equation for $\mathbf{v}_j$ may then be expressed as

$$\frac{\partial \log L_N}{\partial \mathbf{v}_j} = \sum_{l=1}^{s} \frac{r_l}{\overline{P}_l} \int_\theta \frac{\partial \prod_h^{m_j} P_{jh}^{u_{ljhk}}(\theta)\, \partial / \mathbf{v}_j}{\prod_h^{m_j} P_{jh}^{u_{ljhk}}(\theta)} L_l(\theta)\, g(\theta) d\theta = \mathbf{0} \tag{6.15}$$

or applying the rule for differentiating continued products and canceling exponents in the numerator and denominator,

$$\sum_{l=1}^{s} \frac{r_l}{\overline{P}_l} \int_\theta \sum_{h=1}^{m_j} \frac{u_{ljhk}}{P_{jh}(\theta)} \frac{\partial P_{jh}(\theta)}{\partial \mathbf{v}_j} L_l(\theta)\, g(\theta) d\theta = \mathbf{0}. \tag{6.16}$$

The Bock–Aitkin (1981) EM solution of these equations for n items reverses the order of summation and integration

$$\frac{\partial \log L_N}{\partial \mathbf{v}_j} = \int_\theta \sum_{h=1}^{m_j} \frac{\bar{r}_{jh}}{P_{jh}(\boldsymbol{\theta})} \frac{\partial P_{jh}\boldsymbol{\theta}}{\partial \mathbf{v}_j} d\boldsymbol{\theta} = \mathbf{0}, \tag{6.17}$$

where

$$\bar{r}_{jh} = \sum_{l=1}^{s} \frac{r_l\, u_{ljhk}}{\bar{P}_l} L_l(\boldsymbol{\theta})\, g(\boldsymbol{\theta}). \tag{6.18}$$

In the dichotomous case, the likelihood equations simplify to

$$\sum_{l=1}^{s} \frac{r_l}{\bar{P}_l} \int_\theta \left\{ \frac{u_{lj1k}}{P_{j1}(\boldsymbol{\theta})} \frac{\partial P_{j1}(\boldsymbol{\theta})}{\partial \mathbf{v}_l} + \frac{[1 - u_{lj1k}]}{[1 - P_{j1}(\boldsymbol{\theta})]} \frac{\partial [1 - P_{j1}(\boldsymbol{\theta})]}{\partial \mathbf{v}_l} \right\} L_l(\boldsymbol{\theta}) d\boldsymbol{\theta} \tag{6.19}$$

$$= \int_\theta \frac{2\bar{r}_j - \bar{N} P_{j1}}{P_{j1}(\boldsymbol{\theta})[1 - P_{j1}(\boldsymbol{\theta})]} \frac{\partial P_{j1}(\boldsymbol{\theta})}{\partial \mathbf{v}_j} d\boldsymbol{\theta} = \mathbf{0}, \tag{6.20}$$

where

$$\bar{r}_j = \sum_{l=1}^{s} \frac{r_l\, u_{ljhk}}{\bar{P}_l} L_l(\boldsymbol{\theta})\, g(\boldsymbol{\theta}) \tag{6.21}$$

and

$$\bar{N} = \sum_{l=1}^{s} \frac{r_l}{\bar{P}_l} L_l(\boldsymbol{\theta})\, g(\boldsymbol{\theta}). \tag{6.22}$$

See Bock (1989b).

When n, m_j, and d are not too large and $g(\boldsymbol{\theta})$ is d-variate normal, the likelihood equations in both the polytomous and dichotomous case can be evaluated by Gauss–Hermite quadrature or Monte Carlo integration. By using either of these methods for numerical integration, the E-step of the EM algorithm, yield the expected number of responses for each category of every item. They serve as data for the generalized maximum likelihood probit analysis in the M-step that yields the provisional estimates of the parameters of all the item response functions. In practical applications, however, the quadratures of the E-step present the computational problem of requiring, with Q points per dimension, a total of Q^d points in the full $\boldsymbol{\theta}$-space. Even at modern computing speeds, this limits Q to small numbers, 2 or 3, when d is 10 or more. A further complication is that, as the number of items or the number of categories per item increases, the posterior densities become so concentrated that they have almost no probability of including more than one point. Accurate calculation of the expectations then becomes impossible. The same applies to the Monte Carlo method with any practical number of draws.

The key to avoiding this difficulty is realizing that the integrations are not over the full $\boldsymbol{\theta}$-space, but over the posterior density given the observed pattern, $\mathbf{u}_l$. If we then assume that the posterior distribution is sufficiently well approximated by a d-variate normal distribution with mean $\boldsymbol{\mu}_l$ and covariance matrix $\boldsymbol{\Sigma}_l$, we can use the Naylor–Smith Naylor and Smith (1982) procedure to adapt the quadrature to the posterior space rather than the full space. This method of adaptive quadrature conforms well with the EM solution of the likelihood equations because at each cycle, provisional MAP estimates of $\boldsymbol{\mu}_l$ and $\boldsymbol{\Sigma}_l$ are easily calculated (see below). Described briefly, the Naylor–Smith procedure is as follows. The integrals to be evaluated may be expressed as

$$\int_{\psi} f(\boldsymbol{\psi}|\mathbf{v}_j, \boldsymbol{\mu}_l)p_l(\boldsymbol{\psi})d\boldsymbol{\psi}, \tag{6.23}$$

where $p_l(\boldsymbol{\psi})$ is the posterior density $L_l(\boldsymbol{\psi})\, g(\boldsymbol{\psi})$. Dividing and multiplying by the d-variate normal density $\phi(\boldsymbol{\psi}|\boldsymbol{\mu}_l, \boldsymbol{\Sigma}_l)$ gives,

$$\int_{\psi} \frac{f(\boldsymbol{\psi}|\mathbf{v}_j, \boldsymbol{\mu}_l)p_l(\boldsymbol{\psi})}{\phi(\boldsymbol{\psi}|\boldsymbol{\mu}_l, \boldsymbol{\Sigma}_l)} \phi(\boldsymbol{\psi}|\boldsymbol{\mu}_l, \boldsymbol{\Sigma}_l)d\boldsymbol{\psi} = \int_{\psi} F(\boldsymbol{\psi})\phi(\boldsymbol{\psi}|\boldsymbol{\mu}_l, \boldsymbol{\Sigma}_l)d\boldsymbol{\psi}. \tag{6.24}$$

Let

$$\boldsymbol{z} = \boldsymbol{T}_l^{-1}(\boldsymbol{\psi} - \boldsymbol{\mu}_l),$$

where $\boldsymbol{T}_l$ is the Cholesky decomposition of $\sum_l$, such that

$$\boldsymbol{\Sigma}_l = \boldsymbol{T}_l\boldsymbol{T}_l', \tag{6.25}$$

$$\boldsymbol{\Sigma}_l^{-1} = (\boldsymbol{T}_l^{-1})'\boldsymbol{T}_l^{-1}, \tag{6.26}$$

and

$$\boldsymbol{T}_l^{-1}\boldsymbol{\Sigma}_l(\boldsymbol{T}_l^{-1})' = \boldsymbol{I}. \tag{6.27}$$

Then the integral becomes

$$|\boldsymbol{T}_l| \int_{\boldsymbol{z}} F(\boldsymbol{T}_l\boldsymbol{z} + \boldsymbol{\mu}_l)\phi(\boldsymbol{z}|\mathbf{0}, \boldsymbol{I})d\boldsymbol{z}, \tag{6.28}$$

a form suitable for Gauss–Hermite quadrature in which the points in the z-space are well positioned for quadrature (see (Skrondal and Rabe-Hesketh 2004)). Similarly, points for Monte Carlo integration can be drawn for the posterior distribution of $\boldsymbol{\theta}$ given $\mathbf{u}_l$. Simulation studies by Schilling and Bock (2005) in the dichotomous case have shown excellent recovery of generating item parameters in as many as eight dimensions with as few as two quadrature points per dimension.

As exact positioning of the points is not a requirement, at some stage in the EM cycles computing time can be saved by keeping the current points for each pattern and reusing them in all remaining cycles. Because the points are generated separately for each pattern, there is almost no probability that the fixed points will be unfavorable to accurate estimation or calculation of the overall marginal likelihood.

If the values of the parameters of the item response models appropriate to the test or score are known, likelihood-based estimation of the factor scores for any given response pattern is straightforward. Assuming conditional independence of responses, the posterior probability for the pattern of respondent i drawn from a population with density $g(\boldsymbol{\theta})$ is

$$p(\boldsymbol{\theta}|\mathbf{u}_i) = \prod_{j=1}^{n}\prod_{h=1}^{m_j} P_{jh}^{u_{ijhk}} g(\boldsymbol{\theta}). \tag{6.29}$$

Then the d-dimensional gradient of the log likelihood is

$$\mathbf{G}(\boldsymbol{\theta}) = \frac{\partial \log p(\boldsymbol{\theta}|\mathbf{u}_i)}{\partial \boldsymbol{\theta}} = \sum_{j=1}^{n}\sum_{h=1}^{m_j} \frac{u_{ijhk}}{P_{jh}(\boldsymbol{\theta})} \frac{\partial P_{jh}(\boldsymbol{\theta})}{\partial \boldsymbol{\theta}} + \frac{\partial \log g(\boldsymbol{\theta})}{\partial \boldsymbol{\theta}}, \tag{6.30}$$

and the MAP estimate, $\tilde{\boldsymbol{\theta}}$ of $\boldsymbol{\theta}$ given $\mathbf{u}_i$ is the solution of $E(\boldsymbol{\theta}) = \mathbf{0}$. If $g(\boldsymbol{\theta})$ is multivariate normal, $\phi(\boldsymbol{\theta}|\boldsymbol{\mu}, \boldsymbol{\Sigma})$, the so-called"penalty term" in the stationary equations is

$$\frac{\partial \log \phi(\boldsymbol{\theta})}{\partial \boldsymbol{\theta}} = -\boldsymbol{\Sigma}^{-1}(\boldsymbol{\theta} - \boldsymbol{\mu}). \tag{6.31}$$

If $g(\boldsymbol{\theta})$ is standard multivariate normal, the term is simply $-(\boldsymbol{\theta})$. These equations are well-conditioned for solution by d-dimensional Newton–Gauss iterations starting from $\boldsymbol{\theta}_i^{(0)} = \mathbf{0}$,

$$\boldsymbol{\theta}_i^{(t+1)} = \boldsymbol{\theta}_i^{(t)} + \boldsymbol{H}^{-1}(\boldsymbol{\theta}_i^{(t)})\mathbf{G}(\boldsymbol{\theta}_i^{(t)}),$$

where

$$\boldsymbol{H}(\boldsymbol{\theta}) = \sum_{j=1}^{n}\sum_{h}^{m_j} \frac{1}{P_{ij}(\boldsymbol{\theta})} \left[\frac{\partial P_{jh}(\boldsymbol{\theta})}{\partial \boldsymbol{\theta}}\right]' \frac{\partial P_{ij}(\boldsymbol{\theta})}{\partial \boldsymbol{\theta}} + \boldsymbol{\Sigma}^{-1}, \tag{6.32}$$

$H(\boldsymbol{\theta}_i)$ is the provisional Fisher Information matrix evaluated at $\boldsymbol{\theta} = \boldsymbol{\theta}_i^t\, \boldsymbol{I}$, and $g(\boldsymbol{\theta})$ is standard multivariate normal $\boldsymbol{\Sigma}^{-1} = \boldsymbol{I}$. Then, $\boldsymbol{H}^{-1}(\tilde{\boldsymbol{\theta}}_i)$ at convergence is the estimated covariance matrix of the MAP estimator; the estimated standard errors of the estimated factor scores are the square roots of its corresponding diagonal elements.

The expected a posteriori (EAP) (or Bayes) estimate, $\overline{\boldsymbol{\theta}}_i$, is the mean of the posterior distribution of $\boldsymbol{\theta}$, given $\mathbf{u}_i$,

$$\overline{\boldsymbol{\theta}}_i = \int_{\boldsymbol{\theta}} \boldsymbol{\theta} L_i(\boldsymbol{\theta}) g(\boldsymbol{\theta}) d\boldsymbol{\theta}, \tag{6.33}$$

where

$$L_i(\boldsymbol{\theta}) = \prod_{j=1}^{n}\prod_{h=1}^{m_j} P_{jh}^{u_{ijhk}}(\boldsymbol{\theta}). \tag{6.34}$$

If $g(\boldsymbol{\theta})$ is a standard multivariate normal density function, the EAP estimate may be obtained by adaptive Gauss–Hermite quadrature. The square roots of the diagonal elements of the corresponding posterior covariance matrix,

$$\boldsymbol{\Sigma}_{\boldsymbol{\theta}|\mathbf{u}_i} = \int_{\boldsymbol{\theta}} (\boldsymbol{\theta} - \overline{\boldsymbol{\theta}}_i)^2\, L_i(\boldsymbol{\theta}) g(\boldsymbol{\theta}) d\boldsymbol{\theta}, \tag{6.35}$$

serve as standard errors of the EAP estimates.

6.5 Indeterminacies of Item Factor Analysis

6.5.1 Direction of Response

In cognitive testing with dichotomous items, the direction of the response relative to the measurement dimension is clear: A correct response has higher value than an incorrect response. In psychological inventories or social surveys, the direction almost always depends upon the direction wording of the item – positive or negative. For example, "Are you in favor of gun control" versus "Are you opposed to gun control." IRT-based analysis of the data will work correctly in either case, but the slope parameters of items worded in different directions will have different algebraic signs. If a negative slope appears for a *cognitive* multiple choice item, it usually indicates an error in the answer key and is easily corrected. For polytomous Likert-scored items, it indicates that the integer values of the categories are reversed. In those cases, it is advisable to reverse the values for negatively worded items to avoid confusion when interpreting the signs of the factor loadings. The computing procedure should have a direction key indicating which items require reversal.

6.5.2 Indeterminacy of Location and Scale

In the multiple factor models for test scores and item responses, the latent variables are indeterminate both with respect to location (origin) and scale (unit). In the model for test scores, these indeterminacies are resolved by standardizing the variables to mean zero and standard deviation one in the sample. In the model for classical item factor analysis, the response processes and latent variables are assumed to have mean zero and standard deviation one in the population. These choices fix the location and scale of the factor loadings and the item thresholds.

In parameter estimation for IRT item factor analysis, location and scale are set by the assumptions that the *residuals* and latent variables have mean zero and standard deviation one in the population. In estimating factor scores ($\boldsymbol{\theta}$-values) for the respondents, however, it is customary for ease of interpretation to choose other values for the mean and standard deviation. They are typically set so that the scores have effectively no probability of being negative and are expressible to two or three orders of precision by natural numbers. If, say, M is chosen as the mean for the scores and S as the standard deviation, the latent variable v is redefined as

$$\theta_v^* = S\theta_v + M. \tag{6.36}$$

This implies a reparameterization of the IRT model; in the d-dimensional model, for example,

$$\begin{aligned} y_{jh} &= \sum_{v=1}^{d} a_{jv} \frac{(\theta^* - M)}{S} + c_{jh} + \epsilon_{jh} \\ &= \sum_{v=1}^{d} a_{jv}^* \theta^* + c_{jh}^* + \epsilon_{jh}, \end{aligned} \tag{6.37}$$

where the new parameters are

$$a_{jv}^* = a_{jv}/S \quad \text{and} \quad c_{jh}^* = c_{jh} - \sum_{v=1}^{d} a_{jv} M/S.$$

6.5.3 Rotational Indeterminacy of Factor Loadings in Exploratory Factor Analysis

More generally, when there are two or more dimensions, all of the elements in the vector latent variable $\boldsymbol{\theta}$ may be subjected to a one-to-one transformation $\boldsymbol{T}$, which has an inverse $\boldsymbol{T}^{-1}$ that recovers the original values,

$$\boldsymbol{\theta} = \boldsymbol{T}^{-1}\boldsymbol{\theta}^*. \tag{6.38}$$

Substituting the latter in the factor model and transforming the vector of slopes $\mathbf{a}_j$, leaves the response function unchanged. Converting the slopes of the items to factor loadings and applying the $\boldsymbol{T}^{-1}$ transform yields a new pattern for the loadings that may be preferable for purposes of interpretation or scoring. The fit of model is unchanged. If the transformation is orthogonal, the transform $\boldsymbol{T}$ corresponds to a rigid rotation of the coordinate frame and preserves the pairwise zero correlations between the latent variables. If the transformation is *oblique*, the angles between pairs of coordinates may differ from 90°, which implies nonzero correlations among the latent variables. These correlations may be absorbed in so-called *second-order* factors that leave the overall model unchanged (see (Thurstone 1947)).

The choice of a preferred transformation is entirely arbitrary and depends on the objective of the analysis. If the goal is to obtain the most parsimonious representation of association in the data, a transformation that yields a principle factor pattern is clearly preferable in the sense that it accounts for the largest possible common factor variance for a specified number of factors. It also restricts the set of factor loadings to orthogonality; that is, the sum of cross-products of all loadings for any pair of factors is zero. Under this restriction, the number of free parameters for estimating the polytomous model for n items is $nd - d(d-1)/2$ for the loadings, and $\sum_j^n m_j - n$ for the intercepts, or for the rating scale model $n + m - 1$. The principle factor pattern also orders factors with respect to the percent of common factor variance accounted for by each. For these reasons, the principle factor pattern is the best choice for the initial estimation of factor loadings. The other possible patterns are obtained by transformations such as varimax or promax rotation.

6.5.3.1 Varimax Factor Pattern

If the objective of the analysis is to identify mutually exclusive subsets of items each of which account for variance of one and only one factor, the transformation should result in a matrix in which each row has a loading appreciably different from zero in only one column. Thurstone (1947) called this pattern *simple structure.* If the initial estimated pattern can in fact be transformed orthogonally to a simple structure, it can be found by the Kaiser (1958) varimax procedure, which maximizes the total variance of the factor loadings under the ridged rotation. A varimax rotation is a good first step toward identifying the source of the factors by the common features of items that have appreciable loadings on the factor.

6.5.3.2 Promax Factor Pattern

An even more sensitive method of identifying items that best represent each factor is to perform an oblique rotation that moves the coordinates to positions near the clusters of loadings in the factor space that identify the various factors. A convenient procedure for this purpose is Hendrickson and White's (1964) promax rotation. It proceeds by raising the varimax loadings to a higher power and moving the coordinates near the largest of the resulting values. When the loadings are expressed on the obliquely rotated coordinates, the procedure sometimes yields loadings greater than one, but that does not interfere with identifying factors. As a by-product, the procedure gives the cosines of the angles between all pairs of coordinates, which correspond to the correlations between the factors.

6.5.3.3 General and Group Factors

Item factor analysis of responses to cognitive tasks rarely if ever exhibit simple structure. Instead, a general factor with all positive loadings almost always accounts for more of a common variance than any of the item-group factors that may appear in the analysis. In exploratory factor analysis, this general factor will correspond to the first factor of the principle factor pattern. And because of the orthogonality constraint on principal factors, all subsequent factors must have both positive and negative loading – that is, they take the form of bipolar contrasts between item groups. The number of such contrasts cannot be greater than one minus the number of groups. If no significant variance is accounted for by the group factors, the principal factor solution defaults to the one factor case, which is equivalent to conventional one-dimensional IRT analysis.

6.5.3.4 Confirmatory Item Factor Analysis and the Bifactor Pattern

In confirmatory factor analysis, indeterminacy of rotation is resolved by assigning arbitrary fixed values to certain loadings of each factor during maximum likelihood estimation. In general, fixing of loadings will imply nonzero correlations of the latent variables, but this does not invalidate the analysis. (The correlations may also be estimated if desired.) An important example of confirmatory item factor analysis is the bifactor pattern for general and group factors, which applies to tests and scales with item content drawn from several well-defined subareas of the domain in question. Two prominent examples are tests of educational achievement consisting of reading, mathematics and science areas, and self-reports of health status covering physical activity, sleep, nutrition, allergies, worry, etc. The main objective in the use of such instruments is to estimate a single-score measuring, in these examples, general educational achievement or overall health status.

To analyze these kinds of structures for dichotomously scored item responses, Gibbons and Hedeker (1992) developed full-information item bifactor analysis for binary item responses, and Gibbons extended it to the polytomous case Gibbons, Bock, Hedeker, Weiss, Segawa, Bhaumik, Kupfer, Frank, Grochocinski and Stover (2007a). To illustrate, consider a set of n test items for which a d-factor solution exists with one general factor and $d-1$ group or method-related factors. The bifactor solution constrains each item j to a nonzero loading α_{j1} on the primary dimension and a second loading ($\alpha_{jv}, v = 2, \ldots, d$) on not more than one of the $d-1$ group factors. For four items, the bifactor pattern matrix might be

$$\boldsymbol{\alpha} = \begin{bmatrix} \alpha_{11} & \alpha_{12} & 0 \\ \alpha_{21} & \alpha_{22} & 0 \\ \alpha_{31} & 0 & \alpha_{33} \\ \alpha_{41} & 0 & \alpha_{43} \end{bmatrix}. \tag{6.39}$$

This structure, which Holzinger and Swineford (1937) termed the "bifactor" pattern, also appears in the interbattery factor analysis of Tucker (1958) and is one of the confirmatory factor analysis models considered by Jöreskog (1969). In the latter case, the model is restricted to test scores assumed to be continuously distributed. However, the bifactor pattern might also arise at the item level (Muthén 1989). Gibbons and Hedeker (1992) showed that paragraph comprehension tests, where the primary dimension represents the targeted process skill and additional factors describe content area knowledge within paragraphs, were described well by the bifactor model. In this context, they showed that items were conditionally independent between paragraphs, but conditionally dependent within paragraphs. More recently, the bifactor model has been applied to problems in patient-reported outcomes in physical and mental health measurement (Gibbons, Immekus and Bock 2007b, Gibbons et al. 2012, 2016).

The bifactor restriction leads to a major simplification of likelihood equations that (i) permits analysis of models with large numbers of group factors since the integration always simplifies to a two-dimensional problem, (ii) permits conditional dependence among identified subsets of items, and (iii) in many cases, provides more parsimonious factor solutions than an unrestricted full-information item factor analysis.

In the bifactor case, the graded response model is

$$z_{jh}(\boldsymbol{\theta}) = \sum_{v=1}^{d} a_{jv}\theta_v + c_{jh}, \tag{6.40}$$

where only one of the $v = 2, \ldots, d$ values of a_{jv} is nonzero in addition to a_{j1}. Assuming independence of the $\boldsymbol{\theta}$, in the unrestricted case, the multidimensional model above would require a d-fold integral in order to compute the unconditional probability for response pattern $\mathbf{u}$, i.e.

$$P(\mathbf{u} = \mathbf{u}_i) = \int_{-\infty}^{\infty}\int_{-\infty}^{\infty}\cdots\int_{-\infty}^{\infty} L_i(\boldsymbol{\theta})g(\theta_1)g(\theta_2)\cdots g(\theta_d)d\theta_1 d\theta_2\cdots d\theta_d, \tag{6.41}$$

for which numerical approximation is limited as previously described. Gibbons and Hedeker (1992) showed that for the binary response model, the bifactor restriction always results in a two-dimensional integral regardless of the number of dimensions, one for θ_1 and the other for θ_v, $v > 1$. The reduction formula is due to Stuart (1958), who showed that if n variables follow a standardized multivariate normal distribution where the correlation $\rho_{ij} = \sum_{v=1}^{d} \alpha_{iv}\alpha_{jv}$ and α_{iv} is nonzero for only one v, then the probability that respective variables are simultaneously less than γ_j is given by

$$P = \prod_{v=1}^{d}\int_{-\infty}^{\infty}\left\{\prod_{j=1}^{n}\left[\Phi\left(\frac{\gamma_j - \alpha_{jv}\theta}{\sqrt{1-\alpha_{jv}^2}}\right)\right]^{u_{jv}}\right\}g(\theta)d\theta, \tag{6.42}$$

where $\gamma_j = -c_j/y_j$, $\alpha_{jv} = a_{jv}/y_j$, $y_j = (1 + a_{j1}^2 + a_{jv}^2)^{1/2}$, $u_{jv} = 1$ denotes a nonzero loading of item j on dimension v ($v = 1, \ldots, d$), and $u_{jv} = 0$ otherwise. Note that for item j, $u_{jv} = 1$ for only one d. Note also that γ_j and α_{jv} used by Stuart (1958) are equivalent to the item threshold and factor loading and are related to the more traditional IRT parameterization as described above.

This result follows from the fact that if each variate is related only to a single dimension, then the d dimensions are independent and the joint probability is the product of d unidimensional probabilities. In this context, the result applies only to the $d-1$ content dimensions (i.e. $v = 2, \ldots, d$). If a primary dimension exists, it will not be independent of the other $d-1$ dimensions, since each item now loads on each of two dimensions. Gibbons and Hedeker (1992) derived the necessary two-dimensional generalization of Stuart's Stuart (1958) original result as

$$P = \int_{-\infty}^{\infty} \left\{ \prod_{v=2}^{d} \int_{-\infty}^{\infty} \left[\prod_{j=1}^{n} \left(\Phi \left[\frac{\gamma_j - \alpha_{j1}\theta_1 - \alpha_{jv}\theta_v}{\sqrt{1 - \alpha_{j1}^2 - \alpha_{jv}^2}} \right] \right)^{u_{jv}} \right] g(\theta_v) d\theta_v \right\} g(\theta_1) d\theta_1. \tag{6.43}$$

For the graded response model, the probability of a value less than the category threshold $\gamma_{jh} = -c_{jh}/y_j$ can be obtained by substituting γ_{jh} for γ_j in the previous equation. Let $u_{ijhk} = 1$ if the response falls in category k and 0 otherwise. The unconditional probability of a particular response pattern $\mathbf{u}_i$ is, therefore,

$$P(\mathbf{u} = \mathbf{u}_i) = \int_{-\infty}^{\infty} \left\{ \prod_{v=2}^{d} \int_{-\infty}^{\infty} \left[\prod_{j=1}^{n} \prod_{h=1}^{m_j} [\Phi_{jh}(\theta_1, \theta_v) - \Phi_{jh-1}(\theta_1, \theta_v)]^{u_{ijhk}} \right] g(\theta_v) d\theta_v \right\} \times g(\theta_1) d\theta_1, \tag{6.44}$$

which can be approximated to any degree of practical accuracy using two-dimensional Gauss–Hermite quadrature, since for both the binary and graded bifactor response models, the dimensionality of the integral is 2 regardless of the number of subdomains (i.e. $d-1$) that comprised the scale.

6.6 Estimation of Item Parameters and Respondent Scores in Item Bifactor Analysis

Gibbons and Hedeker (1992) showed how parameters of the item bifactor model for binary responses can be estimated by maximum marginal likelihood using a variation of the EM algorithm described by Bock and Aitkin (1981). For the graded case, the likelihood equations are derived as follows.

Denoting the vth subset of the components of $\boldsymbol{\theta}$ as $\boldsymbol{\theta}_v^* = \begin{bmatrix} \theta_1 \\ \theta_v \end{bmatrix}$, we let

$$\begin{aligned} P_i &= P(\mathbf{u} = \mathbf{u}_i) \\ &= \int_{\theta_1} \left\{ \prod_{v=2}^{d} \int_{\theta_v} \left[\prod_{j=1}^{n} \prod_{h=1}^{m_j} (\Phi_{jh}(\boldsymbol{\theta}_v^*) - \Phi_{jh-1}(\boldsymbol{\theta}_v^*))^{u_{ijhk}} \right] g(\theta_v) d\theta_v \right\} g(\theta_1) d\theta_1 \\ &= \int_{\theta_1} \left\{ \prod_{v=2}^{d} \int_{\theta_v} L_{iv}(\boldsymbol{\theta}_v^*) g(\theta_v) d\theta_v \right\} g(\theta_1) d\theta_1, \end{aligned} \tag{6.45}$$

where

$$L_{iv}(\theta_v^*) = \prod_{j=1}^{n}\prod_{h=1}^{m_j}\left(\Phi_{jh}(\theta_v^*) - \Phi_{jh-1}(\theta_v^*)\right)^{u_{ijhk}}. \tag{6.46}$$

Then the log-likelihood is

$$\log L = \sum_{i=1}^{s} r_i \log P_i, \tag{6.47}$$

where s denotes number of unique response patterns, and r_i the frequency of pattern i. As the number of items gets large, s typically is the number of respondents and $r_i = 1$.

In practice, the ultimate objective is to estimate the trait level of person i on the primary trait the instrument was designed to measure. For the bifactor model, the goal is to estimate the latent variable θ_1 for person i. A good choice for this purpose (Bock and Aitkin 1981) is the EAP value (Bayes estimate) of θ_1, given the observed response vector $\mathbf{u}_i$ and levels of the other subdimensions $\theta_2 \cdots \theta_d$. The Bayesian estimate of θ_1 for person i is

$$\hat{\theta}_{1i} = E(\theta_{1i} \mid \mathbf{u}_i, \theta_{2i}, \ldots, \theta_{di}) = \frac{1}{P_i}\int_{\theta_1} \theta_{1i} \left\{ \prod_{v=2}^{d} \int_{\theta_v} L_{iv}(\theta_v^*) g(\theta_v) d\theta_v \right\} g(\theta_1) d\theta_1. \tag{6.48}$$

Similarly, the posterior variance of $\hat{\theta}_{1i}$, which may be used to express the precision of the EAP estimator, is given by

$$V(\theta_{1i} \mid \mathbf{u}_i, \theta_{2i}, \ldots, \theta_{di}) = \frac{1}{P_i}\int_{\theta_1} (\theta_{1i} - \hat{\theta}_{1i})^2 \left\{ \prod_{v=2}^{d} \int_{\theta_v} L_{iv}(\theta_v^*) g(\theta_v) d\theta_v \right\} g(\theta_1) d\theta_1. \tag{6.49}$$

These quantities can be evaluated using Gauss–Hermite quadrature as previously described.

In some applications, we are also interested in estimating a person's location on the secondary domains of interest as well. For the vth subdomain, the EAP estimate and its variance can be written as follows:

$$\hat{\theta}_{vi} = E(\theta_{vi} \mid \mathbf{u}_i, \theta_{1i}) = \frac{1}{P_i}\int_{\theta_v} \theta_{vi} \left\{ \int_{\theta_1} L_{i1}(\theta_1^*) g(\theta_1) d\theta_1 \right\} g(\theta_v) d\theta_v, \tag{6.50}$$

and

$$V(\theta_{vi} \mid \mathbf{u}_i, \theta_{1i}) = \frac{1}{P_i}\int_{\theta_v} (\theta_{vi} - \hat{\theta}_{vi})^2 \left\{ \int_{\theta_1} L_{i1}(\theta_1^*) g(\theta_1) d\theta_1 \right\} g(\theta_v) d\theta_v. \tag{6.51}$$

For the purpose of interpretation, we recommend that the subdomain scores be presented in ipsitized form by expressing them as differences from the general factor score.

6.7 Estimating Factor Scores

The first formulation of IRT estimation of test scores was the maximum likelihood estimator (MLE) derived by Lord (1953) from Lazarsfeld's principle of conditional (or local) independence of item responses (see (Lazarsfeld 1959)). Lord's paper received relatively little attention at the time because the computation required for practical application of the method was then prohibitive. With the coming of readily available electronic computation in the 1960s, the situation changed radically and likelihood-based approaches to test theory began in earnest. Samejima (1969) extended Lord's results to polytomous responses and introduced Bayes modal, or maximum a posteriori (MAP) estimation – more robust than maximum likelihood in this context. Soon after, estimation of the Bayes mean, or EAP, estimation came into use in IRT (see (Bock and Mislevy 1982)). All three of these methods of estimating individual scores depend upon item parameters estimated in data from large samples of respondents drawn from the population in which the test or scale will be used. The parameter estimation step in IRT development of measuring instruments is called item *calibration*.

Conventional test scores for dichotomous items are counts of the number of items correct and for polytomous items are so-called *Likert scores*, i.e. averages of successive integers assigned to the response categories. In contrast, the likelihood-based scores correspond to locations on a scale with fixed origin and unit. They also provide a standard error for each score specifically rather than an overall index of reliability for the test. When Bayes methods of estimation are used, results from the calibration sample provide an empirical index of the reliability.

The estimator of total variance in the population is the variance of the Bayes scores plus the mean of the squared posterior variance for all scores in the sample. The empirical reliability for the test or scale is the Bayes score variance divided by the total score variance. When scoring is based on item parameters of a orthogonal factor solution, these relationships hold approximately for all of the factor scores estimated, although for other solutions this may not be the case. For that reason, it is desirable for consistency among alternative solutions to rescale the factor scores and their standard errors so that the relationship holds exactly in the sample. If unit total variance is fixed at one in the sample, the variance of the Bayes estimates of each factor score equals its empirical reliability.

6.8 Example

As an illustration of the IRT approach to item factor analysis for graded responses, we analyze data obtained with the *Quality of Life Interview for the Chronically Mentally Ill* (Lehman 1988) from 586 chronically mentally ill patients.

The instrument consists of one global life-satisfaction item followed by 34 items in seven subdomains, namely, Family, Finance, Health, Leisure, Living, Safety, and Social, with 4, 4, 6, 6, 5, 5, and 4 items, respectively. The subdomains are identified to the respondents by name and description. Respondents are instructed to rate each item in turn on a seven-point scale consisting of ordered response categories *terrible, unhappy, mostly dissatisfied, mixed- about equally satisfied and dissatisfied, mostly satisfied, pleased, delighted.*

Although rating scale with items of similar content grouped together and labeled descriptively are easier for the respondent to complete, the names of the sections and the descriptive labels encourage responses to the set of items as a whole rather than considered responses to the individual items. This effect creates dependencies between responses within the sets that violate the assumption of conditional independence of response required for conventional one-dimensional IRT analysis. This is an important reason for introducing multidimensional IRT models, especially the bifactor model mentioned above. The effect could be avoided, of course, by presenting the items in random order, but responding to the items would be much more cognitively demanding of the respondent.

6.8.1 Exploratory Item Factor Analysis

The purpose of traditional exploratory factor analysis of quantitative variables is to identify or characterize in some way the sources of common variation that account for the observed pattern of correlation between all pairs of the variables; no prior hypotheses about these sources are required. The same is true of IRT-based methods of item factor analysis, except that these procedures find sources of association of all orders, pairwise and higher, among observed qualitative item responses. In either case, a principal factor solution is the preferred starting point for exploratory analysis, since it accounts for the greatest amount of correlation or association with a given number of factors.

It is up to the investigator, however, to choose that number of factors, usually by performing repeated analyses with increasing numbers of factors. The question is then when to stop adding successive principle factors, which are under the constraint that the loadings of each additional are orthogonal to those that precede. As mentioned above, when MLE of the factor loadings is employed, the amount of decrease of two times the log likelihood provides a statistical criterion of when nothing more can be gained by adding further factors. In large samples, this quantity has a central chi-square distribution on null hypothesis that the additional factor loadings are all zero. The number of degrees of freedom is the number of free parameters represented among the additional loadings. This statistical test is a safeguard against over factoring when the amount of original data are limited. It is of little value, however, when the number of respondents in the sample is very large, for in that case, as few as one or two nonzero loadings will produce a significant decrease in chi-square. To be worth retaining, a factor must show enough loadings of appreciable size to support a plausible interpretation. The traditional rule of thumb is that at least three loadings of the factor should be greater than 0.3. This situation is similar to the choice of number of terms in a polynomial regression analysis. If the sample size is very large, high-order terms may be statistically significant, but they will produce small-scale wiggles in the fitted curve that are likely to be uninterpretable, and they compromise the smoothness of the relationship.

In the present example, our experience that labeled groups of items produce association of responses within the groups above and beyond that contributed by the overall measurement variable gives us a stronger basis for judging the practical significance of additional factors than would a completely unstructured set of items. We use exploratory analysis here to investigate the suitability of the data for a bifactor confirmatory analysis on the expectation that there will be seven factors – one general factor and the six orthogonal bipolar factors possible among seven groups. The first result to be examined in this connection is the chi-square statistics for the factor added in a successive analyses shown in Table 6.1.

Very roughly a chi-square value is significant if it is at least twice as large as its degrees of freedom. By this rule, even the addition of an eighth orthogonal factor shows no danger of over factoring, although its contribution to improved goodness of fit is the smallest of any factor. Notice that the decreases are not monotonic: unlike traditional factor analysis of product-moment correlations, the marginal probabilities of the response patterns (which determined the marginal likelihood) reflect changes in all parameters jointly, including the category parameters and not just the factor loadings. Because our inspection of the signs of the loadings of the first seven factors showed relationships to the item groups and the eighth factor did not, we focused on the seven-factor model, the results for which are displayed in Tables 6.2 and 6.3. These estimates are based on 40 cycles of the adaptive quadrature solution as converged to three or more decimal places of precision.

Table 6.1 Quality of life data ($N = 586$) decrease of −2Log likelihood of solutions with 1 to 8 factors.

Solution	−2Log likelihood	Decrease	Degrees of freedom
1	66 837.1		
2	66 045.0	792.1	34
3	65 089.5	955.5	33
4	64 118.4	971.5	32
5	63 509.1	609.3	31
6	63 063.7	445.4	30
7	62 677.5	386.2	29
8	62 370.5	307.4	28

Table 6.2 Item principal factor loadings.

Item		Factors						
Group	Items	1	2	3	4	5	6	7
0	1	0.769	0.021	0.082	−0.054	0.054	−0.002	0.097
					Family			
	2	0.614	−0.269	0.044	−0.461	0.272	−0.081	0.044
1	3	0.687	−0.181	−0.007	−0.380	0.159	0.007	−0.058
	4	0.703	−0.245	−0.045	−0.522	0.257	0.029	0.008
	5	0.729	−0.214	−0.004	−0.505	0.279	0.046	0.019
					Finance			
	6	0.606	0.468	−0.391	0.116	0.284	0.003	−0.101
2	7	0.515	0.405	−0.300	0.097	0.220	0.021	−0.032
	8	0.647	0.511	−0.342	0.101	0.276	0.048	−0.092
	9	0.632	0.510	−0.305	0.072	0.242	0.023	−0.069
					Health			
	10	0.568	−0.123	0.132	0.201	0.049	−0.236	0.095
	11	0.644	0.007	0.195	0.128	0.038	−0.443	−0.139
3	12	0.627	0.087	0.289	0.074	−0.026	−0.390	−0.167
	13	0.668	−0.052	0.156	0.138	−0.005	−0.383	−0.232
	14	0.678	−0.004	0.154	0.116	0.061	−0.288	0.054
	15	0.701	0.044	0.249	0.045	0.071	−0.154	0.054
					Leisure			
	16	0.741	0.215	0.150	0.030	−0.138	0.156	0.155
	17	0.657	0.149	0.142	−0.017	−0.128	−0.054	0.285
4	18	0.721	0.223	0.101	−0.005	−0.173	0.019	0.331
	19	0.749	0.313	0.144	−0.059	−0.199	0.095	0.301
	20	0.670	0.192	0.078	−0.101	−0.162	−0.030	0.295
	21	0.522	−0.002	−0.056	−0.002	−0.099	−0.049	0.042
					Living			
	22	0.664	−0.241	−0.401	0.038	−0.191	0.048	−0.008
	23	0.549	−0.332	−0.325	0.118	−0.140	−0.013	−0.028
5	24	0.611	−0.253	−0.529	0.006	−0.190	−0.112	0.042
	25	0.626	−0.347	−0.446	0.079	−0.285	−0.127	0.030
	26	0.568	−0.213	−0.439	0.066	−0.177	−0.018	0.034

Table 6.2 (*Continued*).

Item		Factors						
Group	Items	1	2	3	4	5	6	7
					Safety			
	27	0.679	−0.241	0.221	0.299	0.232	0.341	−0.004
	28	0.688	−0.387	0.051	0.317	0.141	0.250	−0.040
6	29	0.594	−0.065	0.231	0.145	0.109	0.123	−0.044
	30	0.670	−0.253	0.181	0.276	0.196	0.223	−0.003
	31	0.702	−0.264	0.140	0.336	0.064	0.197	0.006
					Social			
	32	0.688	0.189	0.169	−0.180	−0.399	0.197	−0.375
7	33	0.696	0.254	0.099	−0.192	−0.317	0.212	−0.218
	34	0.620	0.203	0.149	−0.118	−0.218	0.161	−0.232
	35	0.494	−0.163	0.122	−0.056	−0.179	0.046	−0.202

As expected, all first principal factor loadings are positive, clearly identifying the factor with the overall quality-of-life variable. In fact, the largest loading is that of item number one, which asks for the respondent's rating of overall quality of life. Only one item, the last, has a loading less than 0.5. As for the six bipolar factors, the significant feature is that the sign patterns of loadings of appreciable size conform to the item groups. Factor 2 strongly contrasts the Finance group with Family Living and Safety; to a lesser extent, Leisure is working in the same direction as Finance. In other words, persons who tend to report better financial positions and quality of leisure are distinguish by this factor from those who report better family relationships and safety. Factor 3 then combines Living and Finance and contrasts them primarily with a combination of Health and Safety. Factor 4 contrasts a combination of Family and Social with Finance, Health, and Safety. Factor 5 combines Social, Living, and Safety versus Family, Finance, and Safety. Factor 6 primarily contrasts Health and Safety. Finally, Factor 7 contrasts Social verses Leisure. The fact that the seven-factor solution has the expected all positive first factor and patterns for the remaining bipolar factors that contrast item groups rather than items within groups clearly supports a bifactor model for these data.

Although it is technically possible to estimate respondent scores for all of the principal factors, apart from the first with all positive loadings the resulting scores would not be easy for users to understand. The bifactor model is much more suitable for scoring instruments like the quality-of-life scale that are intended

Table 6.3 Item location and centered category thresholds.

Item	Location	Thresholds 1	2	3	4	5	6
1	−0.107	−1.086	−0.627	−0.361	0.106	0.656	1.313
2	−0.346	−1.066	−0.586	−0.362	0.160	0.642	1.212
3	−0.085	−1.163	−0.632	−0.276	0.075	0.664	1.332
5	−0.029	−1.078	−0.564	−0.292	0.111	0.636	1.187
6	0.283	−1.135	−0.620	−0.261	0.050	0.676	1.289
7	0.009	−1.261	−0.672	−0.332	0.048	0.732	1.486
9	0.318	−1.089	−0.555	−0.200	0.065	0.617	1.162
10	−0.277	−1.179	−0.715	−0.496	−0.114	0.879	1.624
11	−0.410	−1.061	−0.732	−0.450	0.014	0.663	1.566
12	−0.230	−1.230	−0.764	−0.352	0.002	0.755	1.589
13	−0.358	−1.129	−0.794	−0.391	−0.002	0.719	1.598
14	−0.311	−1.231	−0.684	−0.321	0.039	0.671	1.526
15	−0.057	−1.141	−0.695	−0.349	0.053	0.716	1.416
16	0.072	−1.204	−0.673	−0.296	0.104	0.669	1.401
17	−0.081	−1.238	−0.760	−0.321	0.085	0.746	1.488
18	−0.181	−1.139	−0.615	−0.276	0.101	0.632	1.296
19	0.022	−1.116	−0.584	−0.246	0.091	0.639	1.217
20	−0.075	−1.110	−0.692	−0.317	0.039	0.665	1.415
21	−0.581	−1.278	−0.826	−0.451	0.106	0.831	1.617
22	−0.163	−1.009	−0.654	−0.347	0.023	0.680	1.307
23	−0.577	−0.960	−0.663	−0.326	−0.030	0.617	1.361
24	−0.282	−0.814	−0.546	−0.248	0.022	0.468	1.118
25	−0.452	−0.793	−0.483	−0.250	−0.010	0.508	1.028
26	0.087	−0.877	−0.467	−0.224	0.129	0.512	0.928
27	−0.073	−1.101	−0.690	−0.396	−0.017	0.693	1.510
28	−0.271	−0.963	−0.671	−0.393	−0.044	0.638	1.433
29	0.096	−0.941	−0.713	−0.350	0.065	0.618	1.322
30	−0.008	−1.002	−0.695	−0.342	0.018	0.639	1.383
31	−0.217	−1.091	−0.757	−0.450	−0.031	0.701	1.629
32	−0.375	−1.265	−0.880	−0.455	0.062	0.881	1.657
33	−0.249	−1.417	−0.815	−0.427	0.086	0.853	1.721
34	−0.380	−1.219	−0.820	−0.444	0.089	0.780	1.614
35	−0.689	−1.250	−0.847	−0.568	0.045	0.853	1.767

to measure overall performance while allowing for estimation of scores for the separate domains.

Along with the factor loadings, the exploratory analysis also estimates the thresholds representing the six boundaries between rating scale categories of each item. These thresholds are shown in Table 6.3 in the form that is convenient for interpretation: the mean of the thresholds for each item has been subtracted from each, giving what are called "centered thresholds." These means are then shown separately to convey the overall location of the respective centered thresholds. The centered thresholds indicate the spacing of the thresholds, which in turn reflects the sample frequencies of responses in each category.

6.8.2 Confirmatory Item Bifactor Analysis

Turning now to the bifactor analysis of these data, we first examine goodness of fit for three versions of the bifactor model with increasing numbers of free parameters. The first one is the one-dimensional graded rating scale model with no subdomain item groups. The second model assumes seven factors corresponding to the above-defined item groups, plus the general factor, and also assuming a common rating scale for all items. To the 76 parameters of the one-factor model, this model adds 35 slope parameters within the item groups. Finally, the graded model adds $5 \times 35 = 175$ free parameters for the item specific category thresholds.

Because of the simple structure of the item-group bifactor loadings, it is possible to perform MLE with two-dimensional quadrature with many quadrature points without excessive computational burden. For this reason, the numerical integrations required in the solution can be performed with good accuracy by nonadaptive quadrature. For the present example, 21 quadrature points were employed, compared to two or three quadrature points in the seven-dimensional quadrature for the unconstrained principal factor solution above. Considering the radically different computational methods, it is apparent from the comparison of the one-factor case in Tables 6.1 and 6.4 that both procedures reach essentially the same minimum of the $-2\times$ log-likelihood chi-square. Table 6.4 also shows, as expected, that

Table 6.4 Bifactor model fit.

Model	-2 Log L	Decrease	DF
1-Factor	67 278.0		
8-Factor rating scale	65 039.2	2 238.2	35
8-Factor graded	64 233.2	806.0	175

the shift to the eight-factor rating scale bifactor model greatly reduces chi-square, compared to the relevantly small decrease gained by estimating thresholds for the individual items with the graded model. For this reason, we show in Table 6.5, the factor loadings only of the rating-scale solution, which in fact are identical to those the graded solution in the first decimal place. Because the group factor loadings are not constrained to orthogonality with those of the general factor, they are all positive and their magnitudes indicate the strength of the effect of items belonging to common domains. The effects of Family and Finance, for example, are stronger than those of Health and Leisure.

In connection with the information on the common thresholds of the categories shown in Table 6.6, it is important to understand that the common thresholds for different items are the same only up to a linear transformation. The actual thresholds for any given item are obtained by adding the location intercepts shown in the bottom section of the table to the rating-scale intercepts at the top (which are common to all items). The sum is then divided by the constant that converts intercepts to thresholds, i.e. the square root of one plus the communality of the item. The location and spacing of the individual item thresholds reflect the frequency of use of the categories by the respondents in the sample.

Inasmuch as the ultimate goal of this application of the bifactor model is to assign scores to the respondents descriptive of their quality-of-life, the sample statistics in Table 6.7 are of particular interest. They show the measurement properties of empirical Bayes estimation of the factor scores. These figures shown are in the arbitrary scale of the bifactor solution, but they could have been rescaled, for example to mean zero and standard deviation one in the sample, if desired. The RMS error would then change in proportion to the standard deviations, but the empirical reliabilities, which are scale free, would remain the same. If the reporting of scores is limited to the general factor, only the first column is relevant. Although the standard errors of IRT scale scores differ from one respondent to another, the average error for the general factor score is in the neighborhood of one-third of the standard deviation, the overall reliability is close to 0.9. If one were to compute these quantities for the one-dimensional solution – which ignores the conditional dependence of responses within item groups – the values shown in the footnote of the table would result. The mean and standard deviation of the scores would be much the same, but the average error would be appreciably underestimated and the reliability overestimated. Avoiding this type of bias is a major motivation for item bifactor analysis, added to its power to obtain the most reliable score estimates with multiple category items with varying numbers of categories or in combination with binary scored items.

Notice that the bifactor model also provides for estimating scores for the group factors from the responses to items within each group, which can be assumed conditionally independent. In the present example, there are too few items within

Table 6.5 Bifactor model loadings.

Item group	Item	General	Factors 1	2	3	4	5	6	7
0	1	0.789							
					Family				
	2	0.535	0.620						
1	3	0.576	0.509						
	4	0.575	0.586						
	5	0.631	0.547						
					Finance				
	6	0.476		0.634					
2	7	0.437		0.553					
	8	0.544		0.617					
	9	0.535		0.622					
					Health				
	10	0.560			0.256				
	11	0.528			0.504				
3	12	0.486			0.505				
	13	0.529			0.473				
	14	0.650			0.286				
	15	0.714			0.141				
					Leisure				
	16	0.694				0.285			
	17	0.565				0.413			
4	18	0.628				0.451			
	19	0.635				0.506			
	20	0.571				0.473			
	21	0.479				0.208			
					Living				
	22	0.536					0.549		
	23	0.484					0.530		
5	24	0.497					0.668		
	25	0.508					0.688		
	26	0.508					0.672		

Table 6.5 (*Continued*).

	Item		Factors						
group	**Item**	**General**	**1**	**2**	**3**	**4**	**5**	**6**	**7**
						Safety			
	27	0.557						0.517	
	28	0.593						0.474	
6	29	0.533						0.501	
	30	0.558						0.538	
	31	0.591						0.383	
						Social			
7	32	0.545							0.438
	33	0.586							0.351
	34	0.520							0.466
	35	0.446							0.296

Table 6.6 Rating scale centered intercepts and item locations.

Intercepts	**−1.741**	**−1.048**	**−0.533**	**0.077**	**1.065**	**2.180**
Group			**Locations within groups**			
0	−1.288					
1	−2.332	−1.605	−1.772	−1.920		
2	−1.046	−1.156	−1.068	−1.141		
3	−1.311	−1.911	−1.515	−1.745	−1.634	−1.201
4	−1.192	−1.312	−1.853	−1.697	−1.461	−1.470
5	−1.693	−2.170	−2.302	−2.880	−1.637	
6	−1.531	−1.950	−1.150	−1.489	−1.600	
7	−1.722	−1.400	−1.714	−1.617		

groups to allow reliable reporting of group scores, but in larger instruments with as many as 25 items per subdomain, the greater information conveyed by multiple category responses could allow reasonably reliable estimation of subdomain scores as well as highly reliable estimates of the overall score.

Table 6.7 Measurement properties of the bifactor solution.

	General	Item group						
		1	2	3	4	5	6	7
Mean	−0.124	−0.060	−0.060	−0.040	−0.052	−0.068	−0.058	−0.045
SD	0.949	0.650	0.712	0.641	0.565	0.738	0.609	0.518
RMS error	0.322	0.754	0.703	0.782	0.823	0.673	0.791	0.851
Empirical reliability	0.896	0.426	0.507	0.402	0.320	0.546	0.372	0.270

Corresponding properties of the one-dimensional model.
mean = −0.084; SD = 0.969; RMS error = 0.232; reliability = 0.946.

The analyses for this example were performed with an α-version of the POLYFACT program of Bock, Gibbons, and Schilling (Scientific Software International).

6.9 Two-Tier Model

An important generalization of the bifactor model is the two-tier model described by Cai (2010). Unlike the bifactor model that posits a single primary dimension and a series of orthogonal subdomains, the two-tier model extends the bifactor model to a set of p potentially correlated primary dimensions, and S orthogonal subdomains, with the restriction that an item can load on only one of the subdomains as in the bifactor model. The major advantage of the two-tier model is it extends the bifactor model to the case of a series of correlated primary dimensions, making it similar to an unrestricted item-factor analysis model, but still retains the dimension reduction advantage of the bifactor model, reducing the dimensionality from $p + S$ (where p is the number of primary dimensions, and S is the number of subdomains) to $p + 1$. Unlike the bifactor model, however, which always reduces dimensionality of the likelihood to a bivariate integral, the two-tier model is limited in terms of the number of correlated primary dimensions in the same way as an unrestricted item factor analysis model is (i.e. as p gets large numerical evaluation of the likelihood becomes computationally intractable).

To fix ideas, the two-tier model is displayed graphically in Figure 6.1.

Figure 6.1 clearly illustrates that the bifactor model is a special case of the two-tier model when $p = 1$, that is there is a single primary dimension. The likelihood equations are derived as an extension of Eq. (6.45), expanding the single integral for the primary dimension to a p-fold integral as in Eq.(15) in Cai (2010). Cai (2010) describes several possible applications of the two-tier model,

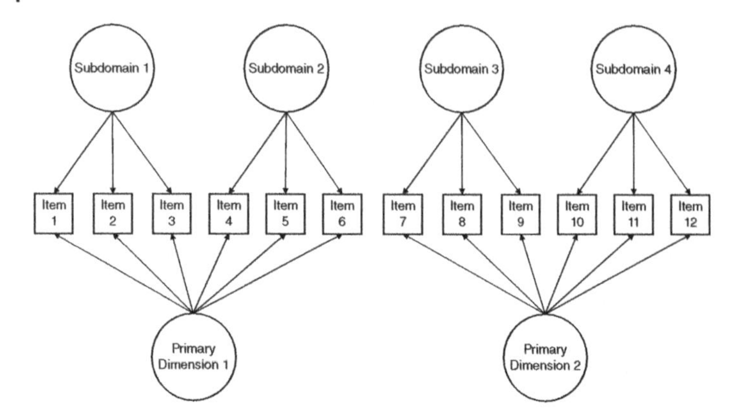

Figure 6.1 Two-tier model with two correlated primary dimensions, four subdomains, each with three items.

including extensions to the analysis of longitudinal item-response data and random intercept models to allow for idiosyncratic response styles in which there is inter-individual variability in the likelihood of selecting different response categories above and beyond the item parameters.

6.10 Summary

In this chapter, we have extended the IRT approach to item factor analysis to the case of graded response data. We have considered both unrestricted item factor analysis and the bifactor model, the first example of a full information confirmatory factor analytic model for item response data. While these models extend traditional applications of IRT to educational measurement problems in many ways, one of their greatest attractions is that they open the door to more widespread applications of item factor analysis in the area of patient reported outcomes in medical research, where graded response data are the rule, not the exception. The bifactor and two-tier models are of particular interest in these new applications because these scales are invariably multidimensional and the item domains are typically well known in advance. The scoring of both primary and subdomains as described here further extend research work in this field and permit new applications of computerized adaptive testing to scales whose multidimensionality previously limited such applications.

7

Analysis of Dimensionality

It has been stated by economists and by other social scientists that affect cannot be measured, and some of the fundamental theory of social science has been written with this explicit reservation. Our studies have shown that affect can be measured. In extending the methods of psychophysics to the measurement of affect we seem to see the possibility of a wide field of application by which it will be possible to apply the methods of quantitative scientific thinking to the study of feeling and emotion, to esthetics, and to social phenomena.[1]

(Source: L.L. Thurstone)

Much of item response theory (IRT) is based on the assumption of unidimensionality; namely, that the associations among the item responses are explained completely by a single underlying latent variable, representing the target construct being measured. While this is often justified in many areas of educational measurement, more recent interest in measuring patient reported outcomes (Gibbons et al. 2008, 2012) involves items that are drawn from multiple uniquely correlated subdomains violating the usual conditional independence assumption inherent in unidimensional IRT models. As alternatives, both unrestricted item factor analytic models (Bock and Aitkin 1981) and restricted or confirmatory item factor analytic models (Gibbons and Hedeker 1992) have been used to accommodate the multidimensionality of constructs for which the unidimensionality assumption is untenable. A concrete example is the measurement of depressive severity, where items are drawn from mood, cognition, and somatic impairment subdomains. While this is a somewhat extreme example, there are many more borderline cases where the choice between a unidimensional model and a multidimensional model is less clear, or the question of how many dimensions is "enough" is of interest. In this chapter, we explore the issue of determining the dimensionality

1 Much of the material presented in this chapter follows from Gibbons and Cai (2017) with permission.

Item Response Theory, First Edition. R. Darrell Bock and Robert D. Gibbons.

of a particular measurement process. We begin by discussing multidimensional item factor analysis models, and then consider the consequences of incorrectly fitting a unidimensional model to multidimensional data. We also discuss non-parametric approaches such as DIMTEST (Stout 1987). We then examine different approaches to testing dimensionality of a given measurement instrument, including approximate or heuristic approaches such as eigenvalue analysis, as well as more statistically rigorous limited-information and full-information alternatives. Finally, we illustrate the use of these various techniques for dimensionality analysis using a relevant example. Much of this material is based on the review of Gibbons and Cai (2017).

7.1 Unidimensional Models and Multidimensional Data

A natural question is whether there is any adverse consequence of applying unidimensional IRT models to multidimensional data. To answer this question, Stout and coworkers (e.g. (Stout 1987, Zhang and Stout 1999)) took a distinctly non-parametric approach to characterize the specific conditions under which multidimensional data may be reasonably well represented by a unidimensional latent variable. They emphasized a core concept that subsequently became the basis of a family of theoretical and practice devices for studying dimensionality, namely, local independence as expressed using conditional covariances. To begin, given n items in a test, the strong form of local independence states that the conditional response pattern probability factors into a product of conditional item response probabilities, i.e.

$$P(\boldsymbol{U} = \mathbf{u}|\theta) = \prod_{j=1}^{n} P(U_j = u_j|\theta). \tag{7.1}$$

Correspondingly, a test is weakly locally independent with respect to θ if the conditional covariance is zero for all item pairs j and j':

$$\mathrm{Cov}((U_j, U_{j'}|\theta) = 0. \tag{7.2}$$

The conditional covariances provide a convenient mechanism to formalize the notion of an *essentially unidimensional* test that possess one essential dimension and (possibly) a number of nuisance dimensions. A test is said to be essentially independent (Stout 1990) with respect to θ if

$$D_n(\theta) = \frac{\sum_{1 \le j < j' \le n} \left|\mathrm{Cov}(U_j, U_{j'}|\theta)\right|}{\binom{n}{2}} \to 0, \text{ as } n \to \infty. \tag{7.3}$$

In other words, essential independence states that the average value of conditional covariances across all item pairs is small as the test length increases. If the minimal dimensionality of $\boldsymbol{\theta}$ necessary for item pool $\boldsymbol{U}$ to satisfy essential independence is equal 1, then the test is said to be essentially unidimensional.

The mathematical condition above suggests a statistical procedure for testing essential unidimensionality (see (Stout 1987)). In brief, the test is split into two subsets called assessment tests (AT1 and AT2) and a longer subset called the partitioning test (PT). The items for AT1 are chosen to be saturated with the same dominant latent trait, but are as dimensionally different as possible from the items in the PT. Then AT2 is selected such that the items have similar difficulty as AT1. Each test taker's total score on the PT is used to group the test takers into several homogeneous subgroups. The PT total score becomes the conditioning score (effectively as a surrogate of θ) to calculate required conditional covariances for statistical hypothesis testing using the AT1 and AT2 item responses. The procedure as formalized by Nandakumar and Stout (1993) is referred to as DIMTEST.

Gibbons, Immekus and Bock (2007b) studied the consequences of fitting unidimensional models to multidimensional data empirically. The question they asked was slightly different. To the extent that the primary dimension of interest can be preserved in a unidimensional model and in the primary factor of a bifactor model or possibly in an exploratory item factor analysis model, does the specific model used make a difference in the results?

They conducted a simulation study to investigate the effects of applying Samejima's (1969) graded response model in unidimensional and bifactor form to multidimensional data. Conditions studied were the following: (i) test length, 50 items or 100 items, (ii) number of dimensions, 5 or 10, (iii) primary loadings, 0.50 or 0.75, and (iv) domain loadings, 0.25 or 0.50. Outcome results include the following: standard deviation of expected *a posteriori* (EAP) estimates of θ, posterior standard deviations (PSDs, or standard errors) of Bayes EAP scores, log-likelihood (model fit), differences between EAP and actual θ, and percentage change between unidimensional and bifactor models of these variables. The generated data were based on a four-point categorical scale, and the examinee distribution was assumed to be normal, $N(0, 1)$, based on 1000 replications. In the following, we summarize the key findings of this study.

Figure 7.1 reports the standard deviations of the θ estimates for the unidimensional and bifactor models across the 12 simulated conditions. Inspection of the figure indicates that the EAP estimates based on the unidimensional model were more varied across all conditions. The magnitude of the difference decreased when the primary and secondary loadings decreased, leading to a more unidimensional solution. As shown, as the number of items increased from 50 to 100, the EAP estimates from both models became more varied, but not as severe for the bifactor model.

Figure 7.2 reports the mean PSD of the Bayes EAP estimates. As shown, the differences in the PSD between the models can be dissected in terms of the

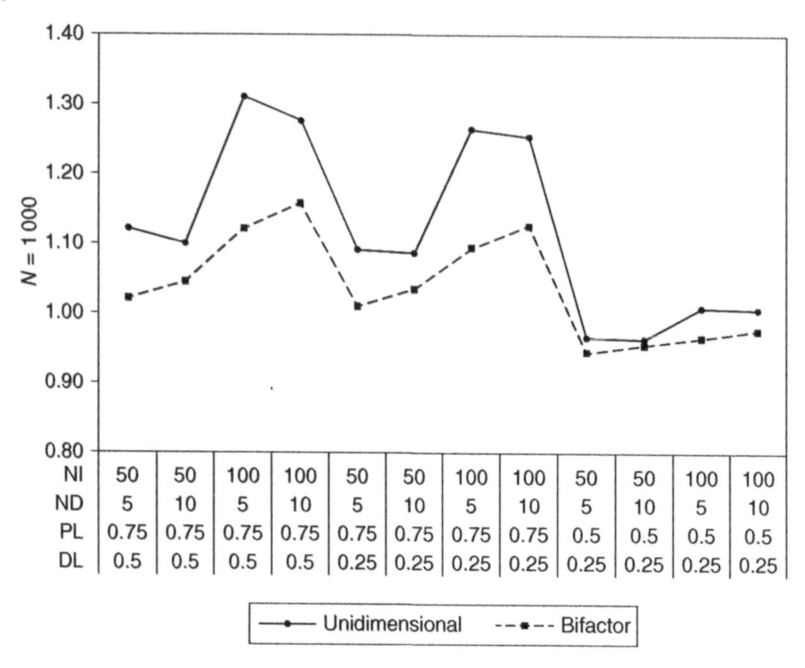

Figure 7.1 Mean standard deviations of theta of the unidimensional and bifactor models based on 1000 Replications per condition (Number Items ([NI]) = 50 or 100, number dimensions [ND] = 5 or 10, primary loadings [PL] = 0.50 or 0.75, domain loadings [DL] = 0.25 or 0.50).

dimensionality of the underlying data. Specifically, in the conditions in which the primary loadings are 0.75 and the domain loadings are 0.50, the PSD of the unidimensional model substantially underestimates the PSDs from the bifactor model. As shown, the largest PSD for the unidimensional model occurs with 100 items and 5 dimensions. The PSD estimated by the bifactor model remains fairly consistent across the conditions in which the underlying structure can be regarded as strongly multidimensional (i.e. primary loadings = 0.75, domain loadings = 0.50). For the conditions in which the primary loadings are 0.75 and the domain loadings are 0.25, the PSD for the unidimensional approaches that for the bifactor model but, nevertheless, continues to underestimate the bifactor result, which is the correct value in this case. The largest discrepancies between the PSD of these models occurs when the number of dimensions is 5 and the number of items is 50 and 100. The smallest difference between the mean PSDs for the unidimensional and bifactor models occurs when the number of dimensions is 10 with 50 items. For the bifactor model, the PSD decreases slightly when the number of items increases from 50 to 100. However, the number of dimensions does not seem to significantly influence the PSD of the bifactor model.

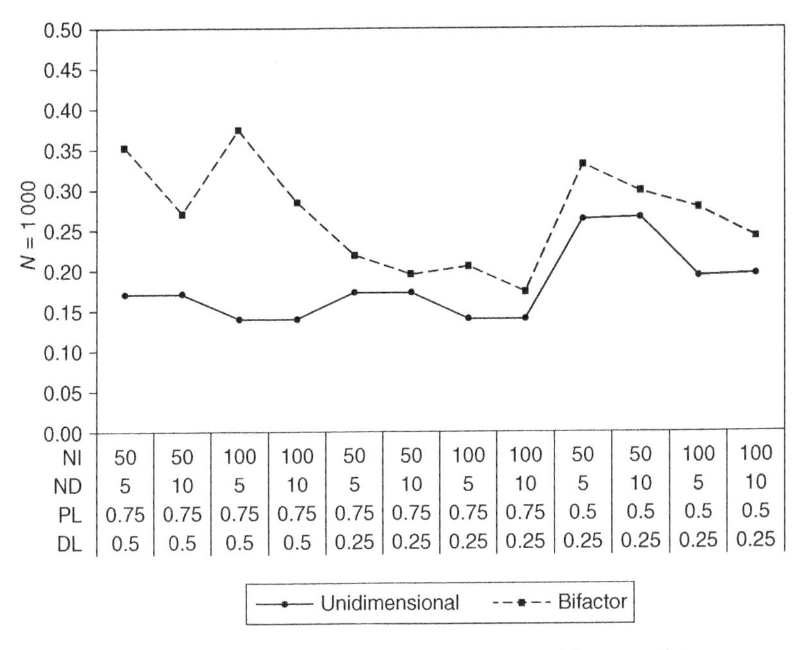

Figure 7.2 Mean posterior standard deviations of Bayes EAP scores of the unidimensional and bifactor models based on 1,000 Replications per condition (number items [NI] = 50 or 100, Number Dimensions [ND] = 5 or 10, primary loadings [PL] = 0.50 or 0.75, domain loadings [DL] = 0.25 or 0.50).

The results of this study illustrate the consequences attached to applying a unidimensional IRT model to data with varying degrees of multidimensionality compared to the bifactor model. The first set of results addressed the variability in estimated θ values, or examinees' standing on the latent trait. Compared to the unidimensional model, the bifactor model yielded θ estimates that were more homogeneous across simulated data structures. As a consequence, studies that are designed to evaluate educational or clinical interventions will have increased statistical power to detect meaningful effects when scores are based on a bifactor model and the underlying data are the result of a multidimensional response process.

PSD estimates were found to be underestimated across all conditions for the unidimensional model. For the bifactor model, PSD values were consistently below 0.20 across conditions, except when the total test length was 50 and the primary loadings were 0.50 and the domain loadings were 0.25. One setting in which the underestimation of PSDs could affect test scores is in computer adaptive testing, in which each item is intentionally selected to provide the most information for estimating of examinee ability in the sense of greatest reduction of PSD. Using PSD estimates based on the unidimensional model may therefore lead to suboptimal estimates of examinee ability. Used as measurement error variance, the inverse squared unidimensional PSDs are not valid for weighting observations in statistical analyses using the scores as data.

7.2 Limited-Information Goodness of Fit Tests

The nonparametric indices based on conditional covariances such as DIMTEST do not explicitly specify a distribution of the θ's. Hence, they require the use of external conditioning subscores such as the partitioning total score. When an item factor analysis model is fitted using standard estimation methods such as marginal maximum likelihood, population distributions of θ are routinely assumed. Therefore, upon finding the maximum likelihood solution, the model yields expected probabilities for each single item, as well as joint probabilities for item pairs, triplets, quadruplets, etc. When contrasted against the observed probabilities, the residuals may be used to derive goodness of fit statistics. Most of the time, univariate, and bivariate association information is used.

In the context of IRT, statistics based on (mostly) univariate and bivariate subtables are referred to as limited-information goodness of fit statistics, in contrast to full-information statistics (e.g. the Pearson's chi-square statistic) that are based on residuals of the full item by item by item cross-classifications. Despite the apparent loss of information due to collapsing the full contingency table into a series of first- and second-order association tables, limited-information test statistics have been suggested as a potential solution to the Achilles' heel of full-information statistics, namely, the sparseness of the underlying multiway contingency table upon which the IRT model is defined (Bartholomew and Tzamourani 1999). The number of cells in the full cross-classification table is exponentially increasing in the number of items, and for tests of realistic length, the table will become extremely sparse for any conceivable sample size. The sparseness invalidates the usual asymptotic chi-square approximations to the distribution of Pearson's statistic or the likelihood ratio statistic, making model fit testing decisions based on full-information statistics untrustworthy in practical situations. On the other hand, test statistics based on univariate and bivariate subtables maintain Type I error rate control and have adequate power (see, e.g. (Cai et al. 2006)). In particular, Maydeu-Olivares and Joe's (2005) M_2 family of test statistics has witnessed increasing popularity.

In the context of multidimensional IRT, Cai and Hansen (2013) extended the dimension reduction technique, already used in parameter estimation of bifactor models, to limited-information goodness of fit testing. For example, for a bifactor model, the probabilities and derivatives for computing limited-information test statistics require at most two-dimensional numerical integration, regardless of the number of factors in the model, making it feasible to test much larger models with many latent variables. In addition, Cai and Hansen (2013) developed a new quadratic form test statistic, which they call M_2^*, that is based on the general limited-information testing principles proposed by Joe and Maydeu-Olivares (2010). The statistic is best understood as a further reduction (or concentration) of the univariate and bivariate subtables. When the item responses are polytomous, this new statistic can be substantially better calibrated and more powerful than M_2. In addition, the chi-square distributed test statistics can be used to calculate fit measures such as the root mean square error of approximation (RMSEA; (Browne and Cudeck 1993)) that are free from the influence of sample size.

The details of limited-information goodness of fit testing is more substantial than can be covered in this chapter. In brief, the development begins with the realization that the IRT model can be written as a function of the (marginal) response pattern probability $\pi_{\mathbf{u}}(\gamma)$ for pattern $\mathbf{u}$, where γ is a notational shorthand for the collection of free and estimable parameters in the model. Suppose there are C possible response patterns. Let us define the $C \times 1$ vector of modeled probabilities as $\boldsymbol{\pi}(\gamma)$ and the corresponding $C \times 1$ vector of observed proportions as $\boldsymbol{p}$. Let the $C \times 1$ population cell probabilities be $\boldsymbol{\pi}$. The null hypothesis being evaluated in the goodness of fit testing situation is $H_0 : \boldsymbol{\pi}(\gamma) = \boldsymbol{\pi}$, for any γ, versus the alternative $H_A : \boldsymbol{\pi}(\gamma) \neq \boldsymbol{\pi}$, for some γ.

Suppose the total sample size is N. Treating $\boldsymbol{p}$ as the fixed observed data, maximizing (for example using the EM algorithm) the multinomial likelihood with cell probabilities given by $\boldsymbol{\pi}(\gamma)$ leads to the marginal maximum likelihood estimator $\hat{\gamma}$. Let the fitted cell probabilities be $\hat{\boldsymbol{\pi}} = \boldsymbol{\pi}(\gamma)$. The cell residuals are $\boldsymbol{e} = \boldsymbol{p} - \hat{\boldsymbol{\pi}}$. Standard discrete multivariate analysis results (Rao 1973) suggest that the cell residuals are asymptotically C-variate normal under the null hypothesis:

$$\sqrt{N}\boldsymbol{e} = \sqrt{N}(\boldsymbol{p} - \hat{\boldsymbol{\pi}}) \xrightarrow{D} \mathcal{N}_C(0, \Xi), \tag{7.4}$$

where $\Xi = \boldsymbol{D} - \boldsymbol{\pi}\boldsymbol{\pi}' - \boldsymbol{\Delta}(\gamma)\mathcal{F}(\gamma)^{-1}\boldsymbol{\Delta}(\gamma)'$, $\boldsymbol{D} = \text{diag}(\boldsymbol{\pi})$ and

$$\boldsymbol{\Delta}(\gamma) = \frac{\partial \boldsymbol{\pi}(\gamma)}{\partial \gamma'} \tag{7.5}$$

is the Jacobian of the model, and $\mathcal{F} = \boldsymbol{\Delta}(\gamma)'\boldsymbol{D}^{-1}\boldsymbol{\Delta}(\gamma)$ is the Fisher information matrix.

Subtable probabilities such as the univariate and bivariate probabilities are linear functions of the cell probabilities (Cai et al. 2006). The relationship can be conveniently expressed using reduction operator matrices (Joe and Maydeu-Olivares 2010). Let $\boldsymbol{T}$ be a particular fixed $q \times C$ matrix with full row rank that achieves the reduction of $\boldsymbol{\pi}$ into lower-order probabilities. The new vector of residuals retains asymptotic normality

$$\sqrt{N}\boldsymbol{r} = \sqrt{N}\boldsymbol{T}\boldsymbol{e} = \sqrt{N}\boldsymbol{T}(\boldsymbol{p} - \hat{\boldsymbol{\pi}}) \xrightarrow{D} \mathcal{N}_q(0, \boldsymbol{\Sigma}), \tag{7.6}$$

where $\boldsymbol{\Sigma} = \boldsymbol{T}\boldsymbol{\Xi}\boldsymbol{T}' = \overline{\boldsymbol{D}} - \overline{\boldsymbol{\pi}}\,\overline{\boldsymbol{\pi}}' - \overline{\boldsymbol{\Delta}}(\boldsymbol{\gamma})\mathcal{F}(\boldsymbol{\gamma})^{-1}\overline{\boldsymbol{\Delta}}(\boldsymbol{\gamma})'$, and $\overline{\boldsymbol{D}} = \boldsymbol{T}\boldsymbol{D}\boldsymbol{T}'$, $\overline{\boldsymbol{\pi}} = \boldsymbol{T}\boldsymbol{\pi}$, with the $q \times \dim(\boldsymbol{\gamma})$ (local) Jacobian matrix given by

$$\overline{\boldsymbol{\Delta}}(\boldsymbol{\gamma}) = \boldsymbol{T}\boldsymbol{\Delta}(\boldsymbol{\gamma}) = \boldsymbol{T}\frac{\partial \boldsymbol{\pi}(\boldsymbol{\gamma})}{\partial \boldsymbol{\gamma}'} = \frac{\partial \overline{\boldsymbol{\pi}}(\boldsymbol{\gamma})}{\partial \boldsymbol{\gamma}'}. \tag{7.7}$$

If the IRT model is locally identified, i.e. $\overline{\boldsymbol{\Delta}}(\boldsymbol{\gamma})$ has full column rank, then there exists a $q \times [q - \dim(\boldsymbol{\gamma})]$ orthogonal complement matrix $\overline{\boldsymbol{\Delta}}_c$ such that $\overline{\boldsymbol{\Delta}}_c'\overline{\boldsymbol{\Delta}}(\boldsymbol{\gamma})$ is a null matrix. This implies

$$\sqrt{N}\overline{\boldsymbol{\Delta}}_c'\boldsymbol{r} \xrightarrow{D} \mathcal{N}_{q-\dim(\boldsymbol{\gamma})}(0, \boldsymbol{\Omega}), \tag{7.8}$$

where $\boldsymbol{\Omega} = \overline{\boldsymbol{\Delta}}_c'(\overline{\boldsymbol{D}} - \overline{\boldsymbol{\pi}}\,\overline{\boldsymbol{\pi}}')\overline{\boldsymbol{\Delta}}_c'$. Evaluating the model-implied probabilities and the Jacobian elements at the maximum likelihood estimate, the following limited information statistic is asymptotically centrally chi-square distributed with $q - \dim(\boldsymbol{\gamma})$ degrees of freedom:

$$M = N\boldsymbol{r}'\overline{\boldsymbol{\Delta}}_c\boldsymbol{\Omega}^{-1}\overline{\boldsymbol{\Delta}}_c'\boldsymbol{r}. \tag{7.9}$$

7.3 Example

As an illustration, we return to the previously described *Quality of Life Interview for the Chronically Mentally Ill* (Lehman 1988) from 586 chronically mentally ill patients. Both the multiple content areas of the subdomains and their labeling as such encourage responses to the set of items as a whole rather than considered responses to the individual items. This effect creates dependencies between responses within the sets that violate the assumption of conditional independence of response required for conventional one-dimensional IRT analysis.

7.3.1 Exploratory Item Factor Analysis

Given that the items are clustered within seven content domains, for the purpose of dimensionality assessment, we considered models containing one through eight factors, to determine if an additional factor explained any significant additional variation in item responses over the seven specified subdomains. Chi-square statistics for the addition of each successively added factor are shown in Table 6.1. Very roughly, a chi-square value is significant if it is at least twice as large as its degrees of freedom. By this rule, even the addition of an eighth orthogonal factor shows no danger of over factoring, although its contribution to improved goodness of fit is the smallest of any factor. Notice that the decreases are not monotonic: unlike traditional factor analysis of product-moment correlations, the marginal probabilities of the response patterns (which determined the marginal likelihood) reflect changes in all parameters jointly, including the category parameters and not just the factor loadings. Because our inspection of the signs of the loadings of the first seven factors showed relationships to the item groups and the eighth factor did not the seven factor model is likely the most parsimonious choice.

As expected, all first principal factor loadings are positive, clearly identifying the factor with the overall quality of life variable (see Table 6.2). In fact, the largest loading is that of item number one, which asks for the respondent's rating of overall quality of life. Only one item, the last, has a loading less than 0.5. As for the six bipolar factors, the significant feature is that the sign patterns of loadings of appreciable size conform to the item groups. Factor 2 strongly contrasts the Finance group with Family Living and Safety; to a lesser extent, Leisure is working in the same direction as Finance. In other words, persons who tend to report better financial positions and quality of leisure are distinguished by this factor from those who report better family relationships and safety. Factor 3 then combines Living and Finance and contrasts them primarily with a combination of Health and Safety. Factor 4 contrasts a combination of Family and Social with Finance, Health, and Safety. Factor 5 combines Social, Living, and Safety versus Family, Finance, and Safety. Factor 6 primarily contrasts Health with Safety. Finally, Factor 7 contrasts Social versus Leisure. The seven factor solutions have a bifactor pattern where there are positive first factor loadings, and secondary factors which contrast item groups rather than items within groups. Estimated thresholds for this analysis are reported in Table 6.3.

7.3.2 Confirmatory Item Bifactor Analysis

The bifactor model produced a value of $-2 \text{ Log } L = 64{,}233.3$ which is similar to that obtained for a four-factor model. While the seven-factor unrestricted model provides significant improvement in fit, inspection of the estimated factor loading in Table 6.5 shows that the bifactor model provides the most parsimonious and easily interpretable results. Because the group factor loadings are not constrained to orthogonality with those of the general factor, they are all positive and their magnitudes indicate the strength of the effect of items belonging to common domains. The effects of Family and Finance, for example, are stronger than those of Health and Leisure. It is interesting to note that the empirical reliability for the primary dimension for the bifactor model is 0.9, but is overestimated as 0.95 for a unidimensional model applied to these same data (standard errors of 0.322 and 0.232, respectively). As expected, reliability is overestimated and uncertainty in estimated scale scores is underestimated when the conditional dependencies are ignored. Avoiding this type of bias is a major motivation for item bifactor analysis.

Limited information goodness of fit testing lends additional support for the appropriateness of the bifactor solution. Take the unidimensional model for instance, the Cai–Hansen modified M_2^* statistic is equal to 2674.85 on 385 degrees of freedom, $p < 0.0001$. The statistic uses both univariate and bivariate residual tables. Since there are 35 items, there are $35 + 35 \times (35 - 1)/2 = 630$ residuals available for testing model fit. The unidimensional graded model contains $35 \times 7 = 245$ free item parameters, resulting in $630 - 245 = 385$ degrees of freedom. The null hypothesis of exact unidimensionality is rejected and the unidimensional model is untenable for this data set. We may compute the RMSEA index, which is a widely used measure of fit in factor analysis and structural equation modeling, and it is equal to 0.08 and a 90% confidence interval of RMSEA is (0.076, 0.084). Adopting established conventions in factor analysis (Browne and Cudeck 1993), an RMSEA that exceeds 0.05 cannot be taken as indication of good fit. On the other hand, the M_2^* statistic is equal to 546.88 on 351 degrees of freedom. While it remains significant at the 0.05 level, the RMSEA index for the bifactor model is equal to 0.03 and the 90% confidence interval is $(0, 0.035)$, indicating substantially improved fit.

7.4 Discussion

We have shown that for many applications of IRT, multidimensionality rather than unidimensionality should represent the null hypothesis. There are a variety of limited information and full information methods for determining the goodness of fit and the underlying dimensionality of a particular test. As it turns out, the bifactor model produces excellent results for a variety of different IRT applications because it (i) uses expert judgment to define the underlying factor structure and (ii) evaluates the likelihood in an always computationally tractable way because it reduces it to a two-dimensional integral, which is relatively easy to evaluate in practice. Traditional methods based on eigenvalues have a tendency to identify factors that are not indicative of the underlying trait of interest. By contrast, goodness of fit statistics which compare various nested and nonnested statistical models can be used efficiently to evaluate the dimensionality of a particular test. In general, unrestricted item factor analysis should not be used to evaluate multidimensionality. It is poorly specified and is subject to considerable rotational variance leading to a plethora of different conclusions regarding the latent variables. This is not the case for the bifactor model which provides essentially the same answer regardless of small changes in the model specification. Finally, one should be extremely cautious regarding the fitting of a unidimensional model to what are inherently multidimensional data. The net result is an underestimate of the point at which adaptive testing should terminate (i.e. underestimates of the posterior variance of the latent variable estimate), and increases in the empirical variance of the resulting test score. Neither of these conditions is good. As a consequence, the best possible approaches to determining dimensionality should always be used.

8

Computerized Adaptive Testing

In our lust for measurement, we frequently measure that which we can rather than that which we wish to measure... and forget that there is a difference.

(Source: George Udny Yule)

8.1 What Is Computerized Adaptive Testing?

Imagine a 1000-item mathematics test with items ranging in difficulty from basic arithmetic through advanced calculus. Now consider two examinees, a fourth-grader and a graduate student in mathematics. Most questions will be uninformative for both examinees (too difficult for the first and too easy for the second). To decrease examinee burden, we could create a short test of 10 items, equally spaced along the mathematics difficulty continuum. Although this test would be quick to administer, it would provide very imprecise estimates of their abilities because only an item or two would be appropriate for either examinee. A better approach would be to begin by administering an item of intermediate difficulty, and based on the response scored as correct or incorrect, select the next item at a level of difficulty either lower or higher. This process would continue until the uncertainty in the estimated ability is smaller than a predefined threshold. This process is called computerized adaptive testing (CAT). To use CAT, we must first calibrate a bank of test items using an item response theory (IRT) model that relates properties of the test items (e.g. their difficulty and discrimination) to the ability (or other trait) of the examinee. The paradigm shift is that rather than administering a fixed number of items that provide limited information for any

Item Response Theory, First Edition. R. Darrell Bock and Robert D. Gibbons.

given subject, we adaptively administer a small but varying number of items (from a much larger item bank) that are optimal for the subject's specific level of severity.

8.2 Computerized Adaptive Testing – An Overview

The field of psychological testing has depended almost exclusively on conventional psychological tests since its inception about 100 years ago. In a conventional test, a set of test items is selected in advance to comprise an instrument designed to measure a particular psychological trait. All questions in that instrument are administered to every individual who takes that test. Only a few exceptions, such as the intelligence tests based on Alfred Binet's test model and a few other individually administered tests of that type, have not used the conventional testing approach.

The Binet types of intelligence tests are adaptive tests (Weiss 1985). In an adaptive test, items are selected during the process of test administration for each individual being tested. Adaptive tests are designed to allow the test administrator to control the precision of a given measurement and to maximize the efficiency of the testing process. In the Binet test, items are classified during the process of development with respect to "mental age" levels. These levels correspond to increasing levels of item difficulty. When a test administrator administers a Binet-type test, he/she begins test administration at whatever "mental age" level the examinee appears to be functioning. Items are scored as they are administered, and when all items at a given mental age level have been administered, the test administrator determines whether additional items are needed.

If the examinee answered some of the items at a given mental age level correctly, testing continues with items of either a higher or a lower mental age level. If none or a few of the items are answered correctly, easier items are administered to that examinee. If all or most of the items at a given mental age level have been answered correctly, more difficult items are administered. Test administration continues until two mental age levels are identified: One at which the examinee answers all items incorrectly (the ceiling level) and one at which the examinee answers all items correctly (the basal level). In between the ceiling and basal level is the effective range of measurement for that individual. The result of this adaptive item-selection process is that individuals with different trait levels will be administered items at different difficulty levels.

The Binet-type tests have all the characteristics of an adaptive test, including:

1. A precalibrated bank of test items. To create an adaptive test, items must previously be administered to a group of individuals, and item difficulty and other data must be obtained on the items. An adaptive test based on IRT, for example,

will use an item bank in which items are precalibrated on item difficulty, discrimination, and (if appropriate) the pseudoguessing parameter.
2. A procedure for item selection. Because items are selected based on an examinee's previous answers, items must be scored as they are administered. The next item (or item subset) to be administered is then based on how the examinee answered all previously administered items.
3. A method of scoring the test. Because the purpose of test administration is to obtain a test score for the examinee, the procedure for adaptive testing requires not only that items be scored as they are administered but also that a test score of some type be determined at multiple points during the process of test administration.
4. A procedure for terminating the test. In contrast to a conventional test, the number of test items is not fixed in an adaptive test. Thus, in a Binet-type test, an individual may receive test items from as few as two mental age levels to as many as eight or nine, depending on how he/she performs on the test.

Research since the 1970s has shown that adaptive testing procedures are most effective when combined with IRT procedures (Kingsbury and Weiss 1980, 1983, McBride and Martin 1983). Thus, an item bank for use in adaptive testing can be calibrated according to an IRT model. The point at which a test is to be started (frequently referred to as the entry point) can be determined by taking into account individual status variables or other data about an individual (e.g. previous test scores, age, gender, clinical evaluations). Explicit procedures for estimating an entry point for an adaptive test are available in conjunction with IRT using Bayesian statistical methods (Baker 1992);(Weiss and McBride 1984).

IRT procedures for estimating an individual's trait level are applicable to the adaptive testing process. Procedures of maximum likelihood or Bayesian estimation permit estimation of trait, based on one or more responses made by a single individual in an adaptive test. Thus, a continuous updating of the trait level can be accomplished after each item is administered in an adaptive test, and the next item to be administered can be based on the trait estimate derived from all previous items administered. In addition, maximum likelihood and Bayesian estimation procedures also provide individualized standard errors of measurement (SEM), for each trait level.

8.3 Item Selection

Item selection rules derived from IRT and adaptive testing can explicitly use concepts of item information (Hambleton and Swaminathan 1985, Weiss 1985). Thus, at a given current trait estimate, the most informative item not yet administered can be chosen for administration. When items are selected using this maximum-information item selection rule, the net effect is an extremely efficient procedure for reducing the error of measurement at each successive stage in the administration of an adaptive test (Weiss 1985). Item information describes the information contained in a given item for a specific trait estimate. Our goal is to administer the item with maximum item information at each step in the adaptive process.

8.3.1 Unidimensional Computerized Adaptive Testing (UCAT)

According to Lord (1968), an examinee is measured most effectively when the test items are neither too hard nor too easy for her. An examinee's latent trait level is fit precisely by selecting test items sequentially from the item pool on the basis of her responses to the items already administered. In other words, the test is tailored to each examinee's θ level, thus matching the difficulties of the items to the examinee (Chang 2004).

Research in tailored testing at that time (see (Linn et al. 1969)) were typically built on the following branching rule: If the examinee answers an item correctly, the next item administered should be harder; if she answers it incorrectly, the next item should be easier. To implement the branching rule, Lord proposed an item selection algorithm as an extension of Robbins–Monro process where: "Start with large changes in item difficulty (b), then use smaller and smaller steps, so as to zero in on the difficulty level at which the examinee answers correctly 50% of the time." (Lord 1968) This final difficulty level, b^{n+1}, is a measure of the examinee's standing of the trait measured by the test. Also he demonstrated that conditions for convergence to the desired level are not difficult in practice.

However, the speed of convergence may not be fast (Chang 2004). In other words, it may need many items for $\hat{\theta}_n$ to be close to θ. Furthermore, there is an efficiency issue encountered in CAT and for efficiency comparison we may need more information on item response in a CAT design. Theory given in Lord (1968) assumes that items differ from each other only on difficulty level. In practice, they differ on a_i (discrimination parameter) and c_i (guessing parameter) of the IRT model. Thus, Lord's theory should be modified to deal with more general conditions. For contents of his tailored testing procedure and its limitations, see Lord (1968) and Chang (2004).

8.3.1.1 Fisher Information in IRT Model

Because most CAT theory is built on the IRT model, we assume the two parameter logistic (2PL) model described below by which the items in the item pool are calibrated. Let X_j be the score for randomly selected examinee on the jth item, with $X_j = 1$ if the answer is correct and $X_j = 0$ if incorrect. In this model, the probability of a correct answer to item j by an examinee with trait parameter θ, valued in $(-\infty, \infty)$, is given by

$$P_j(\theta) = \text{Prob}(X_j = 1|\theta) = \frac{1}{1 + e^{-a_j(\theta - b_j)}}, \tag{8.1}$$

where $b_j \in (-\infty, \infty)$ and $a_j \in (0, \infty)$ represent the difficulty and discrimination parameters on item j, respectively. Suppose n test items are given to an examinee with θ and assume that they are locally independent with respect to θ. Then, the likelihood function can be written as:

$$L_n(\theta; X_1, \ldots, X_n) = \prod_{j=1}^{n} P_j(\theta)^{X_j} Q_j(\theta)^{1-X_j}, \tag{8.2}$$

where $Q_j = 1 - P_j$. The maximum likelihood estimator is the θ which maximizes (8.2). Let $\hat{\theta}_n$ denote the resulting estimator. Under general IRT modeling assumptions, $\hat{\theta}_n$ is consistent and asymptotically normally distributed as $N(\theta_0, I^{-1}(\theta_0))$ where θ_0 is the true θ and $I^{-1}(\theta_0)$ is the inverse of the Fisher information evaluated at θ_0. The Fisher information is given by:

$$I(\theta) = -E\left(\frac{\partial^2 \log L_n(\theta; X_1, \ldots, X_n)}{\partial \theta^2}\right) = E\left(\frac{\partial \log L_n(\theta; X_1, \ldots, X_n)}{\partial \theta}\right)^2. \tag{8.3}$$

The third expression in (8.3) is the expected squared relative rate at which the density changes at $X_1, \ldots, X_n$ as θ changes. The greater this expectation is at a given θ_0, the easier it distinguishes θ_0 from neighboring θ, and therefore the more accurate θ can be estimated at $\theta = \theta_0$ (Lehmann and Casella 1998).

In IRT models $I(\theta)$, consists of independent and additive contribution from each of n separate items in a test, that is,

$$I(\theta) = \sum_{j=1}^{n} I_j(\theta), \tag{8.4}$$

where

$$I_j(\theta) = \frac{\left(\frac{\partial P_j(\theta)}{\partial \theta}\right)^2}{P_j(\theta) Q_j(\theta)}. \tag{8.5}$$

Because of this additivity property of $I(\theta)$, we can separately calculate $I_j(\theta)$ and add them to form updated test information $I(\theta)$ at each stage. Notable is that the item information function $I_j(\theta)$ in (8.5) of practical interest is the maximum amount of information obtainable. For IRT model (8.1), computation gives us:

$$I_j(\theta) = a_j^2 P_j(\theta) Q_j(\theta) = \frac{a_j^2 e^{z_j}}{(1 + e^{z_j})^2} \tag{8.6}$$

and

$$\frac{\partial I_j(\theta)}{\partial \theta} = \frac{a_j^3 e^{z_j}(1 - e^{z_j})}{(1 + e^{z_j})^3}, \tag{8.7}$$

where $z_j = a_j(\theta - b_j)$. Equations (8.6) and (8.7) show that the maximum of $I_j(\theta)$ occurs at $\theta = b_j$ and the maximum is $(1/4)a_j^2$. Thus, under the IRT model (8.1), $I_j(\theta)$ is the largest when the difficulty level of item j, b_j, is matched to the examinee's θ, and that maximum of $I_j(\theta)$, when $\theta = b_j$, rises with high a_j parameter.

8.3.1.2 Maximizing Fisher Information (MFI) and Its Limitations

An adaptive test differs from a conventional test in that the assignment of the test items is sequential with selection of each item depending on the responses of the examinee to the preceding items. Item selection criterion in CAT used most frequently now is the maximizing Fisher information (MFI; Birnbaum (1968); (Lord 1980)). Under MFI, the next item to be selected from item pool should have the highest value of item information at the current θ estimate, based on prior item responses. The motivation for maximizing the Fisher information is to make $\hat{\theta}_n$ the most efficient. Because the variance of the maximum likelihood estimator is inversely related to the Fisher information as n goes to infinity, to minimize the variance of the estimator, we have to maximize the Fisher information at the current estimate.

For a long time, the use of the adaptive maximum Fisher information in CAT missed the asymptotic motivation that existed under the paper-and-pencil test framework (van der Linden and Pashley 2010). Chang and Ying (2009) showed that under the adaptive design for the 2PL model in which "after $\hat{\theta}_k$ is defined, set $b_{k+1} = \hat{\theta}_k$ and $\hat{a}_{k+1} \in [n, M]$," the maximum likelihood estimator of θ converges to true θ with a sampling variance approaching the reciprocal of (8.4). Their study suggests that, for large n, MFI is more efficient than the Robbins–Monro process for tailored testing because $I(\theta_0)^{-1}$ is the asymptotic lower bound of the sample variance of the maximum likelihood estimator $\hat{\theta}$.

Unfortunately, under MFI, however, the θ estimation is unstable at the early stages of the test. To see this in some detail, let true θ be θ_0 and choose an item such that $b = \hat{\theta}_n$. Then the true item information is

$$I_j(\theta_0 | a, b = \hat{\theta}_n) = \frac{a^2 e^{a(\theta_0 - \hat{\theta}_n)}}{(1 + e^{a(\theta_0 - \hat{\theta}_n)})^2}, \tag{8.8}$$

which reaches the maximum value $a^2/4$ at $\hat{\theta}_n = \theta_0$. But at the initial stage of CAT when θ estimates have relatively low precision, $\hat{\theta}_n$ is often far from θ_0. If that is the case then the true information (8.8) is much smaller than the expected $a^2/4$. Hence, when using MFI during the early stage of CAT, MFI tends to have the optimal properties at a highly biased θ, which is known as the attenuation paradox (Lord and Novick 1968).

Furthermore, under IRT circumstances, items with larger discrimination parameters are more likely to be selected by MFI than items with smaller ones (Lima Passos et al. Lima Passos, Berger, and Tan, 2007). To see why, let us look at (8.6) again. For a fixed a, if θ is known, we get maximum $a^2/4$ by setting $b = \theta_0$, a true θ value. This maximum increases as far as the a-value increases. However, θ_0 is unknown in practice. One close approximation of such an approach for maximizing I_j is to select items with b near the current estimate of $\hat{\theta}$ and with a as large as possible. Thus, the MFI rule selects an item with a large a-value from the item pool. At the initial stage of CAT when θ estimates have relatively low precision, selecting high-discrimination items actually provides little information toward θ estimation. If a gets larger with $\theta_0 \neq \hat{\theta}_n$, the true item information (8.8) goes down to zero as can be seen in:

$$\lim_{a\to\infty} I_j(\theta_0|a, b = \hat{\theta}_n) = 0, \quad \text{if } \theta_0 \neq \hat{\theta}_n. \tag{8.9}$$

8.3.1.3 Modifications to MFI

To deal with this early stage instability problem, some modifications to the MFI approach have been proposed.

(1) Veerkamp and Berger (1997) introduced an interval information criterion for item selection to account for the uncertainty of θ estimation at the early stage of CAT. Instead of maximizing the Fisher information function at a θ estimate, they proposed to integrate the function over an interval around the estimate,

$$\text{FII}_j(\theta) = \int_{\hat{\theta}-\frac{z}{\sqrt{I(\theta)}}}^{\hat{\theta}+\frac{z}{\sqrt{I(\theta)}}} I_j(\theta)d\theta, \tag{8.10}$$

where $z = \Phi^{-1}(\frac{1+\delta}{2})$, and δ is a confidence coefficient for the standard normal cumulative distribution Φ. FII_j is the basis for selecting the item that provides maximum area value in a confidence interval of $\hat{\theta}$. As the test progresses, $I(\theta)$ increases and the confidence interval narrows (see (8.10)). As the estimated θ approaches true θ, there will be no difference between FII and MFI.

(2) In the same paper, Veerkamp and Berger proposed the other alternative approach in which the Fisher information function is integrated throughout the θ scale, weighted by the likelihood function. They call this criterion likelihood weighted information (LWI). With the LWI criterion, the item to be selected is the item j maximizing:

$$\int_{\theta=-\infty}^{\theta=+\infty} L(\theta; X_1, \dots, X_n)I_j(\theta)d\theta, \tag{8.11}$$

where $L(\theta; X_1, \dots, X_n)$ is the likelihood function of responses after the nth item administration. LWI seems to be superior to FII. van der Linden and Reese (1998) pointed out, however, that for the few items, the likelihood function is still flat and high-discrimination items might be over-used again. One solution to improve LWI is to use the posterior distribution of θ to weight information.

(3) An alternative solution to the problem inherent in FII or LWI is to impose a posterior distribution as a weight function. Call this criterion Fisher information with a posterior distribution (FIP). Under FIP an item selection base is

$$\mathrm{FIP}_j(\theta) = \int_{\theta=-\infty}^{\theta=+\infty} p(\theta|X_1, \dots, X_n) I_j(\theta) d\theta, \tag{8.12}$$

where $p(\theta|X_1, \dots, X_n)$ is the posterior θ distribution after n items have been administered. Because the posterior distribution $p(\theta|X_1, \dots, X_n)$ is the product of a prior distribution and likelihood function, the posterior distribution is strongly influenced by the prior distribution at the early stage of CAT. After many items have been administered, the posterior distribution is dominated by the likelihood function and degenerates to a point distribution at true θ (Chen et al., 2000). For large n, FIP will perform similarly to LWI or MFI.

(4) Chang and Ying (1996) pointed out that the Fisher information is a local concept around the true value of θ and its usefulness is questionable at the early stage when θ estimate is often far from its true value. They proposed to select items based on a global definition of information called the Kullback–Leibler (KL) information measure. Generally, the KL information measures the distance between two likelihoods over the same parameter space. When the KL information is applied in CAT, the purpose is to select items that maximize the distance between true $\theta = \theta_0$ and the current estimate $\hat{\theta}$. The KL information for item j can be written (Chang and Ying 1996), Eq. (8.10) as:

$$\mathrm{KL}_j(\hat{\theta}, \theta_0) = E_{\theta_0} \log \left(\frac{L_j(\theta_0|X_j)}{L_j(\hat{\theta}|X_j)} \right). \tag{8.13}$$

Because of the conditional independence between responses, the information in the responses for the first n items can be written as:

$$\mathrm{KL}^{(n)}(\hat{\theta}, \theta_0) = E_{\theta_0} \log \left(\frac{L_j(\theta_0|X_1, \dots, X_n)}{L_j(\hat{\theta}|X_1, \dots, X_n)} \right) = \sum_{j=1}^{n} \mathrm{KL}_i(\hat{\theta}, \theta_0). \tag{8.14}$$

Hence, the next item to be selected has to maximize the KL item information $\mathrm{KL}_j(\hat{\theta}, \theta_0)$. Since true θ_0 is generally unknown, Chang and Ying (1996) further proposed the KL index by integrating $\mathrm{KL}_j(\hat{\theta}, \theta_0)$ of (8.13) over an interval centered with $\hat{\theta}$

$$\mathrm{KL}_j(\hat{\theta}) = \int_{\hat{\theta}-\frac{r}{\sqrt{k}}}^{\hat{\theta}+\frac{r}{\sqrt{k}}} \mathrm{KL}_j(\hat{\theta}, \theta) d\theta, \tag{8.15}$$

where k is the number of the items that have been administered so far, and r is a constant. Then the item selected in the test is the one which maximizes (8.15) among the items that have not been administered yet. As shown in (8.15), the KL index is an overall information of an item to differentiate $\hat{\theta}$ and neighboring θ values. Chang and Ying (1996) also showed that when k gets large and hence the integration interval of (8.15) shrinks, maximizing the KL index will be equivalent to maximizing the Fisher information.

(5) FII, LWI, and KL methods cope with the standard errors of the MFI approach by replacing a point estimate with a range of estimates. However, these item selection methods were not intended to resolve the problem with MFI in its excessive use of items with higher a parameter values and low utilization of items with lower a parameter values. To prevent this problem, Chang and Ying (1999) proposed stratifying items in the item pool by a-parameter value. In their method, items for the lowest a-value would be administered at the early stage of the test, and those from the highest level would be administered at the last stage of the test. At each stage, an item with a b-parameter value that is closest to the current estimate $\hat{\theta}$ is selected from the item stratum with a similar level of a-parameter values. According to their simulation results, the a-stratification method reduced the uneven use of items and also managed the increase in estimation errors and biases, compared to MFI.

However, some drawbacks with this a-stratification design were pointed out (see (Han 2012)). For example, correlations between a- and b-parameters are commonly observed in practice, and hence if items are specified by a-parameter values, the item stratum will not likely be equivalent to each other in terms of b-parameters. Chang et al. (Chang, Qian, and Ying) developed a modification, "a-stratification with b-blocking," that addressed the $a - b$ correlation issue with the original a-stratification design. Alternative methods that compensate for the defects of the a-stratification method include a multiple stratification variant of the a-stratified design that incorporates content balancing (Yi and Chang 2003), and the enhanced a-stratified design (Leung, Chang and Han, 2003).

8.3.2 Multidimensional Computerized Adaptive Testing (MCAT)

We assume that the items are calibrated with a multidimensional version of (8.1):

$$P_j(\boldsymbol{\theta}) = \text{Prob}\left(X_j = 1|\boldsymbol{\theta}\right) = \frac{1}{1 + e^{-\boldsymbol{a}_j^T\boldsymbol{\theta}+b_j}}. \tag{8.16}$$

Here $\boldsymbol{\theta}^T = (\theta_1, \dots, \theta_p)$ denotes a set of p trait coordinates, $\boldsymbol{a}_j$ is the vector of discrimination parameters for item j, b_j is the scalar parameter for the difficulty of the item. Let $Q_j = 1 - P_j$. In order to estimate $\boldsymbol{\theta}$, we can use either a maximum likelihood estimator or a Bayesian estimator. For details of the $\boldsymbol{\theta}$ estimation methods in MIRT, see Segall (1996) and Mulder and van der Linden (2009) and Section 10.2.5.

8.3.2.1 Two Conceptualizations of the Information Function in Multidimensional Space

The first concept is the Fisher information function that is generalized to multidimensional space. The Fisher information for item j is defined as:

$$I_j(\boldsymbol{\theta}) = -E\left(\frac{\partial^2 \log L(\boldsymbol{\theta}; X_j)}{\partial\boldsymbol{\theta}\, \partial\boldsymbol{\theta}^T}\right). \tag{8.17}$$

If the 2PL model in (8.16) is used, the Fisher information for item j becomes a $p \times p$ matrix given by:

$$I_j(\boldsymbol{\theta}) = P_j(\boldsymbol{\theta})Q_j(\boldsymbol{\theta})\, \boldsymbol{a}_j \cdot \boldsymbol{a}_j^T = P_j(\boldsymbol{\theta})Q_j(\boldsymbol{\theta}) \begin{pmatrix} a_{j1}^2 & a_{j1}a_{j2} & \cdots & a_{j1}a_{jp} \\ a_{j2}a_{j1} & a_{j2}^2 & \cdots & a_{j2}a_{jp} \\ \cdots & \cdots & \cdots & \cdots \\ a_{jp}a_{j1} & a_{jp}a_{j2} & \cdots & a_{jp}^2 \end{pmatrix}, \tag{8.18}$$

where $\boldsymbol{a}^T = (a_{j1}, a_{j2}, \dots, a_{jp})$.

The second concept was proposed by Reckase and McKinley (1991):

$$I_j^{\gamma}(\boldsymbol{\theta}) = \frac{\left(\nabla_{\gamma} P_j(\boldsymbol{\theta})\right)^2}{P_j(\boldsymbol{\theta})Q_j(\boldsymbol{\theta})} = \frac{\left(P_j(\boldsymbol{\theta})Q_j(\boldsymbol{\theta}) \sum_{\ell=1}^{p} a_{j\ell} \cos\gamma_{j\ell}\right)^2}{P_j(\boldsymbol{\theta})Q_j(\boldsymbol{\theta})} \tag{8.19}$$

where $\boldsymbol{\gamma}^T = (\gamma_{j1}, \dots, \gamma_{jp})$ is the vector of angles with coordinate axes that define the direction taken from the $\boldsymbol{\theta}$ point, and ∇_{γ} is the directional derivative in the direction $\boldsymbol{\gamma}$. $I_j^{\gamma}(\boldsymbol{\theta})$ is essentially a multidimensional critical ratio ((Lord 1980), p. 69). $I_j^{\gamma}(\boldsymbol{\theta})$ is a measure of how effective test score X is at discriminating between $\boldsymbol{\theta}$ level of $(\theta_1, \dots, \theta_p)^T$ and a level "closed by" $\boldsymbol{\theta}' = (\theta_1', \dots, \theta_p')^T$ along a line through $\boldsymbol{\theta}$ at angle $\boldsymbol{\gamma}$ (Ackerman 1994). Equation (8.19) shows that $I_j^{\gamma}(\boldsymbol{\theta})$ is related to $I_j(\boldsymbol{\theta})$ by:

$$I_j^{\gamma}(\boldsymbol{\theta}) = (\cos\boldsymbol{\gamma}_j)^T I_j(\boldsymbol{\theta})(\cos\boldsymbol{\gamma}_j), \tag{8.20}$$

where $(\cos\boldsymbol{\gamma}_j)^T = (\cos\gamma_{j1}, \dots, \cos\gamma_{jp})$. This relation implies that $I_j^{\gamma}(\boldsymbol{\theta})$ is the directional information in the direction $\boldsymbol{\gamma}$ in the multidimensional $\boldsymbol{\theta}$ space. At each point in the $\boldsymbol{\theta}$ space, the shape of the multidimensional item response surface (or hyper space) differs on the direction of the movement from the point. For example, if we take the direction $\boldsymbol{\gamma}_j$ such that $\gamma_{j\ell} = \pi/2$ for all ℓ except ℓ' and $\gamma_{j\ell'} = 0$, that is, if we move along the direction parallel to the ℓ'th coordinate axis then

$$I_j^{\gamma}(\boldsymbol{\theta}) = P_j(\boldsymbol{\theta})Q_j(\boldsymbol{\theta})\, a_{j\ell'}^2, \tag{8.21}$$

which becomes the same form as (8.6) in the unidimensional IRT model.

8.3.2.2 Selection Methods in MCAT

Based on the information concepts described above, several methods have been proposed to deliver multidimensional computerized adaptive testing (MCAT). Three of them are presented here.

(1) In unidimensional computerized adaptive testing (UCAT), to minimize the variance of $\hat{\theta}$ we selected an item which maximizes the Fisher information at the current θ estimate. Because in MIRT the Fisher information (8.18) is a $p \times p$ matrix, not a scalar, it cannot be maximized over the real-valued line and hence some real-valued functionals of (8.18) have to be considered as a maximand. Segall (1996) proposed the test item selection criterion based on the relationship between the Fisher information matrix and the confidence region around the estimates of θ specified by Anderson (1984). Because the current estimate of $\hat{\theta}_k$ (obtained from the first k responses) is asymptotically distributed as $N(\theta_0, \Omega_k)$, we have

$$\Pr\left(\hat{\theta}_k^T \Omega_k^{-1} \hat{\theta}_k \leq \chi_p^2(\alpha)\right) = 1 - \alpha, \tag{8.22}$$

that is, the probability that $\hat{\theta}_k$ falls inside the ellipsoid

$$\boldsymbol{x}^T \Omega_k^{-1} \boldsymbol{x} = \chi_p^2(\alpha) \tag{8.23}$$

is $1 - \alpha$. The volume of this ellipsoid is $\delta|\Omega_k|^{1/2}$ where δ is a coefficient relying only on p and α. Here we want small, rather than large, volume of the confidence ellipsoid since the same probability mass, $1 - \alpha = 0.95$ for $\alpha = 0.05$, for example, is contained inside of small, rather than large, volume of the ellipsoid, implying small variance of the distribution of $\boldsymbol{x}$. Segall showed that when

$$W = |\mathbf{I}_k(\hat{\theta}_k) + I_{k+1}(\hat{\theta}_k)| \tag{8.24}$$

is maximized, the volume of the confidence ellipsoid around the θ estimate is minimized, where $\mathbf{I}_k(\hat{\theta}_k)$ is the information obtained from the already selected k items at $\hat{\theta}_k$, and $I_{k+1}(\hat{\theta}_k)$ is the added information from the next adaptively selected item conditional on the current estimate $\hat{\theta}_k$. The candidate item $k + 1$, which maximizes (8.24) provides the largest decrement in the volume of the confidence ellipsoid.

Use of the determinant of the information matrix or generalized variance as a criterion of optimality is known as D-optimality in the optimal design literature (Silvey 1980). D_s-optimality, as a variant of D-optimality, reflects the optimal item selection for MCAT if the first set of elements of θ are "intentional" and others are "nuisance" (Mulder and van der Linden 2009).

The following algorithm can be used for implementing Segall's method.

1. For each item m in the item pool, compute $W_m = |\mathbf{I}_k(\theta_k) + P_m(\theta)Q_m(\theta)\mathbf{a}_m \cdot \mathbf{a}_m^T|$, where $\mathbf{I}_k(\theta_k) = \sum_{j=1}^{k} P_j(\theta_k)Q_j(\theta_k)\mathbf{a}_j \cdot \mathbf{a}_j^T$;
2. Select item $= m$ such that W_m has the maximum value;
3. Update θ_k based on the information contained in the k previously selected items and the new item.

Segall (1996); Segall (2000) also proposed a Bayesian version of the method described above. Assuming a multivariate normal prior we select the next item $k+1$ by maximizing the determinant of the posterior information:

$$W^B = |\mathbf{I}_k(\hat{\theta}_k) + I_{k+1}(\hat{\theta}_k) + \Sigma^{-1}|, \tag{8.25}$$

where Σ is the covariance matrix of the prior distribution of $\boldsymbol{\theta}$ and $\hat{\boldsymbol{\theta}}_k$ is a Bayesian estimator, such as expected a priori (EAP) or maximum a posteriori (MAP). Note that two criteria by (8.24) and (8.25) differ only by the term Σ^{-1}.

(2) van der Linden (1999) derived an algorithm that minimizes the asymptotic error variance when the linear combination of different $\boldsymbol{\theta}$ dimensions is of interest. Ackerman (1994); Ackerman (1996) proposed an index of multidimensional test information as well as a general approach for determining a linear composite in the trait space that maximizes the test information. Ackerman's approach is equivalent to determining a linear composite whose ML estimate has minimal standard error of measurement. Yao (2012); Yao (2013) applied Ackerman's idea to MCAT and proposed the method of selecting an item that has the minimum error variance for the composite score of optimized weight.

In order to see how the weight is derived, we let $\mathbf{I}_n(\boldsymbol{\theta})$ be the test information for a test with n items with known item parameters and a given $\boldsymbol{\theta}$. The composite score $\theta_C = \sum_{\ell=1}^{p} \theta_\ell w_\ell$ has measurement variance

$$\text{Var}(\theta_C) = w^T V(\boldsymbol{\theta}) w, \tag{8.26}$$

where $\boldsymbol{w} = (w_1, \dots, w_p) = (\cos^2\gamma_1, \dots, \gamma^2\gamma_p)$, and $V(\boldsymbol{\theta})$ can be approximated by $I(\boldsymbol{\theta})^{-1}$. Then finding the optimized weight becomes solving the minimization problem:

$$\text{Minimize } \boldsymbol{w}^T V(\boldsymbol{\theta})\boldsymbol{w} \quad \text{subject to } \mathbf{1}_p^T \cdot \boldsymbol{w} = 1, \tag{8.27}$$

where $\mathbf{1}_p^T = (1, \dots, 1)$. Forming Lagrangian $\mathcal{L}$ for this problem,

$$\mathcal{L} = \boldsymbol{w}^T V(\boldsymbol{\theta})\boldsymbol{w} + \lambda(1 - \mathbf{1}_p^T \boldsymbol{w}). \tag{8.28}$$

From the necessary condition for a minimum we get

$$\boldsymbol{w} = (1/2)\lambda V^{-1}\mathbf{1}_p. \tag{8.29}$$

Premultiplying both sides of (8.29) by $\mathbf{1}_p^T$, we obtain $(1/2)\lambda \mathbf{1}_p^T V^{-1}\mathbf{1}_p = 1$ and hence $\lambda = 2/(\mathbf{1}_p^T V^{-1}\, \mathbf{1}_p)$. Substituting this into (8.29) yields the optimized weight $\boldsymbol{w}^*$

$$\boldsymbol{w}^* = \frac{V^{-1}\,\mathbf{1}_p}{\mathbf{1}_p^T V^{-1}\mathbf{1}_p}, \tag{8.30}$$

and at $\boldsymbol{w} = \boldsymbol{w}^*$ the optimized variance of (8.26) is:

$$\boldsymbol{w}^T V \boldsymbol{w} = \frac{1}{\mathbf{1}_p^T V^{-1}\mathbf{1}_p}. \tag{8.31}$$

Assuming that V is a symmetric positive definite matrix, (8.31) can be shown to be the global minimum (Fletcher 1987). The weighted overall score has the smallest standard error of measurement $1/\sqrt{\mathbf{1}_p^T V^{-1}\mathbf{1}_p}$ among all the weighted scores.

The following steps may be used for implementing Yao's method (2013). Let $M < J$ be an integer.

1. For $k \leq M$, the weight is prefixed as an equal value, that is, $\boldsymbol{w}_k = (w_{k,1}, \dots, w_{k,p})$, $w_{k,\ell} = 1/p$ for $\ell = 1, 2, \dots, p$;
2. For $j > M$, compute the optimized weight $\boldsymbol{w}_k^*$ based on the k selected items (see (8.30));
3. Select item $k + 1 = m$ such that $(\boldsymbol{w}_k^*)^T(\mathbf{I}_{k+1}^m(\boldsymbol{\theta}_k))^{-1}\boldsymbol{w}_k^*$ has a minimum value, where $\mathbf{I}_{k+1}^m(\boldsymbol{\theta}_k) = \mathbf{I}_k(\boldsymbol{\theta}_k) + P_m(\boldsymbol{\theta})Q_m(\boldsymbol{\theta})a_m \cdot a_m^T$, and $\mathbf{I}_k(\boldsymbol{\theta}_k) = \sum_{j=1}^k P_j(\boldsymbol{\theta}_k)Q_j(\boldsymbol{\theta}_k)\mathbf{a}_j \cdot \mathbf{a}_j^T$.

(3) Reckase (2009) and Yao (2012) proposed the minimum angle method, which is to select an item that has the maximum information in the direction that has the minimum information for previously selected items.

The direction angle $\boldsymbol{\gamma} = (\gamma_1, \dots, \gamma_p)$ is determined as follows. For this let $\mathbf{I} = (t_{ij})$ be the $p \times p$ information matrix at $\boldsymbol{\theta}$ for a set of items. Determining the direction angle that has the minimum information for previously selected items is equivalent to solving the minimization problem:

$$\text{Minimize } (\cos\boldsymbol{\gamma})^T\mathbf{I}(\cos\boldsymbol{\gamma}), \text{ subject to } \|\cos\boldsymbol{\gamma}\|^2 = 1. \tag{8.32}$$

This is a standard problem of finding the extrema of a quadratic form on the unit sphere (Schott 1997). The solution vector of weights $\cos\boldsymbol{\gamma}$ to the problem (8.32) is the eigenvector of $\mathbf{I}$ associated with the smallest eigenvalue of $\mathbf{I}$. If $\mathbf{I}$ is the information matrix for items of simple structure, that is, $\mathbf{I}$ is a diagonal matrix $\text{diag}(t_{11}, \dots, t_{pp})$, then the eigenvalues of $\mathbf{I}$ are its diagonal elements. Because $\mathbf{I}$ is the information matrix, it is a nonnegative definite matrix (Fedorov and Hackl 1997) and thus its eigenvalues are all nonnegative. Let t_{ii} be the smallest eigenvalue. Then one of the eigenvectors associated with t_{ii} is $(0, \dots, 1, \dots, 0)^T$, where 1 is in the ith place of the vector. Equating $(\cos\gamma_1, \dots, \cos\gamma_i, \dots, \cos\gamma_p) = (0, \dots, 1, \dots, 0)$ yields $(\gamma_1^*, \dots, \gamma_i^*, \dots, \gamma_p^*) = (\pi/2, \dots, 0, \dots, \pi/2)$. Thus, the minimum direction angle is the direction parallel to ith coordinate axis.

The following steps may be used for implementing Reckase's method (Yao 2012).

1. At $\boldsymbol{\theta}^k$, let $\boldsymbol{\gamma}^*$ be the minimum direction angle as obtained above;
2. Select item $k + 1 = m$ such that $(\cos\boldsymbol{\gamma}^*)^T\mathbf{I}k + 1(\boldsymbol{\theta}_k)(\cos\boldsymbol{\gamma}^*)$ has a maximum value (among all the items not administered yet in the pool), where $\mathbf{I}_{k+1}^m(\boldsymbol{\theta}_k) = \mathbf{I}_k(\boldsymbol{\theta}_k) + P_m(\boldsymbol{\theta})Q_m(\boldsymbol{\theta})\mathbf{a}_m \cdot \mathbf{a}_m^T$, and $\mathbf{I}_k(\boldsymbol{\theta}_k) = \sum_{j=1}^k P_j(\boldsymbol{\theta})Q_j(\boldsymbol{\theta})\mathbf{a}_j \cdot \mathbf{a}_j^T$.

8.3.3 Bifactor IRT

Suppose there are $i = 1, 2, \ldots, N$ examinees, and $j = 1, 2, \ldots, n$ items. Let the probability of a response in category $h = 1, 2, \ldots, m_j$ to graded response item j for examinee i with factor $\boldsymbol{\theta}$ be denoted by $P_{ijh}(\boldsymbol{\theta})$. We call $P_{ijh}(\boldsymbol{\theta})$ a category probability. $P_{ijh}(\boldsymbol{\theta})$ is given by the difference between two adjacent boundaries,

$$P_{ijh}(\boldsymbol{\theta}) = P(x_{ij} = h \mid \boldsymbol{\theta}) = P^*_{ijh}(\boldsymbol{\theta}) - P^*_{ijh-1}(\boldsymbol{\theta}), \tag{8.33}$$

where $P^*_{ijh}(\boldsymbol{\theta})$ is the boundary probability. Under the normal ogive model, the boundary probability is given by

$$\begin{aligned} P^*_{ijh}(\boldsymbol{\theta}) &= \Phi(z_{jh}) \\ &= \int_{-\inf}^{z_{jh}} \frac{1}{\sqrt{2\pi}} e^{\frac{t^2}{2}}\, dt, \end{aligned}$$

where

$$z_{jh} = a_{j1}\theta_1 + a_{j2}\theta_2 + c_{jh}.$$

In the bifactor model, the item information function for the graded response model can be written as:

$$I_j(\theta) = \sum_{h=1}^{m_j} \left(\frac{1}{P_{jh}(\theta)}\right)\left(\left(\frac{\partial P_{jh}}{\partial \theta_1}, \frac{\partial P_{jh}}{\partial \theta_2}\right) \cdot (\cos\alpha_{j1}, \cos\alpha_{j2})'\right)^2, \tag{8.34}$$

where θ_1 and θ_2 are primary and secondary factors, respectively. The item information function has a specific form:

$$\begin{aligned} I_j(\theta) &= \sum_{h=1}^{m_j}\left(\frac{1}{P_{jh}}\right) \\ &\times\left[(a_{j1}\phi(z_{jh}) - a_{j1}\phi(z_{jh-1}), a_{j2}\phi(z_{jh}) - a_{j2}\phi(z_{jh-1}))\cdot(\cos\alpha_{j1}, \cos\alpha_{j2})'\right]^2 \\ &= \left[\sum_{h=1}^{m_j} \frac{[\phi(z_{jh}) - \phi(z_{jh-1})]^2}{\Phi(z_{jh}) - \Phi(z_{jh-1})}\right][a_{j1}\cos\alpha_{j1} + a_{j2}\cos\alpha_{j2}]^2, \end{aligned} \tag{8.35}$$

where

$$\phi(z_{jh}) = \frac{\partial\Phi(z_{jh})}{\partial z_{jh}} = \frac{1}{\sqrt{2\pi}} e^{-\frac{z_{jh}^2}{2}}.$$

Given θ and a_{j1}, a_{j2}, and c_{jh}, the first term of the right-hand side of (8.35)

$$\sum_{h=1}^{m_j} \frac{(\phi(z_{jh}) - \phi(z_{jh-1}))^2}{\Phi(z_{jh}) - \Phi(z_{jh-1})}$$

is constant; so at a fixed point of θ, $I_j(\theta)$ depends only on the direction α. Two special but important cases are

$$I_j(\theta) = \sum_{h=1}^{m_j} \frac{\left(a_{j1}[\phi(z_{jh}) - \phi(z_{jh-1})]\right)^2}{\Phi(z_{jh}) - \Phi(z_{jh-1})}, \tag{8.36}$$

which is the item information associated with a change in a primary factor, but no change in a secondary factor, and

$$I_j(\theta) = \sum_{h=1}^{m_j} \frac{(a_{j2}[\phi(z_{jh}) - \phi(z_{jh-1})])^2}{\Phi(z_{jh}) - \Phi(z_{jh-1})}, \tag{8.37}$$

which is the item information associated with a change in a secondary factor, but no change in a primary factor. It should be noted that the previous two equations depend on both the primary θ_1 and a_{j1}, and secondary θ_2 and a_{j2} factors and loadings.

When we are interested in estimating the item information function for θ_1 in the presence of other subdomains, the subdomains can be integrated out of the objective function. For the purpose of CAT administration, θ_1 is typically our focus; however, θ_2 is also present in a bifactor model. In this case, we are interested in obtaining $I_j(\theta_1)$, which is a function only of θ_1. To get $I_j(\theta_1)$, we integrate the previous bifactor item information function expression with the conditional distribution $h(\theta_2 \mid \theta_1)$ of θ_2 and obtain

$$I_j(\theta_1) = \sum_{h=1}^{m_j} \int \frac{\left[\phi(z_{jh}) - \phi(z_{jh-1})\right]^2}{\Phi(z_{jh}) - \Phi(z_{jh-1})} h(\theta_2 \mid \theta_1) d\theta_2, \tag{8.38}$$

which provides an estimate of the information associated with θ_1 averaged over the θ_2 distribution for item j. Given a set of precalibrated item parameters and a provisional estimate of $\theta = \hat{\theta}_k$ at that point in the adaptive test, we select the next item $(k + 1)$ with maximum information $\max(I_j \mid \theta_k)$.

8.4 Terminating an Adaptive Test

Finally, adaptive testing procedures developed in accordance with IRT can take advantage of a number of different procedures for terminating an adaptive test. One procedure frequently applied is to reduce the individualized SEM to a prespecified level before a test is terminated (Weiss and Kingsbury 1984). An individualized SEM allows the number of test items administered to an individual to vary, but it also results in control of the subsequent level of SEM for individuals tested. Thus, for the individual who responds essentially in accordance with the IRT model, a given level of SEM will be achieved more quickly than for the individual for whom the responses are not in accordance with the IRT model, resulting in a slower reduction of the individualized SEM.

8.5 Additional Considerations

Although individually administered tests are efficient and effective, they are labor intensive, requiring a highly trained test administrator to achieve necessary levels of standardization. When adaptive testing uses IRT, however, the calculations required at each stage of item selection eliminate the possibility of using a human test administrator. Under these circumstances, the adaptive test must be administered by interactive computers. CAT procedures administer items on an individual basis by presenting them on a computer screen. Responses are entered on the keyboard or by a touch screen and are immediately scored by the computer. Various algorithms for selecting items according to maximum information or other criteria are then implemented using the computational capabilities of the computer ((Vale and Weiss 1977); (Weiss 1985)), and typically in less than one second another item is selected for administration and presented on the screen. Meanwhile, the computer continually updates the person's estimated latent trait level and its SEM, again using IRT methods, and constantly monitors the appropriate termination criterion. Once the termination criterion is reached, the test is ended. Tests such as the Graduate Record Examination (GRE) and the Graduate Management Admission Test (GMAT) have become CATs, although the GRE was recently changed to a multi-stage test, which is adaptive by sections rather than by items.

Research shows that adaptive tests are more efficient than conventional tests (Brown and Weiss 1977, McBride and Martin 1983). That is, in an adaptive test a given level of measurement precision can be reached much more quickly than in a test in which all examinees are administered the same items. This results from selecting items that are most informative for an individual at each stage of test administration in the adaptive test. Typical adaptive tests result in a 50% average reduction in number of items administered, and some reductions in the range of 80–90% have been reported, with no decrease in measurement quality (Brown and Weiss 1977). In addition, as has been indicated, adaptive tests allow control over measurement precision. Thus, adaptive tests result in measurements that are both efficient and effective.

Although IRT was developed originally in the context of measuring ability and achievement, the family of IRT measurement models also includes numerous models that are applicable to personality instruments that are not dichotomously scored (Andrich 1978, 1988, Muraki 1990, Tucker 1958). Research has demonstrated that the IRT family of models can be meaningfully applied to the measurement of attitudes and personality variables (Reise and Waller 1993) and that benefits that result from this application are similar to those observed for measuring ability and achievement. Research has also begun into improving the measurement of personality constructs using CAT (Baek 1997, Dodd et al. 1995).

The bifactor model is extremely useful for MCATs. The conditional dependencies produced by the subdomains can be directly incorporated in trait estimation and item information functions as shown in Section 8.3.3, leading to improved estimates of uncertainty and elimination of premature termination of the CAT and potential bias in the estimated trait score. After each item administration, the primary ability estimate θ_1 and posterior standard deviation are recomputed, and based on the estimate of θ_1, the item with maximal information is selected as the next item to be administered. This process continues until the posterior SEM is less than a threshold value (e.g. 0.3). Once the primary dimension has been estimated via CAT, subdomain scores can be estimated by adding items from the subdomain that have not been previously administered, until the subdomain score is estimated with similar precision. Seo and Weiss (2015) provide and evaluate a fully MCAT algorithm for the dichotomous case of the bifactor model.

When the trait score is at a boundary (i.e. either the low or high extreme of the trait distribution), it may take a large number of items to reach the intended posterior SEM convergence criterion (e.g. SEM < 0.3). In such extreme cases, we generally do not require such high levels of precision because we know that e.g. in mental health testing, the subject either does not suffer from the condition of interest or is among the most severely impaired. A simple solution to this problem is to add a second termination condition based on item information at the current estimate of the trait score, and if there is less information than the threshold, the CAT terminates. The choice of the threshold is application specific and can be selected based on simulated CATs. A good value will affect only a small percentage of cases (e.g. $<20\%$) and only be used in extreme (i.e. high or low) cases.

Large item banks may contain items that are too similar to be administered within a given session. These can be declared as "enemy items" and not co-administered. The idea of enemy items can be extended to the longitudinal case to ensure that the same respondent is not repeatedly administered the same items on adjacent testing sessions. CAT will often result in a subset of the entire item bank being used exclusively, because these items have the highest loadings on primary domains and subdomains. Often the difference between the loadings of items that are selected by the CAT and those that are not is quite small and the items have similar information. To ensure that the majority of the items in the item bank are administered, we can add a probabilistic component in which a selected item is administered only if a uniform random number exceeds a threshold. Typically, a threshold of 0.5 works well (for a uniform random number), but again, the exact choice can be based on simulated adaptive testing, in which the largest set of unique items is used without compromising the other characteristics of the measurement process (i.e. average number of items administered and correlation with the total bank score).

8.6 An Example from Mental Health Measurement

The importance of performing research in real-world clinical settings is widely recognized, as is the need for measurement-based care outside the bounds of clinical research. However, in busy medical and psychiatric practices and clinics, the feasibility of conducting the kind of extensive evaluations typical of clinical research is questionable. Therefore, any strategy that reduces the burden of empirically based assessment has the potential to improve outcomes through measurement-based clinical decision making. Traditional mental health measurement has been based on classical test theory, in which a patient's impairment level is estimated by a total score, which requires that the same items be administered to all respondents. These items are weighted equally, so that the response to the question "I am sad" is weighted of equal importance as the response to the question "I feel that those around me would be better off if I were dead." In an effort to decrease patient burden, mental health instruments are often restricted to a small number of symptom items (e.g. the 9-item Patient Health Questionnaire [PHQ-9] or the 17-item Hamilton Depression Rating Scale [HAM-D]). For a patient with a given level of depressive severity, only a few of the items will be discriminating. As a consequence, CAT is immediately applicable to mental health measurement. For example, a depression inventory can be administered adaptively, such that an individual responds only to items that are most informative for assessing his or her level of depression. The net result is that a small, optimal number of items are administered to the individual without loss (and frequently with gains) of measurement precision.

The paradigm shift is from traditional measurement, which fixes the number of items administered and allows measurement uncertainty to vary, to IRT-based CAT, which fixes measurement uncertainty and allows the number of items to vary. The results are a dramatic reduction in the number of items needed to measure mental health constructs and an increased precision of measurement. Inexpensive, efficient, and accurate screening of depression in medical and behavioral health settings is a direct application of the general theory and related methodology. For longitudinal assessments using traditional instruments, each testing session begins anew and is not informed by the results of prior testing sessions. This is not true for CAT administration, in which the next testing session can begin with the estimated severity score from the previous testing session.

Although there have been some applications of IRT-based CAT in mental health measurement (Fliege et al. 2005, Gardner et al. 2004, Pilkonis et al. 2011), this work has been based on the assumption of unidimensionality, an assumption that is generally inconsistent with the multidimensional nature of mental health constructs. Mental health questions (items) are traditionally drawn from content domains (e.g. mood, cognition, behavior), within which the items are more highly correlated than items from different content domains. This leads to a violation of the conditional independence assumption of the unidimensional IRT model, underestimation of the SEM, and greater variability in the estimated scale scores (Gibbons, Immekus and Bock 2007b). The net result is that we overestimate the precision of measurement and prematurely conclude adaptive testing sessions. Resulting test scores are more variable, less valid, and lead to the need for larger sample sizes in research studies.

8.6.1 The CAT-Mental Health

The computerized adaptive testing-Mental Health (CAT-MH) study (Achtyes et al. 2015, Gibbons et al. 2012, 2014) developed a bifactor-based CAT for ordinal response data (Gibbons, Bock, Hedeker, Weiss, Segawa, Bhaumik, Kupfer, Frank, Grochocinski and Stover 2007a) and applied it to a 1008-item bank consisting of 452 depression, 467 anxiety, and 89 bipolar items. We review the results only for depression.

The total depression item bank consisted of 452 items. The items were organized into conceptually meaningful categories using a hierarchical approach. The hierarchy included domains (depression), subdomains (e.g. mood, cognition, behavior), and factors (e.g. within depressed mood, factors included both increased negative affect and decreased positive affect). The items were selected based on a review of more than 100 existing depression or depression-related rating scales. Subjects for this study were male and female treatment-seeking outpatients between 18 and 80 years of age. Patients were recruited from the Western Psychiatric Institute and Clinic (WPIC) at the University of Pittsburgh, a community clinic (Dubois Regional Medical Center), and community controls. A total of 798 subjects (WPIC) were used to calibrate the IRT model, and 816 subjects (414 WPIC and 402 Dubois) received the live CAT-Depression Inventory (CAT-DI). To study the validity of the CAT-DI, 292 consecutive subjects received a full clinician-based DSM-IV (Am. Psychiatr. Assoc. 1994) diagnostic interview (First et al. 1996) and the live CAT-DI. To examine convergent validity, data were also obtained for the HAM-D, PHQ-9, and Center for Epidemiologic Studies Depression scale (CES-D). The HAM-D was administered by a trained clinician, and the PHQ-9 and CES-D were self-reports.

Results of the calibration study revealed that the bifactor model with five subdomains (mood, cognition, behavior, somatic, and suicide) dramatically improved fit over a unidimensional IRT model ($\chi^2_{389} = 6825$, $p < 0.0001$). A total of 389 items with a primary factor loading of 0.3 or greater (96% > 0.4 and 79% > 0.5) were retained in the model. Results of simulated CAT revealed that for SEM < 0.3 (approximately 5 points on a 100-point scale), an average of 12.31 items per subject (range 7–22) were required. The correlation between the 12-item average-length CAT and the total 389 item score was $r = 0.95$. For SEM < 0.4 (less precise), an average of 5.94 items were required (range 4–16), but a strong correlation with the 389-item total score ($r = 0.92$) was maintained. The average length of time required to complete the 12-item (average) CAT was 2.69 minutes in comparison with 51.66 minutes for the 389-item test.

Figure 8.1 reveals the existence of two discrete distributions of depressive severity, with the lower component representing the absence of clinical depression and the higher component representing severity levels associated with clinical depression.

Figure 8.2 displays the distributions of CAT-DI scores for patients with minor depression (including dysthymia), major depressive disorder (MDD), and those not meeting criteria for depression. There is a clear linear progression between CAT-DI depression severity scores and the diagnostic categories from the Structured Clinical Interview for the DSM. Statistically significant differences were

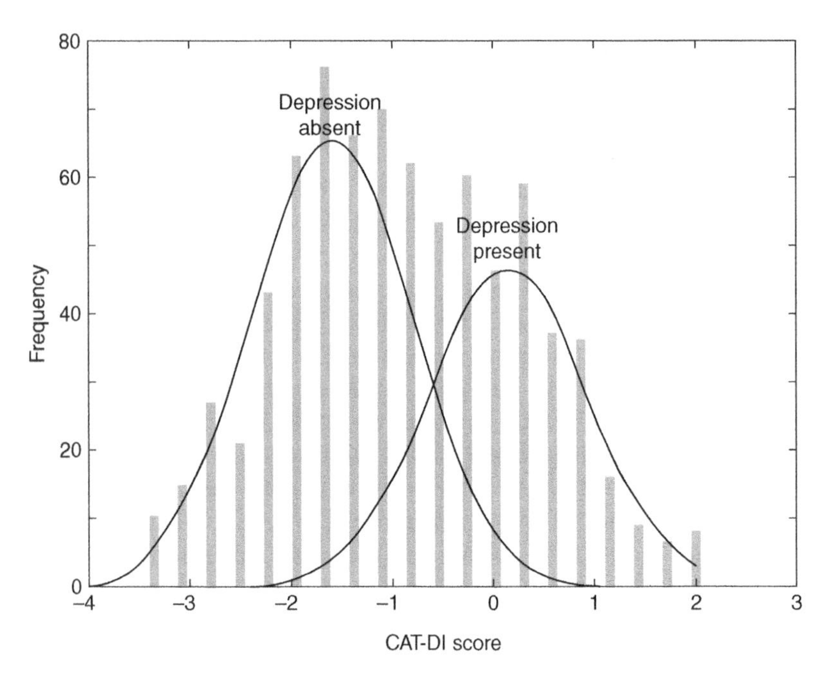

Figure 8.1 Observed and estimated frequency distributions using the computerized adaptive testing-Depression Inventory (CAT-DI) depression scale.

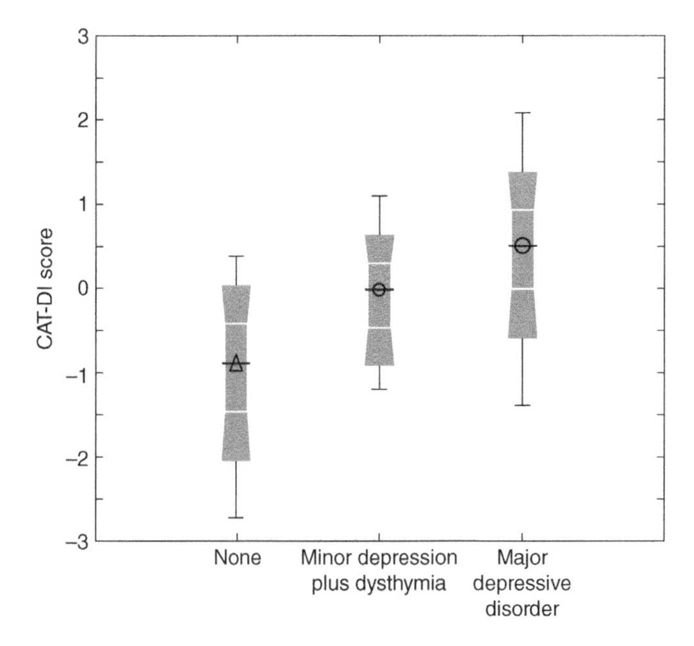

Figure 8.2 Box-and-whiskers plot for computerized adaptive testing-Depression Inventory (CAT-DI) depression scores.

found between none and minor ($p < 0.000\ 01$), none and MDD ($p < 0.000\ 01$), and minor and MDD ($p < 0.000\ 01$), with corresponding effect sizes of 1.271, 1.952, and 0.724 SD units, respectively.

Convergent validity of the CAT-DI was assessed by comparing results of the CAT-DI to the PHQ-9, HAM-D, and CES-D. Correlations were $r = 0.81$ with the PHQ-9, $r = 0.75$ with the HAM-D, and $r = 0.84$ with the CES-D. In general, the distribution of scores between the diagnostic categories showed greater overlap (i.e. less diagnostic specificity), greater variability, and greater skewness for these other scales relative to the CAT-DI. Using the 100 healthy controls as a comparator, sensitivity and specificity for predicting MDD were 0.92 and 0.88, respectively (threshold based on Figure 8.1). CAT-DI scores were significantly related to MDD diagnosis (odds ratio [OR] = 24.19, 95% CI 10.51–55.67, $p < 0.0001$). A unit increase in CAT-DI score has an associated 24-fold increase in the probability of meeting criteria for MDD (Figure 8.3). Figure 8.3 also presents the CAT-DI score percentile ranking for patients with MDD. A patient with a CAT-DI score of −0.6 has a 0.5 probability of meeting criteria for MDD but would be at the lower 7th percentile of the distribution of confirmed cases, whereas a score of 0.5 would have a 0.97 probability of MDD and would be at the 50th percentile of cases.

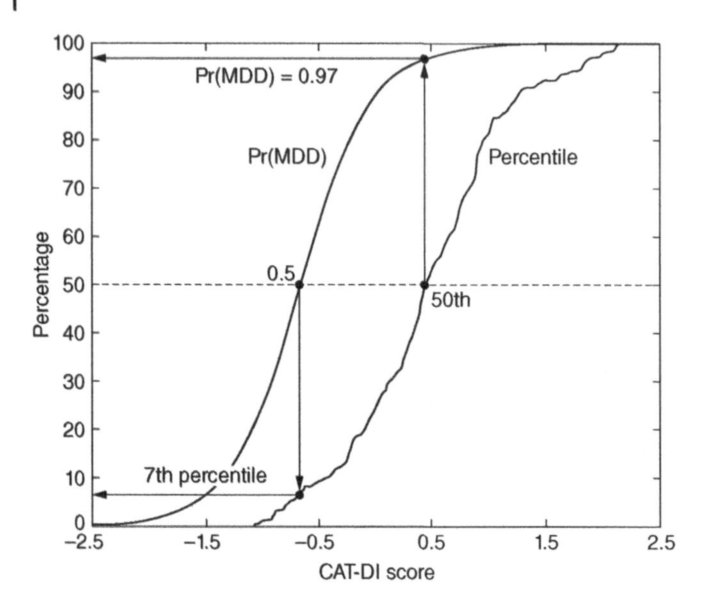

Figure 8.3 Percentile rank (PR) among patients with major depressive disorder (MDD) and probability (expressed as %) of MDD diagnosis.

Results for the CAT inventories for anxiety (CAT-ANX) (Gibbons et al. 2014) and mania (CAT-MANIA) (Achtyes et al. 2015) closely paralleled those for the CAT-DI. Using an average of 12 adaptively administered items, they found correlations of $r = 0.94$ and $r = 0.92$ for the total anxiety and mania item bank scores. For both anxiety and mania, there was a 12-fold increase in the likelihood of the corresponding DSM-5 (Am. Psychiatr. Assoc. 2013) disorder (generalized anxiety disorder or current bipolar disorder) from the low end to the high end of each scale.

8.6.2 Discussion

Beyond the academic appeal of building a better and more efficient system of measurement, CAT of mental health constructs is important for our nation's public health. Approximately 1 in 10 primary care patients has MDD, and the presence of MDD is associated with poor health outcomes in numerous medical conditions. Rates of depression in hospitalized patients are even higher (10–20%), partially because depression increases hospitalization and re-hospitalization rates (Whooley 2012). Unfortunately, clinicians often fail to identify depression in hospitalized patients. This is despite the existence of brief screening tools for depression, such as the (older) Hospital Anxiety and Depression Scale and the more widely used PHQ-9. To increase the likelihood that clinicians will perform some screening, even simpler approaches have been considered. For example, the American Heart

Association recommends a two-stage screening method consisting of a two-item PHQ followed by the PHQ-9 for identifying depression in cardiovascular patients (Elderon et al. 2011). The method yields high specificity (0.91) but low sensitivity (0.52), indicating that it misses almost half of the patients with MDD.

By 2030, MDD is projected to be the number one cause of disability in developed nations and the second leading cause of disability in the world after human immunodeficiency virus and AIDS (Mathers and Loncar 2006, Mitchell et al. 2009, Whooley 2012). Depression affects approximately 19 million Americans per year, or 10% of the adult US population (Natl. Inst. Mental Health 2017). Depression has human costs such as suicide, which ends 35 000 lives per year in the United States (Joiner 2010). Depressed people are 30 times more likely to kill themselves and 5 times more likely to abuse drugs (Hawton and Fagg 1992). As discussed in a 2012 editorial in the American Journal of Public Health (Gibbons et al. 2012), veterans have four times the risk of suicide relative to the general population during the first four years following military service. The ability to screen veterans for depression and suicide risk during this period of high risk and to refer them for appropriate treatment could be lifesaving. Depression is the leading cause of medical disability for people ages 14–44 (Stewart et al. 2003). Depression is a lifelong vulnerability for millions of people.

The information obtained in only two minutes during administration of the CAT-DI would take hours to obtain using traditional fixed-length tests and clinician DSM interviews. In contrast to traditional fixed tests, adaptive tests can be repeatedly administered to the same patient over time without response set bias because the questions adapt to the changing level of depressive severity. For the clinician, CAT provides a feedback loop that informs the treatment process by providing real-time outcomes measurement. For organizations, CAT provides the foundation for a performance-based behavioral health system and can detect those previously unidentified patients in primary care who are in need of behavioral health care and would otherwise be among the highest consumers of physical health-care resources. From a technological perspective, these methods can be delivered globally through the Internet and therefore do not require the patient to be in a clinic or doctor's office to be tested; rather, secure testing can be performed anywhere using any Internet-capable device (e.g. computer, tablet, smart phone). The testing results can be interfaced to an electronic medical record and/or easily maintained in clinical portals that are accessible by clinicians from any Internet-capable device.

The future direction of this body of research is immense. Screening patients in primary care for depression and other mental health disorders including risk of suicide is of enormous importance, as is monitoring their progress in terms of changes in severity during behavioral health treatment. Of critical importance is diagnostic screening of mental health disorders such as depression, anxiety, and mania and attention-deficit/hyperactivity disorder, oppositional defiant disorder, and conduct disorder in children based on adaptive self-ratings and parent ratings. A major priority should be applications of mental health CAT in military settings and among veterans who are at high risk of depression, post-traumatic stress disorder, and suicide. In genetic studies, the ability to obtain phenotypic information in large populations that can in turn be studied in terms of their genetic basis can and should be pursued. Global mental health applications should also be more rigorously understood and evaluated. Differential item functioning can be used to identify items that are good discriminators of high and low levels of depression in one language and culture but may not be effective differentiators in another. The same is true of patients identified for different indications: Somatic items are good discriminators of high and low levels of depression in a psychiatric population but may not be effective in a perinatal population or in patients presenting in an emergency department with comorbid and possibly severe physical impairments. A model of measurement based on multidimensional IRT allows for a much more sensitive understanding of these population-level differences in the measurement process and can provide rapid, efficient, and precise adaptive measurement tools that are insulated from real differences between cultures, languages, and diseases. This is the future of mental health measurement.

9

Differential Item Functioning

An approximate answer to the right problem is worth a good deal more than an exact answer to an approximate problem.

(Source: John Tukey)

9.1 Introduction

Differential item functioning (DIF) occurs when individuals who are at the same level on the construct(s) being assessed, but are from different subpopulations, have unequal probabilities of attaining a given score on a given item. Methods for investigating DIF have been developed for both dichotomously and polytomously scored items. These methods may be classified by whether they condition on an unobserved or observed variable. Item response theory (IRT), logistic regression, and Mantel–Haenszel procedures for dichotomously scored responses and their extensions to polytomous responses are currently the most widely used methods for detecting DIF. Thissen et al. (1993) introduced the IRT approach to DIF detection. The IRT approach usually involves the comparison of two models, a compact model (with common parameters between the different subpopulations) and an augmented model where a subset of the parameters are allowed to vary across the subgroups. Most procedures involve fitting the model twice per hypothesis, once for the compact model and once for the augmented model, resulting in considerable computational complexity. Woods et al. (2013) have described an approach suitable for assessing DIF for multidimensional IRT models that can be adapted to the bifactor model. We generally refer to the two groups as the "reference group" and the "focal group" although they have also been termed the "original calibration sample" and the "target sample."

Item Response Theory, First Edition. R. Darrell Bock and Robert D. Gibbons.

9.2 Types of DIF

DIF can occur for several reasons. An item can be more difficult (or a symptom more prevalent) in one population or another. This is referred to as uniform DIF, where the item gives an advantage to one group for all levels of ability or impairment. Here the item is equally discriminating in both groups but there is increased likelihood of getting the item correct or endorsing it more positively in one of the two groups. Figure 9.1 illustrates the item characteristic curves for uniform DIF.

Nonuniform DIF occurs when the discrimination parameters differ between two groups. When we think of differences in the experience of some construct (e.g. depression) in two different groups (e.g. cultures), we are generally considering differences in the ability of the item to discriminate high and low levels of the underlying construct in the two cultures. This is an example of nonuniform DIF (see Figure 9.2).

Figure 9.2 reveals that at low levels of the latent variable the reference group has a lower probability of correctly answering the question, whereas the reverse is true at higher levels of the latent variable. Often we are interested in determining whether there are items in the focal group that are simply poor discriminators of

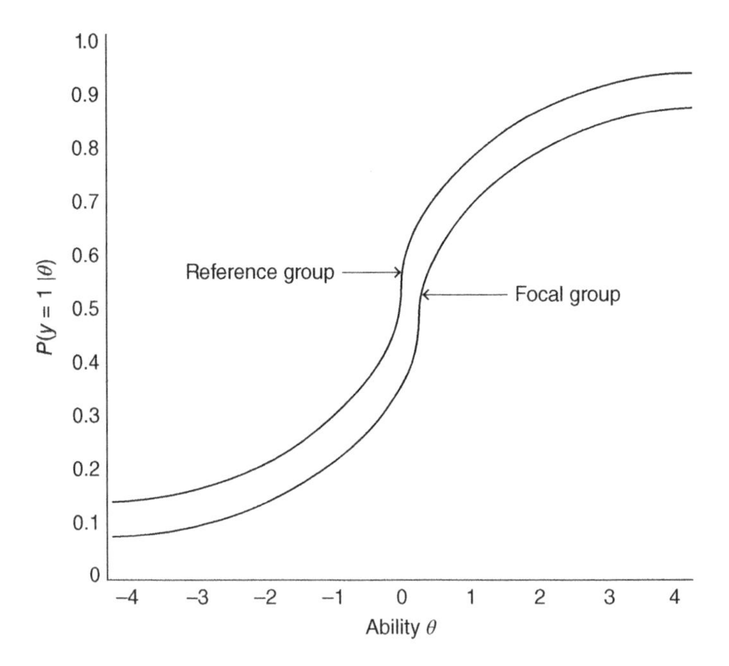

Figure 9.1 Example of uniform DIF.

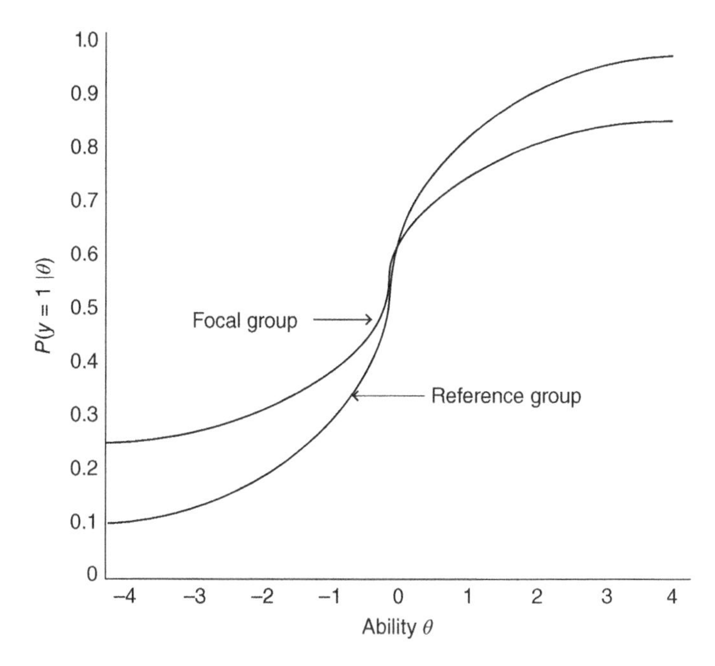

Figure 9.2 Example of nonuniform DIF.

high and low levels of the construct of interest that are already known to function well in the reference group.

9.3 The Mantel–Haenszel Procedure

There are many statistical problems that involve a cross-classification (e.g. a 2×2 contingency table) that is stratified by a grouping variable. For example, in meta-analysis we have treated and control conditions for which we observe a binary outcome variable (presence or absence of success, response to treatment, side effect) for K trials that we want to synthesize and determine if there is an overall treatment-related effect (treatment versus control difference). This same idea has been used to assess DIF where the contingency table describes the cross-tabulation of the two groups and correct and incorrect responses and the stratification variable represents different levels of ability. The outcome of the ith ability level with binary outcome can be summarized in the following 2×2 contingency table.

In Table 9.1, x_{Ri} and x_{Fi} are the number of individuals correctly answering the item in the reference and the focal groups, respectively.

Table 9.1 Contingency table of group by correct or incorrect response in the *i*th ability level.

	Response		
	Correct	**Incorrect**	**Total**
Reference	x_{Ri}	$n_{Ri} - x_{Ri}$	n_{Ri}
Focal	x_{Fi}	$n_{Fi} - x_{Fi}$	n_{Fi}
Total	m_{Ci}	m_{Ii}	N_i

The Mantel–Haenszel method is a heuristic estimate of the assumed common risk ratio across the levels of the stratification variable. The Mantel–Haenszel method estimate of the common risk ratio is

$$\widehat{RR}_{\text{MH}} = \frac{\sum_{i=1}^{K} x_{Ri} n_{Fi} / N_i}{\sum_{i=1}^{K} x_{Fi} n_{Ri} / N_i}, \tag{9.1}$$

which can be expressed as a weighted average estimate as follows:

$$\widehat{RR}_{\text{MH}} = \frac{\sum_{i=1}^{K} w_i \widehat{RR}_i}{\sum_{i=1}^{K} w_i}, \tag{9.2}$$

where

$$w_i = x_{Fi} n_{Ri} / N_i \,. \tag{9.3}$$

The Robins–Breslow–Greenland estimate of variance for $\ln(\widehat{RR}_{\text{MH}})$ is:

$$\widehat{\text{Var}}[\ln(\widehat{RR}_{\text{MH}})] = \frac{\sum_{i=1}^{K} \frac{n_{Ri} n_{Fi}}{N_i^2} m_{Ci}}{\left(\sum_{i=1}^{K} \frac{x_{Ri} n_{Fi}}{N_i}\right)\left(\sum_{i=1}^{K} \frac{x_{Fi} n_{Ri}}{N_i}\right)}. \tag{9.4}$$

The null hypothesis of equal likelihood of getting the item correct in reference and focal groups across the ability continuum, i.e. $RR_{\text{MH}} = 1$, may be tested by the following 1 degree of freedom chi-square test, which is obtained based on the assumption of a conditional binomial distribution for the events x_{Ri} in Table 9.1:

$$X_{\text{MH}}^2 = \frac{\left[\sum_{i=1}^{K} (x_{Ri} - n_{Ri} m_{Ci}) / N_i\right]^2}{\sum_{i=1}^{K} \frac{n_{Ri} n_{Fi} m_{Ci} m_{Ii}}{N_i (N_i - 1)}}. \tag{9.5}$$

Careful inspection of this test reveals that it is a test of uniform DIF in that the overall OR is pooled across the ability levels. Generally the ability levels are defined by a simple total score of the items on the test and then stratified. Referring to Figure 9.2, the overall OR = 1.0 (suggesting the absence of DIF) despite the fact that there is considerable DIF. The problem is that the pooled estimator of the OR assumes a homogeneous effect when nonuniform DIF produces a heterogeneous effect. Although there are tests for the homogeneity assumption (e.g. Cochran's Q-statistic – see (Gibbons and Amatya 2015)), these tests generally have quite limited power to detect heterogeneity.

9.4 Lord's Wald Test

Lord's Wald test for DIF detection compares vectors of IRT item parameters between groups. If the item parameters differ significantly between groups, then trace lines differ across the groups and thus the item functions differently for the groups studied. Lord's Wald Test is asymptotically equivalent to the likelihood ratio test (LRT). Consider the 2-parameter logistic (2PL) model for item j

$$P_j(\theta) = \text{Prob}(X_j = 1|\theta) = \frac{1}{1 + e^{-a_j(\theta - b_j)}}, \tag{9.6}$$

where θ is the latent variable, X_j is the response to item j, a_j is item j's discrimination parameter, and b_j is item j's difficulty. Lord's statistic for item j is

$$\chi^2 = \boldsymbol{v}_j' \Sigma_j^{-1} \boldsymbol{v}_j, \tag{9.7}$$

where $\boldsymbol{v}' = [\hat{a}_{Fi} - \hat{a}_{Ri}, \hat{b}_{Fi} - \hat{b}_{Ri}]$, and Σ_j is the covariance matrix for differences in the item parameters between the focal and reference groups, and df equal to the number of parameters compared for that item (e.g. df = 2 for the 2PL model illustrated here). Item parameters can be estimated using the previously described EM MML methods.

The fundamental problem with Lord's Wald test for DIF detection is the difficulty in estimating the standard errors. However, new computational approaches (Woods et al. 2013) may provide a viable solution to this problem.

9.5 Lagrange Multiplier Test

Glas (1998) argued that the LR and Wald test approach to IRT detection are not efficient because they require estimation of the parameters of IRT model under the alternative hypothesis of DIF for every single item. Therefore, Glas proposed using the Lagrange multiplier (LM) test by Aitchison and Silvey (1958), which does not require estimation of the parameters of the alternative model. The LM test is similar in setup to the LRT and the Wald test, where interest is in testing a special model against a more general alternative. We consider a null-hypothesis about a model with parameters ϕ_0, which is a special case of a more general model with parameters fixed at ϕ. For assessing DIF, the special model is a special case of the more general model obtained by fixing one or more parameters to known constants, not on the boundary of the parameter space. To this end, we partition ϕ_0 as $\phi_0' = (\phi_{01}', \phi_{02}') = (\phi_{01}', c)$, where c is a vector of known constants. Denoting the partial derivatives of the log-likelihood of the general model $h(\theta)$, we have $h(\theta) = (\partial/\partial\phi) \ln \text{ L}(\phi)$ and the Hessian $H(\phi, \phi) = -(\partial^2/\partial\phi\partial\phi') \ln \text{ L}(\phi)$. The *LM* statistic is therefore

$$\text{LM} = h(\phi_0)' H(\phi_0 \phi_0)^{-1} h(\phi_0). \tag{9.8}$$

When evaluated using the maximum likelihood estimate (MLE) of ϕ_{01} and hypothesized values of c, the LM statistic has an asymptotic chi-square distribution with degrees of freedom equal to the number of fixed parameters. Note that at the MLE, the partial derivatives of the free parameters are zero and therefore we only require that the parameters of the special model be estimated, providing a major computational advantage over LRT and Wald tests. In this case (9.8) becomes

$$LM(c) = h(c)' W^{-1} h(c), \tag{9.9}$$

where

$$W = H(c, c) - H(c, \hat{\phi}_{01}) H(\hat{\phi}_{01}, \hat{\phi}_{01})^{-1} H(\hat{\phi}_{01}, c). \tag{9.10}$$

To use the LM test to detect DIF, let $z = 0$ denote the reference population and $z = 1$ denote the focal population. Then, the 2PL model becomes

$$P_j(\theta) = \text{Prob}(X_j = 1|\theta) = \frac{1}{1 + e^{-a_j(\theta - b_j) + z_i(-g_j(\theta - d_j))}}. \tag{9.11}$$

The model is a generalization of (9.6) where the responses of the reference population are adequately described by (9.6), but the focal population requires the additional parameters g_j and d_j. When g_j is nonzero, this is an example of "uniform DIF" where the ICC is shifted for the focal group, whereas nonzero d_j is an example of "nonuniform DIF" where the slope of the ICC differs between the focal and reference group (potentially crossing at some point). The latter is much more of a concern because it indicates that an item that is a good discriminator of high and low levels of θ in the reference group may be a poor discriminator in the focal group. For example, an item related to fatigue in a depression inventory is a good somatic sign of depression in a psychiatric population, but not in a perinatal population, where fatigue is related to pregnancy or giving birth regardless of whether or not the mother is experiencing perinatal depression. Under the null hypothesis, g_j and d_j are fixed at zero. Glas (1998) illustrates computation of the LM statistic for DIF for a variety of unidimensional IRT models.

9.6 Logistic Regression

An alternative approach which extends the previously described Mantel–Haenszel procedure to tests of both uniform and nonuniform DIF can be obtained using logistic regression (Swaminathan and Rogers 1990). The basic idea is to model the likelihood of either a binary or ordinal item response as a function of group, ability level and their interaction. The main effect of group describes the difference in likelihood of a positive response when the ability level is at its mean of zero (uniform DIF) and the interaction between group and ability describes how the difference between groups in likelihood of a positive response changes with ability level (nonuniform DIF). Parameter estimates which can be expressed as ORs, Wald tests for the significance of the main effect of group, and the group by ability interaction can be computed to test for significance and/or likelihood ratio chi-square statistics can be computed to compare nested models including the no effect of group or ability, group only (uniform DIF) and the saturated model including both main effects and the group by ability interaction (nonuniform DIF).

Let $\{y_i, \boldsymbol{x}_i\}$ denote an item response and associated group membership and ability level for subject i, and p_i be the probability of a positive response to the item for the ith subject. Thus, $\boldsymbol{Y}$ is an $N \times 1$ vector, $\boldsymbol{x}_i$ is a 1×4 vector including constant term, group, ability and group by ability interaction, and $\boldsymbol{X}$ is an $N \times 4$ matrix. The logistic regression is a common choice for modeling the odds of the binary (or ordinal) outcome y as a function of $\boldsymbol{x}$.

$$\ln\left(\frac{\boldsymbol{p}}{1-\boldsymbol{p}}\right) = \boldsymbol{X\beta}. \tag{9.12}$$

Since the ith outcome variable y_i is a Bernoulli variable, the log-likelihood function in terms of parameter vector $\boldsymbol{\beta}$ is

$$\ln\ \mathrm{L}(\boldsymbol{\beta}) = \boldsymbol{Y}^T\boldsymbol{X\beta} - \boldsymbol{1}^T \ln\left(1 + e^{\boldsymbol{X\beta}}\right). \tag{9.13}$$

Let $\boldsymbol{p}$ be a column vector of length N with elements $\frac{e^{x_i^T\beta}}{1+e^{x_i^T\beta}}$ and $\boldsymbol{W}$ be an $N \times N$ diagonal matrix with elements $\frac{e^{x_i^T\beta}}{(1+e^{x_i^T\beta})^2}$. Then the score equation (i.e. first derivatives) and the Hessian matrix is given by the following two sets of equations:

$$\frac{d\ln\ \mathrm{L}(\boldsymbol{\beta})}{d\boldsymbol{\beta}} = \boldsymbol{X}^T(\boldsymbol{Y} - \boldsymbol{p}), \tag{9.14}$$

$$\frac{d^2\ \ln\ \mathrm{L}(\boldsymbol{\beta})}{d\boldsymbol{\beta} d\boldsymbol{\beta}^T} = -\boldsymbol{X}^T\boldsymbol{WX}. \tag{9.15}$$

The MLE for $\boldsymbol{\beta}$ satisfies the following estimating equation:

$$\boldsymbol{X}^T(\boldsymbol{Y} - \boldsymbol{p}) = \boldsymbol{0}. \tag{9.16}$$

Equation (9.16) is in this case a system of 4 nonlinear simultaneous equations without a closed form solution. The Newton–Raphson algorithm is commonly used to solve (9.16). The kth step of Newton–Raphson is as follows:

$$\boldsymbol{\beta}^k = \boldsymbol{\beta}^{k-1} - [\boldsymbol{X}^T\boldsymbol{W}\boldsymbol{X}]^{-1}\boldsymbol{X}^T(\boldsymbol{Y} - \boldsymbol{p}). \tag{9.17}$$

The algorithm starts with the initial value $\boldsymbol{\beta}^0$ and iteration continues until convergence. $\boldsymbol{\beta}^0$ is the best guess of $\boldsymbol{\beta}$. For this example, $\boldsymbol{\beta}^0$ can be found by linear regression of $\boldsymbol{Y}$ on $\boldsymbol{X}$.

Significance of the terms in the model can be assessed using Wald's test which in this case is simply the ratio of the MLE to its standard error and is distributed normally. Significance is determined by comparing the absolute value of the ratio to the corresponding upper percentage point of the normal distribution. Alternatively, the likelihood ratio chi-square statistic can be computed as

$$G^2 = -2\ \ln(L_1/L_2), \tag{9.18}$$

where L_1 is without the term of interest and L_2 is for the model that contains the term. G^2 is distributed chi-square with degrees of freedom equal to the difference in the number of terms included in the more complete model relative to the restricted model.

9.7 Assessing DIF for the Bifactor Model

To investigate DIF in the bifactor IRT model, let the multiple-group model of Jeon et al. (2013) be reformulated as a two-group model. Suppose there are $s = S$ specific dimensions in addition to one general dimension (denoted θ_0); the bifactor model is then expressed as:

$$\text{logit}(\pi_{jhi}|\theta_{jh0}, \theta_{jhs}) = G_{jh}\left[a_{jh}(\theta_{jh0}\sigma_{h0} + \mu_{h0}) + b^s_{jh}(\theta_{jhs}\sigma_{hs} + \mu_{hs}) + c_{jh}\right], \tag{9.19}$$

where π_{jhi} is the probability of correct response to item j with binary response category for person i in group $h = 1, 2$; a_{jh}, b^s_{jh}, and c_{jh} are the item discrimination parameters on the general and the sth specific dimensions, and difficulty parameter, respectively for item j in group h; G_{ih} is an indicator variable for the group membership h for subject i. A necessary preliminary step of DIF analysis is to place the subgroups on a common scale of measurement. For this, group differences in latent variables are adjusted for by specifying the group specific mean and standard deviation. Assume that θ's are latent variables that follow

$$(\theta_{ih0}, \theta_{ih1}, \theta_{ih2}, \ldots, \theta_{ihS}) \sim N(0, I_{S+1}) \quad \text{for group} \;\; h = 1, 2. \tag{9.20}$$

A latent variable in the general dimension has mean $= \mu_{h0}$ and $sd = \sigma_{h0}$ for group $h = 1, 2$, and a latent variable on specific dimension s has mean $= \mu_{hs}$ and $sd = \sigma_{hs}$ for group $h = 1, 2$. To identify the model, the means and standard deviations of general and specific dimensions in the reference group ($h = 1$) are set to zero and one, respectively; i.e. $\mu_{W0} = 0$, and $\mu_{Ws} = 0$ for all s, and $\sigma_{W0} = 1$, and $\sigma_{Ws} = 1$ for all s.

Let item parameters for comparison be denoted as $\xi_j' = (a_{jW}, a_{jB}, b_{jW}, b_{jB}, c_{jW}, c_{jB})$. The null hypothesis takes the form

$$H_0 : R\xi_j = \begin{pmatrix} 1 & -1 & 0 & 0 & 0 & 0 \\ 0 & 0 & 1 & -1 & 0 & 0 \\ 0 & 0 & 0 & 0 & 1 & -1 \end{pmatrix} \begin{pmatrix} a_{jW} \\ a_{jB} \\ b_{jW} \\ b_{jB} \\ c_{jW} \\ c_{jB} \end{pmatrix} = \begin{pmatrix} 0 \\ 0 \\ 0 \end{pmatrix}. \tag{9.21}$$

To test this null we can use the Wald statistic

$$Z^2 = (R\hat{\xi}_j)' \left(R(\text{Est.Asy.Var}(\hat{\xi}_j))R'\right)^{-1} (R\hat{\xi}_j) \tag{9.22}$$

as the test statistic, which, under null, is chi-square distributed with df $= 3$ for a given item. Tests for the individual item parameters can be constructed as well.

9.8 Assessing DIF from CAT Data

While the general multiple-group IRT model is the most rigorous approach to this problem, it is not without limitations. Most applications are based on a unidimensional IRT and often data are clearly multidimensional. As shown above, it is possible to generalize the approach to the bifactor model, it further increases the complexity. It also is not possible to use this approach unless the sample size in the new population is quite large, both overall and at the item level. With recent interest in CAT for multidimensional mental health constructs (Gibbons et al. 2016), and concern that there may be DIF in the experience of mental health constructs such as depression in different cultures (Alegría et al. 2016), different indications such as postpartum depression (Kim et al. 2016), and different treatment settings (e.g. a mental health clinic versus a hospital emergency department – (Beiser et al. 2016)), simpler procedures for the assessment of DIF at the item level are needed for continuous quality improvement. For example, Kim et al. (2016) studied 500 women in a perinatal clinic using a multidimensional CAT-based inventory for depression, anxiety and mania to determine the extent to which calibration parameters derived from a psychiatric clinic sample applied to a women in perinatal care, where the possibility of depression (e.g. postpartum depression) is relatively high. There is concern that somatic symptoms of depression may be uninformative in pregnant women where these symptoms are related to their pregnancy rather than their depression. However, the sample size in this target population is small ($N = 500$) and the data are based on CAT administration of an item bank consisting of 1000 items, where each individual will take an average of approximately 36 items (12 for each of the three mental health constructs). Here, all we can do is focus on the most frequently administered items (based on CAT administration) on an item by item basis where for example attention is focused on only those items taken by 50 or more subjects (see (Kim et al. 2016)). The general procedure for assessment of DIF for unidimensional or multidimensional CAT-based tests is as follows.

First, based on the original IRT model calibration (e.g. a bifactor model) in the control sample score the target sample subject's response patterns for the primary domain of interest (e.g. depression). The binary or ordinal response data are then regressed on the estimated scores for each item using a logistic regression model (see Section 9.6). A slope of 1.0 is considered to represent the lower bound on good discrimination (factor loading equivalent of 0.5). The beta coefficient for the estimated severity score based on the control sample calibration in the logistic regression describes the strength of association between the original calibration based severity estimate and the probability of a category increase in the response scale for the target sample subjects. This estimate is equivalent to the slope in an IRT model where the latent variable is known (e.g. the bifactor IRT model for the primary dimension) and can also be expressed as an odds ratio (OR) of 2.72 for slope = 1.0 (i.e. $e^1 = 2.72$). As such, items with ORs < 2.72 have evidence of DIF and do not discriminate well in the target population. Differences in the intercepts of the logistic regression between the two populations can be produced by either differences in the underlying means between the two populations or differences in the amount of severity it takes to shift between categories between the two populations. In this analysis, our focus is on the key question of differences between the two populations in terms of the items' ability to discriminate between high and low levels of the underlying construct of interest (e.g. depression), adjusting for differences in overall mean at both the item and population levels which are absorbed in the intercept of the regression.

As a test of the overall procedure, we can score the target sample subjects using both the original control sample calibration and a new calibration based on the target sample data only. The correlation between the two estimates using all of the items and then eliminating the items exhibiting DIF allow us to determine the extent of the DIF on our estimates of the latent construct of interest and the extent to which we can eliminate bias due to the detection of DIF and elimination of those items from the scoring algorithm. As an example Kim et al. (2016) found that for perinatal women, 15 of the 53 commonly administered mania items from the CAT-MH exhibited DIF as defined here. These items were generally related to symptoms that would be either commonly expected (craving sweets and carbohydrates; diminished interest in sex) or not expected (risk-taking behavior; sexual promiscuity) among perinatal women whether they are manic or not. The correlation between the mania/hypomania severity scores based on the original calibration and the perinatal calibration was $r = 0.73$. Removal of the items that exhibited DIF increased the correlation to $r = 0.93$, validating this simple DIF procedure.

Unlike other simple tests such as the Mantel–Haenszel procedure, the logistic regression model preserves the multidimensional nature of the IRT model used in the calibration and unlike IRT-based procedures, can be used effectively in much smaller sample sizes since it is applied individually to each item. As such it is similar to traditional methods for DIF analyses such as the Mantel–Haenszel test with the improvement that it relies upon more sophisticated estimates of severity scores for the target group based on the original calibration data and can accommodate both unidimensional and multidimensional IRT model parameterizations.

10

Estimating Respondent Attributes

I couldn't claim that I was smarter than sixty-five other guys–but the average of sixty-five other guys, certainly

(Source: Richard Feynman)

10.1 Introduction

For most item response theory (IRT) models, the estimation of a person's ability or abilities and estimation of the parameters associated with the items are two separate operations. Joint estimation is complicated by the fact that for each individual that is added there is at least one parameter that needs to be estimated. This is the so-called nuisance parameter problem in which the parameter space grows linearly with the number of subjects. It is important to note that this is the analogous problem in generalized mixed-effects regression models where the unit-specific parameters are estimated separately from the structural model parameters (both fixed and random effects variance components) as Bayes or empirical Bayes estimates of the unit (e.g. subject) specific trend parameters (Hedeker and Gibbons 2006). In this chapter, we present maximum likelihood, Bayes maximum *a posteriori* (MAP), and Bayes expected *a posteriori* (EAP) ability estimation procedures. We also draw a connection between classical test theory and IRT in a discussion of domain scores, and finally, we discuss nonparametric estimation of the form of the ability distribution. Finally, we also describe ability estimation for multidimensional IRT models, both unrestricted and bifactor models.

Item Response Theory, First Edition. R. Darrell Bock and Robert D. Gibbons.

10.2 Ability Estimation

In Sections 10.2.1–10.2.3, we highlight the three general approaches to IRT ability estimation (Bock and Aitkin 1981). We present results for binary outcomes and then briefly describe the more general case of polytomous data.

10.2.1 Maximum Likelihood

We assume that the item parameters are known based on a previous calibration and for subject i we have the response pattern $\boldsymbol{x}_i = \left[x_{ij}\right], j = 1, 2, \dots, n$. The maximum likelihood estimator (MLE) $\hat{\theta}_i$ is the solution to

$$\sum_{j}^{n} a_j B_{ij} = 0, \tag{10.1}$$

which can be obtained using the Newton–Raphson algorithm with second derivative

$$\sum_{j}^{n} a_j^2 \left(B_{ij} C_{ij} - W_j(\theta)\right)_{\theta=\hat{\theta}_i}, \tag{10.2}$$

where

$$B_{ij} = \frac{\left[x_{ij} - \Phi_j(\theta)\right] \phi_j(\theta)}{\Phi_j(\theta)\left[1 - \Phi_j(\theta)\right]}, \tag{10.3}$$

$$C_{ij} = -z_j(\theta) - \phi_j(\theta)\left(\frac{1}{\Phi_j(\theta)} - \frac{1}{1 - \Phi_j(\theta)}\right), \tag{10.4}$$

and

$$W_j(\theta) = \frac{\phi_j^2(\theta)}{\Phi_j(\theta)\left[1 - \Phi_j(\theta)\right]}. \tag{10.5}$$

For starting values, we can use

$$\hat{\theta}_i^{(0)} = \Phi^{-1}\left(\frac{\sum_j^n r_j}{n}\right). \tag{10.6}$$

The large sample variance of the ability estimate is

$$V(\hat{\theta}_i) = \left[\sum_{j}^{n} a_j^2 W_j(\theta)\right]^{-1}_{\theta=\hat{\theta}_i}. \tag{10.7}$$

A distinct advantage of the IRT-based ability estimates over traditional number correct or simple total scores is the variance estimate that expresses the precision of the test score. In longitudinal studies, this permits the assessment of change at the individual respondent level, where changes in the ability estimate of less than twice the standard deviation of the ability estimate do not represent statistically significant change for that individual; whereas larger changes are statistically significant.

10.2.2 Bayes MAP

The Bayes modal or MAP estimator of ability $\hat{\theta}_i$ is the solution of

$$\left[\sum_j^n a_j B_{ij} + \frac{\partial g(\theta)}{\partial \theta}\right]_{\theta=\hat{\theta}_i} = 0. \tag{10.8}$$

for any $g(\theta)$ with finite mean and variance. The Newton–Raphson algorithm is easy to implement where the required second derivative is now:

$$\left[\sum_j^n a_j^2 \left(B_{ij}C_{ij} - W_{ij}\right) + \frac{\partial^2 g(\theta)}{\partial^2 \theta}\right]_{\theta=\hat{\theta}_i}, \tag{10.9}$$

beginning from $\theta_i^{(0)}$. The large sample variance of the MAP estimator is:

$$V(\hat{\theta}_i) = \left[\sum_j^n a_j^2 W_{ij} + \frac{\partial^2 g(\theta)}{\partial^2 \theta}\right]^{-1}_{\theta=\hat{\theta}_i}. \tag{10.10}$$

10.2.3 Bayes EAP

To derive the EAP ability estimator and its variance, recall that under the assumption of conditional independence, the probability of subject i responding in pattern $\boldsymbol{x}_i = \left[x_{ij}\right], j = 1, 2, \ldots, n$ conditional on ability θ_i is

$$P(\boldsymbol{x} = \boldsymbol{x}_i \mid \theta_i) = \sum_j^n \left[\Phi_j(\theta_i)\right]^{x_{ij}} \left[1 - \Phi_j(\theta_i)\right]^{1-x_{ij}}. \tag{10.11}$$

The unconditional probability for a randomly selected subject sampled from a population with continuous ability distribution $g(\theta)$ is

$$P(\boldsymbol{x} = \boldsymbol{x}_i) = \int_{-\infty}^{\infty} P(\boldsymbol{x} = \boldsymbol{x}_i \mid \theta) g(\theta) d\theta. \tag{10.12}$$

Assuming normality of $g(\theta)$, the unconditional probability can be approximated using Gauss–Hermite quadrature (Stroud and Secrest 1966) as

$$\sum_k^q P(\boldsymbol{x} = \boldsymbol{x}_i \mid X_k) A(X_k), \tag{10.13}$$

where X_k is a quadrature node and $A(X_k)$ is the corresponding quadrature weight. The EAP ability estimator is the conditional expectation of θ given $\boldsymbol{x} = \boldsymbol{x}_i$,

$$\hat{\theta} = E(\theta \mid \boldsymbol{x}_i) = \frac{\int_{-\infty}^{\infty} \theta g(\theta) \prod_j^n [\Phi_j(\theta_i)]^{x_{ij}} [1 - \Phi_j(\theta_i)]^{1-x_{ij}} d\theta}{\int_{-\infty}^{\infty} g(\theta) \prod_j^n [\Phi_j(\theta_i)]^{x_{ij}} [1 - \Phi_j(\theta_i)]^{1-x_{ij}} d\theta}, \tag{10.14}$$

which can be approximated as

$$\hat{\theta} \approx \frac{\sum_k^q X_k L_i(X_k) A(X_k)}{P_i}, \tag{10.15}$$

where

$$\begin{aligned} P_i &= \sum_k^q \left[\prod_j^n [\Phi_j(X_k)]^{x_{ij}} [1 - \Phi_j(X_k)]^{1-x_{ij}} \right] A(X_k) \\ &= \sum_k^q L_i(X_k) A(X_k). \end{aligned}$$

The variance of the EAP ability estimator is approximated as

$$V(\hat{\theta}_i) = \sum_k^q \frac{(X_k - \hat{\theta}_i)^2 L_i(X_k) A(X_k)}{P_i}. \tag{10.16}$$

As noted by Bock and Aitkin (1981), the maximum likelihood estimator has the disadvantage of not providing finite values for constant response patterns and unlikely patterns may fail to converge. The two Bayes estimators are free from these problems; however, the MAP estimator requires a proper continuous density function for the prior distribution. The EAP estimator implemented by quadrature can use both continuous and discrete prior distributions. In practical applications, the EAP estimator is generally used.

10.2.4 Ability Estimation for Polytomous Data

In the unidimensional case, the ability estimate for polytomous data is obtained by substituting the marginal probability for the polytomous model for the marginal probability for a binary response model in Eqs. (10.11) and (10.12). The marginal probability for subject i is

$$\begin{aligned} P_i &= \int_\theta \left(\prod_{j=1}^n \prod_h^{m_j} \left(\Phi_{jh}(\theta) - \Phi_{jh-1}(\theta) \right)^{\delta_{ijh}} \right) g(\theta) d\theta \\ &= \int_\theta L_i g(\theta) d\theta, \end{aligned} \tag{10.17}$$

where $\delta_{ijh} = 1$ if person i responds in category h for item j and $\delta_{ijh} = 0$ otherwise. For computation, we approximate the marginal probability as

$$P_i = \sum_k^q \left(\prod_j^n \prod_h^{m_j} (\Phi_{jh}(X_k) - \Phi_{jh-1}(X_k))^{\delta_{ijh}} \right) A(X_k)$$
$$= \sum_k^q L_i(X_k)A(X_k). \quad (10.18)$$

10.2.5 Ability Estimation for Multidimensional IRT Models

In the multidimensional case, there are d latent variables or dimensions, and the underlying response strength for item j and subject i is

$$y_{ij} = \alpha_{j1}\theta_{i1} + \alpha_{j2}\theta_{i2} + \cdots + \alpha_{jd}\theta_{id} + \varepsilon_j. \quad (10.19)$$

The vector of latent variables is denoted

$$\boldsymbol{\theta}_i = (\theta_{i1}, \theta_{i2}, \ldots, \theta_{id}). \quad (10.20)$$

Letting $\mathbf{v}_j$ denote the factor loadings and threshold for item j, the probability of a response in category h of an m category item conditional on $\boldsymbol{\theta}$ is

$$P_{jh}(\boldsymbol{\theta}_i) = P(u_j \mid \mathbf{v}_j, \boldsymbol{\theta}_i) = \prod_h^{m_j} P_{jh}^{u_{jh}}(\boldsymbol{\theta}_i), \quad (10.21)$$

where $u_{jh} = 1$ if the response is in category h and 0 otherwise. Assuming conditional independence on $\boldsymbol{\theta}$, the conditional probability for the n-item response pattern $\boldsymbol{u}_i$ conditional on $\boldsymbol{\theta}$ is

$$P(\boldsymbol{u}_i \mid \boldsymbol{\theta}) = \prod_j^n \prod_h^{m_j} P_{jh}^{u_{ijh}}(\boldsymbol{\theta}). \quad (10.22)$$

The unconditional or marginal probability of $\boldsymbol{u}_i$ in the population is the d-fold integral

$$P_i = \int_{\boldsymbol{\theta}} L_i(\boldsymbol{\theta})g(\boldsymbol{\theta})d\boldsymbol{\theta}, \quad (10.23)$$

where

$$L_i(\boldsymbol{\theta}) = \prod_j^n \prod_h^{m_j} P_{jh}^{u_{ijh}}(\boldsymbol{\theta}). \quad (10.24)$$

In the multidimensional case, the EAP estimator is the mean vector of the posterior distribution of $\boldsymbol{\theta}$ given $\boldsymbol{u}_i$

$$\hat{\boldsymbol{\theta}}_i = \frac{1}{P_i} \int_{\boldsymbol{\theta}} \boldsymbol{\theta} L_i(\boldsymbol{\theta})g(\boldsymbol{\theta})d\boldsymbol{\theta}, \quad (10.25)$$

which can be evaluated numerically using adaptive Gauss–Hermite quadrature under the assumption of multivariate normality. The posterior covariance matrix is

$$\hat{\boldsymbol{\Sigma}}_{\theta|\boldsymbol{u}_i} = \frac{1}{P_i}\int_{\theta}(\theta - \hat{\theta}_i)^2 L_i(\theta)g(\theta)d\theta, \tag{10.26}$$

where the square roots of the diagonal elements are the standard errors of the EAP estimates.

10.2.6 Ability Estimation for the Bifactor Model

Under the bifactor restriction, each item loads on the primary dimension and only one of the $d - 1$ sub-domains $\boldsymbol{\theta}_v^* = \left[\theta_1\ \theta_v\right]'$. Here we have different forms of the EAP estimators for the primary and secondary dimensions. For the primary dimension, the EAP estimator is

$$\hat{\theta}_{1i} = E(\theta_{1i} \mid \boldsymbol{u}_i, \theta_{2i}, \dots, \theta_{di}) = \sum \frac{1}{P_i}\int_{\theta_1}\theta_{1i}\left[\prod_{v=2}^{d}\int_{\theta_v} L_{vi}(\boldsymbol{\theta}_v^*)g(\theta_v)d\theta_v\right]g(\theta_1)d\theta_1. \tag{10.27}$$

The posterior variance of $\hat{\theta}_{1i}$ is

$$V(\theta_{1i} \mid \boldsymbol{u}_i, \theta_{2i}, \dots, \theta_{di}) = \sum \frac{1}{P_i}\int_{\theta_1}(\theta_{1i} - \hat{\theta}_{1i})^2\left[\prod_{v=2}^{d}\int_{\theta_v} L_{vi}(\boldsymbol{\theta}_v^*)g(\theta_v)d\theta_v\right]g(\theta_1)d\theta_1. \tag{10.28}$$

These estimators can be computed using Gauss–Hermite quadrature under the assumption of multivariate normality of the latent variables. In some cases, we are also interested in scoring the secondary domains of interest as well. The EAP estimator for sub-domain v ($v = 2, \dots, d$) is

$$\hat{\theta}_{vi} = E(\theta_{vi} \mid \boldsymbol{u}_i, \theta_{1i}) = \sum \frac{1}{P_i}\int_{\theta_v}\theta_{vi}\left[\int_{\theta_1} L_{vi}(\boldsymbol{\theta}_v^*)g(\theta_1)d\theta_1\right]g(\theta_v)d\theta_v. \tag{10.29}$$

The corresponding posterior variance is

$$V(\theta_{vi} \mid \boldsymbol{u}_i, \theta_{1i}) = \sum \frac{1}{P_i}\int_{\theta_v}(\theta_{vi} - \hat{\theta}_{vi})^2\left[\int_{\theta_1} L_{vi}(\boldsymbol{\theta}_v^*)g(\theta_1)d\theta_1\right]g(\theta_v)d\theta_v. \tag{10.30}$$

10.2.7 Estimation of the Ability Distribution

Bock and Aitkin (1981) suggested a novel approach to empirically estimating the prior distribution of the underlying latent ability or impairment variable, thereby relaxing the assumption of univariate or multivariate normality. Their approach involves characterizing the continuous ability distribution as a discrete distribution on a finite number of equally spaced points, essentially a histogram. More specifically, an estimate of the density of the ability distribution at point X_k is the posterior density given the data. This is a much more general solution than estimating the parameters of the underlying assumed normal distribution originally described by Andersen and Madsen (1977) and Sanathanan and Blumenthal (1978).

Beginning from a normal prior distribution, we can estimate the posterior density as

$$g(X_k) \approx \frac{\sum_i^N L_i(X_k)A(X_k)}{\sum_t^q \sum_i^N L_i(X_t)A(X_t)}. \tag{10.31}$$

$g(X_k)$ can then be substituted for $A(X_k)$ in the estimating equations. $g(X_k)$ can be updated on each iteration until joint convergence of the item parameters and the quadrature weights. At convergence, $g(X_k)$ represents a histogram of the empirical prior distributions, the precision of which can be increased by expanding the number of quadrature points. The approach is similar to the nonparametric maximum likelihood estimation of a mixing distribution described by Laird (1978).

10.2.8 Domain Scores

In classical test theory, a test is made up of a sample of items from a specific domain. If the items are random samples from the population of all items in the domain then the percent correct or total score provide an estimate of the domain score. However, this is rarely if ever the case. The same is not true however, for IRT. In IRT if the domain is represented as a large set of calibrated items, then IRT provides an alternative estimator of the domain score that does not require a random sample of items from the domain. This has profound implications for reporting test scores for ability domains in educational measurement or patient reported outcomes in health measurement. Bock et al. (1997) discuss how IRT scale scores can be transformed into an estimate of the domain score. The necessary conditions are (i) there is a bank of n items that can be regarded as a probability sample from the domain consisting of arbitrary allocations of items to strata with corresponding sampling weights which reflect the domain proportions of items per strata; (ii) the item bank can be calibrated using an appropriate unidimensional or multidimensional IRT model; (iii) item parameters are available from a large sample of respondents from a specified population; and (iv) a test has been created with items distributed across the underlying dimensions of the domain.

The transformation from IRT scale score $\hat{\theta}_i$ to domain expected number correct score is

$$d(\hat{\theta}_i) = \sum_j^n w_j P_j(\hat{\theta}_i) \bigg/ \sum_j^n w_j, \tag{10.32}$$

and $D(\hat{\theta}_i) = 100d(\hat{\theta}_i)/n$ is the domain percent correct score, where w_j is the sampling weight and $P_j(\hat{\theta}_i)$ is the response function probability for item j conditional on the underlying ability θ or abilities $\boldsymbol{\theta}$. The uncertainty (standard error) in the domain percentage score is

$$\text{SE}(D) = \frac{100\text{SE}(\hat{\theta}_i)}{n \sum_j^n w_j} \sqrt{\sum_j^n w_j^2 \left[\frac{\partial P_j(\theta_i)}{\partial \theta_i} \right]_{\hat{\theta}_i}^2}, \tag{10.33}$$

where $[\partial P_j(\theta)/\partial \theta]_{\hat{\theta}}$ is the derivative of the response function j with respect to θ evaluated at $\hat{\theta}_i$, and $\text{SE}(\hat{\theta}_i)$ is the standard error of the IRT scale score.

As noted by (Bock 1997), there are several advantages of the IRT estimated domain score relative to the traditional classical test theory estimator. First, the items do not have to be a random probability sample from the domain. Relaxing this requirement opens the door to adaptive testing in domain-referenced testing, where an optimal set of items are administered to each subject, targeted to their ability or impairment that is learned through the adaptive testing process (see Chapter 8). Second, the IRT approach to deriving the domain score generalizes directly to multidimensional tests or tests that have a bifactor structure (see Chapter 6) and/or to graded response items (see Chapter 5). Third, as illustrated by Bock et al. (1997), the IRT estimator has greater accuracy than the traditional test score for predicting domain scores.

For a multidimensional model, assume a normal item response function $P_j(\theta) = \Phi(y_j)$ where y is the standard normal deviate

$$y_j = \gamma_j + \alpha_{1j}\theta_{1i} + \alpha_{2j}\theta_{2i} + \cdots + \alpha_{dj}\theta_{di}, \tag{10.34}$$

where γ_j is the item threshold and the α are the factor loadings. The estimated domain score will represent a composite of the underlying dimensions. A better choice is to use a bifactor model and use the primary dimension (θ_1) to estimate the domain score. The bifactor model preserves the multidimensionality of the item bank, but at the same time provides a single valued index of the primary domain of interest.

11

Multiple Group Item Response Models

We may at once admit that any inference from the particular to the general must be attended with some degree of uncertainty, but this is not the same as to admit that such inference cannot be absolutely rigorous, for the nature and degree of the uncertainty may itself be capable of rigorous expression.

(Source: Sir Ronald Fisher)

11.1 Introduction

To this point, our primary focus has been on the analysis of data arising from a single population, with two notable exceptions. In Chapter 9, we discuss differential item functioning (DIF) where the item parameters are compared between a reference group and a focal group, and in Section 10.2.7 where we accommodate non-normality of the underlying latent distribution by empirically estimating the prior distribution. This non-normality could be produced by the mixing of different populations in the calibration sample. As we move beyond the single sample case, there are numerous different applications of item response theory (IRT) where the data are clustered within different manifest or latent subpopulations, or cases in which the sample may be stratified in terms of characteristics of the respondents. In these cases, there may be several different goals. In the case of DIF, we are interested in determining whether the item parameters may be different in the different populations. In the case of an empirical prior distribution, we are interested in obtaining unbiased estimates of item parameters that accommodate the mixing of different populations in the ability distribution. These are two very different goals and lead to very different statistical solutions. Multiple group IRT also draws connections between IRT and multivariate generalizations of probit analysis of multivariate binary data that has grown

Item Response Theory, First Edition. R. Darrell Bock and Robert D. Gibbons.

out of statistical and biostatistical applications. Similarly, there are equivalences between IRT and multi-level models that permit IRT to be used in the context of more general mixed-effects regression models. Different versions of these models result depending on whether the group effects are related to the items or to the subject-level parameters (i.e. group effects on the parameters of the ability distribution). In the following, we consider several different approaches to the treatment of multiple groups in IRT models and their applications.

11.2 IRT Estimation When the Grouping Structure Is Known: Traditional Multiple Group IRT

Bock et al. (1982) suggested a method by which IRT could be applied to multiple-matrix sampling (MMS) designs (Lord 1962). MMS designs provide estimation of population parameters without the need to estimate individual level parameters. Their suggested approach applies an IRT model to groups of individuals defined by relevant educational, demographic, or other characteristics that stratify the population into meaningful subpopulations. Mislevy (1983) expanded the idea of group-level IRT models in considerable detail, providing a better understanding of the relationship between individual level and group level IRT models and a general approach to parameter estimation for group level IRT models.

Following Mislevy (1983), we let γ_j be the fixed threshold for item j, and that subject i has fixed ability θ_i. Let ε be a continuous random variable with mean zero and density function f_j. The manifest binary response x_i is assumed to depend on θ_i and γ_j as:

$$\begin{aligned} x_{ij} &= 1 \quad \text{if } \theta_i + \varepsilon_j > \gamma_j, \text{ or equivalently, } \varepsilon_j > \gamma_j - \theta_i \\ x_{ij} &= 0 \quad \text{otherwise.} \end{aligned} \tag{11.1}$$

The probability of observing a correct response is then

$$\begin{aligned} P(x_{ij} = 1 \mid \theta_i, \gamma_j) &= \int_{-\infty}^{\infty} f_j(\varepsilon) d\varepsilon \\ &= F_j(\theta_i - \gamma_j). \end{aligned} \tag{11.2}$$

For the normal ogive model

$$\begin{aligned} F_j(\theta_i - \gamma_j) &= \frac{1}{\sigma_j \sqrt{2\pi}} \int_{-\infty}^{\theta_i - \gamma_j} \exp\left(-\frac{\varepsilon^2}{2\sigma_j^2}\right) d\varepsilon \\ &= \frac{1}{2\pi} \int_{-\infty}^{(\theta_i - \gamma_j)/\sigma_j} \exp\left(-t^2/2\right) dt \\ &= \Phi\left[\left(\theta_i - \gamma_j\right)/\sigma_j\right]. \end{aligned} \tag{11.3}$$

In the multiple group case, there are a number of subpopulations indexed k in which θ is distributed with potentially different continuous density functions g_i. The g_i are typically assumed to be identical in shape and dispersion, but differ in location according to their mean $\overline{\theta}_k$, where g_0 denotes the density function with a mean of zero. In this case, (11.1) becomes

$$\begin{aligned} x_{ij} &= 1 \quad \text{if } \theta_{ik} + \varepsilon_j > \gamma_j, \text{ or equivalently, if, } \eta_j \equiv \varepsilon_j + (\theta_{ik} - \overline{\theta}_k) > \gamma_j - \overline{\theta}_k \\ x_{ij} &= 0 \quad \text{otherwise.} \end{aligned} \tag{11.4}$$

Let h_j denote the density function of η_j, then the probability of a correct response to item j from a randomly selected subject from subpopulation k is

$$\begin{aligned} P(x_{kj} = 1 \mid \overline{\theta}_i, \gamma_j) &= \int_{\gamma_j - \overline{\theta}_k}^{\infty} h_j(\eta) d\eta \\ &\equiv H_j(\overline{\theta}_k - \gamma_j). \end{aligned} \tag{11.5}$$

Note that F_j represents the subject-level response curve for item j and H_j represents the group-level response curve for item j, which is the expected value of the subject-level model for subjects in group k.

For the two-parameter normal model with normal ability density functions g_i, the group-level item-response density functions are also normal. Let γ_j and σ_j be the threshold and standard deviation parameters for item j in the subject-level model, and σ be the common standard deviation in the ability distributions in the group-level model. η_j is then normally distributed with mean zero and standard deviation $\sqrt{\sigma_j^2 + \sigma^2}$. The probability of a correct response to item j from a randomly selected subject from group k is

$$\begin{aligned} P(x_{kj} = 1 \mid \overline{\theta}_k, \gamma_j) &= \int_{-\infty}^{\overline{\theta}_k - \gamma_j} h_j(\eta) d\eta \\ &= \frac{1}{\sqrt{2\pi}} \frac{1}{\sqrt{\sigma_j^2 + \sigma^2}} \int_{-\infty}^{\overline{\theta}_k - \gamma_j} \exp\left[-\frac{1}{2} \left(\frac{\eta}{\sqrt{\sigma_j^2 + \sigma^2}} \right)^2 \right] d\eta \\ &= \frac{1}{2\pi} \int_{-\infty}^{z} \exp\left(-t^2/2\right) dt \\ &= \Phi(z), \end{aligned} \tag{11.6}$$

where

$$z = \frac{\overline{\theta}_k - \gamma_j}{\sqrt{\sigma_j^2 + \sigma^2}} . \tag{11.7}$$

γ_j and σ_j are the threshold and standard deviation of item j in the subject-level model and γ_j and $\sqrt{\sigma_j^2 + \sigma^2}$ are the item threshold and standard deviation in the group-level model.

Bock and Zimowski (1997) provide a general overview of parameter estimation for the multiple group multiple category model (also see Section 5.7). The likelihood equations are in a form similar to Eq. (5.9); however, there is an additional summation over groups and the conditional likelihood L_{ki} and marginal probability P_{ki} now carry the group subscript k. The likelihood equations for the population parameters under the assumption of normality are

$$\frac{\partial \log L_N}{\partial \mu_k} = \sigma_k^{-2} \sum_i^N \int_{-\infty}^{\infty} (\theta - \mu_k) \frac{L_{ki}(\theta)}{\overline{P}_{ki}} g(\theta) d\theta = 0, \tag{11.8}$$

and

$$\frac{\partial \log L_N}{\partial \sigma_k^2} = -\frac{1}{2} \sum_i^N \int_{-\infty}^{\infty} \left[\sigma_k^2 - (\theta - \mu_k)^2\right] \frac{L_{ki}(\theta)}{\overline{P}_{ki}} g(\theta) d\theta = 0, \tag{11.9}$$

where

$$g(\theta) = \frac{1}{\sqrt{2\pi}\sigma} \exp\left(\frac{-(\theta - \mu_k)^2}{2\sigma^2}\right). \tag{11.10}$$

Given provisional values of the item parameters, the likelihood equation for μ_k is

$$\hat{\mu}_k = \frac{1}{N_k} \sum_i^N \overline{\theta}_{ki}, \tag{11.11}$$

where

$$\overline{\theta}_{ki} = \frac{1}{\overline{P}_{ki}} \int_{-\infty}^{\infty} \theta L_{ki}(\theta) g_k(\theta) d\theta \tag{11.12}$$

is the posterior mean of θ, given $\boldsymbol{x}_{ki}$. The likelihood equation for the pooled variance σ^2 when the item parameters are known is

$$\hat{\sigma}^2 = \frac{1}{N_k} \sum_i^N \left[(\overline{\theta}_{ki} - \hat{\mu}_k)^2 + \sigma^2_{\theta|\boldsymbol{x}_{ki}}\right], \tag{11.13}$$

where

$$\sigma^2_{\theta|\boldsymbol{x}_{ki}} = \frac{1}{\overline{P}_{ki}} \int_{-\infty}^{\infty} (\theta - \overline{\theta}_{ki})^2 L_{ki}(\theta) g_k(\theta) d\theta \tag{11.14}$$

is the posterior variance of θ, given $\boldsymbol{x}_{ki}$.

11.2.1 Example

Cai et al. (2011) present a multiple group IRT example using data from the Program for International Student Assessment (PISA). PISA is a worldwide evaluation of 15-year-old school pupils' scholastic performance, performed first in 2000 and repeated every three years. It is coordinated by the Organization for Economic Co-operation and Development (OECD), with a view to improving educational policies and outcomes. The dataset contains responses by a subset of students to 14 items from math booklet 1. The grouping variable is Country, United States (US) versus United Kingdom (UK). There are 358 students in the US group and 889 in the UK group. Ten of the items are binary and four are graded, three with 3 graded categories and one with 4 graded categories. Cai et al. (2011) used a mixture of 2PL and general partial credit (GPC) models to analyze these 14 items. Table 11.1 displays the 2PL parameter estimates that are fixed to be the same between the two groups (US and UK). Table 11.2 displays the GPC parameter estimates that are fixed to be the same between the two groups (US and UK).

Using the US as the comparator and fixing $\mu_1 = 0$ and $\sigma_1^2 = 1$, the UK has lower mean $\mu_2 = -0.13$ (SE = 0.09), and lower variance $\sigma_2^2 = 0.85$ (SE = 0.18).

Table 11.1 2PL model item parameter estimates common for US and UK logit: $a\theta + c$ or $a(\theta - b)$.

Item	a	SE	c	SE	b	SE
Cube1	1.11	0.17	0.47	0.12	−0.42	0.12
Cube3	1.57	0.22	1.48	0.17	−0.94	0.14
Cube4	1.12	0.18	−1.04	0.13	0.93	0.20
Farms1	2.14	0.33	−0.02	0.22	0.01	0.10
Farms4	0.79	0.14	0.19	0.10	−0.24	0.13
Walking1	2.68	0.44	−2.18	0.30	0.81	0.18
Apples1	1.46	0.22	0.21	0.15	−0.14	0.10
Apples2	2.79	0.50	−3.05	0.33	1.09	0.22
Grow1	1.24	0.20	−0.06	0.13	0.05	0.11
Grow3	1.46	0.22	0.26	0.15	−0.18	0.10

Table 11.2 GPC model item parameter estimates (SE) common for US and UK logit: $a[k(\theta - b) + \sum dk]$.

Item	a	b	d_1	d_2	d_3	d_4
Walking3	1.65 (0.25)	1.52 (0.27)	0.00	0.69 (0.13)	−0.30 (0.09)	−0.39 (0.12)
Apples3	2.24 (0.38)	1.50 (0.27)	0.00	0.07 (0.07)	−0.07 (0.07)	
Continent	1.64 (0.32)	1.32 (0.25)	0.00	0.55 (0.08)	−0.55 (0.08)	
Grow2	0.69 (0.10)	−0.66 (0.14)	0.00	0.94 (0.19)	−0.94 (0.19)	

11.3 IRT Estimation When the Grouping Structure Is Unknown: Mixtures of Gaussian Components

Mislevy (1984) introduced the idea of estimating the population parameters of the distribution of θ without estimating θ for each individual. He considered four cases, (i) a nonparametric approximation, (ii) a normal model, (iii) a resolution into Gaussian components, and (iv) a beta-binomial model. We discuss the nonparametric approximation in Section 10.2.7, and the normal model in Section 11.2. Here we further explore the model in which the population distribution of θ is resolved into a mixture of Gaussian components and describe estimation of the parameters of the mixing distribution.

Following Bock and Aitkin (1981), consider binary responses to a test with n items for each of N examinees. Let $x_{ij} = 1$ if examinee i responds correctly to item j, and $x_{ij} = 0$ otherwise. Assume:

$$\begin{aligned} P(x_{ij} = 1|\theta_i) &= \Phi_j(\theta_i) = \int_{-\infty}^{z_{ij}(\theta_i)} \frac{1}{\sqrt{2\pi}} e^{-\frac{t^2}{2}} dt, \\ P(x_{ij} = 0|\theta_i) &= 1 - \Phi_j(\theta_i). \end{aligned} \tag{11.15}$$

In (11.15), $z_{ij}(\theta_i)$ is a linear function of a one-dimensional ability θ_i; therefore, $z_{ij}(\theta_i) = c_j + a_j\theta_i$. Let the unique response patterns observed in N examinees be indexed by $l = 1, 2, \ldots, s$ and let r_l be the number of subjects responding in pattern l, and $\sum_{l=1}^{s} r_l = N$. Under the assumption of conditional independence for items, the probability of response pattern $x_l = (x_{l1}, \ldots, x_{ln})$ conditional on ability θ_i, is given by

$$p_l(\theta_i) = P(x = x_l|\theta_i) = \prod_{j=1}^{n} \Phi_j(\theta_i)^{x_{lj}}(1 - \Phi_j(\theta_i))^{1-x_{lj}}. \tag{11.16}$$

Following Mislevy (1984), we include parameters for the distribution of θ into the model. Here, we assume that the distribution of θ is a mixture of t normal components, with means $\mu_1, \ldots, \mu_t$ and variances $\sigma_1^2, \ldots, \sigma_t^2$. Let $\pi_1, \ldots, \pi_t$ be the unknown proportions of the mixture. The marginal probability of x_l is expressed as:

$$p_l = P(x = x_l) = \sum_{c=1}^{t} \pi_c \int L_l(\theta) g_c(\theta) d\theta, \tag{11.17}$$

where

$$L_l(\theta) = p_l(\theta) \text{ and } g_c(\theta) = \frac{1}{\sqrt{2\pi\sigma_c^2}} \exp\left(-\frac{(\theta - \mu_c)^2}{2\sigma_c^2}\right). \tag{11.18}$$

The likelihood of a random sample of size N, consisting of r_l examinees responding in pattern $l = 1, \ldots, s$, is given by

$$\log L = \sum_{l=1}^{s} r_l \log p_l. \tag{11.19}$$

Now, subject to a constraint $\sum_{c=1}^{t} \pi_c = 1$ we want to maximize $\log L$ with respect to the IRT parameters contained in the likelihood and parameters of the mixture distribution. Computationally, we proceed by using a two-stage maximization. To obtain the maximizers for the parameters of the mixture distribution, assume that the IRT parameters are fixed and maximize $\log L$ with respect to the mixture distribution parameters, and to obtain the maximizers for the IRT parameters, assume that the mixture distribution parameters are fixed and maximize $\log L$ with respect to the IRT parameters. In the following, we describe the maximization steps.

11.3.1 The Mixture Distribution

Let

$$\log L' = \sum_{l=1}^{s} r_l \log\left(\sum_{c=1}^{t} \pi_c \int L_l(\theta) g_c(\theta) d\theta \right) + \lambda \left(1 - \sum_{c=1}^{t} \pi_c \right). \tag{11.20}$$

Assume that the IRT parameters contained in $L_l(\theta)$ are fixed and maximize $\log L'$ with respect to the parameters of mixture distribution $g_c(\theta)$. Taking derivatives with respect to π_c, μ_c, σ_c^2 for $c = 1, 2, \ldots, t$, respectively, and setting them to zero yields,

$$\frac{\partial \log L'}{\partial \pi_c} = \sum_{l=1}^{s} \frac{r_l}{p_l} \int L_l(\theta) g_c(\theta) d\theta - \lambda = 0, \tag{11.21}$$

$$\frac{\partial \log L'}{\partial \mu_c} = \sum_{l=1}^{s} \frac{r_l}{p_l} \pi_c \int L_l(\theta) g_c(\theta) \frac{\partial \log g_c(\theta)}{\partial \mu_c} d\theta = 0, \tag{11.22}$$

$$\frac{\partial \log L'}{\partial \sigma_c^2} = \sum_{l=1}^{s} \frac{r_l}{p_l} \pi_c \int L_l(\theta) g_c(\theta) \frac{\partial \log g_c(\theta)}{\partial \sigma_c^2} d\theta = 0. \tag{11.23}$$

If $\sigma_c^2 = \sigma^2, c = 1, 2, \ldots, t$, then

$$\frac{\partial \log L'}{\partial \sigma^2} = \sum_{l=1}^{s} \frac{r_l}{p_l} \sum_{c=1}^{t} \pi_c \int L_l(\theta) g_c(\theta) \frac{\partial \log g_c(\theta)}{\partial \sigma^2} d\theta = 0. \tag{11.24}$$

Multiplying (11.21) by π_c and summing over c we find that $\lambda = N$. Using the expressions:

$$\frac{\partial \log g_c(\theta)}{\partial \mu_c} = \frac{\theta - \mu_c}{\sigma_c^2} \quad \text{and} \quad \frac{\partial \log g_c(\theta)}{\partial \sigma_c^2} = -\frac{1}{2\sigma_c^2} + \frac{(\theta - \mu_c)^2}{2\sigma_c^4} \tag{11.25}$$

and substituting them into (11.22) and (11.23) we obtain:

$$\hat{\pi}_c = \frac{1}{N}\sum_{l=1}^{s} r_l \int \left(\frac{\hat{\pi}_c L_l(\theta) g_c(\theta)}{p_l}\right) d\theta, \tag{11.26}$$

$$\hat{\mu}_c = \frac{1}{N\hat{\pi}_c}\sum_{l=1}^{s} r_l \int \theta \left(\frac{\hat{\pi}_c L_l(\theta) g_c(\theta)}{p_l}\right) d\theta, \tag{11.27}$$

$$\hat{\sigma}_c^2 = \frac{1}{N\hat{\pi}_c}\sum_{l=1}^{s} r_l \int (\theta - \hat{\mu}_c)^2 \left(\frac{\hat{\pi}_c L_l(\theta) g_c(\theta)}{p_l}\right) d\theta. \tag{11.28}$$

If $\sigma_c^2 = \sigma^2, c = 1, 2, \ldots, t$, then

$$\hat{\sigma}^2 = \frac{1}{N}\sum_{l=1}^{s}\sum_{c=1}^{t} r_l \int (\theta - \hat{\mu}_c)^2 \left(\frac{\hat{\pi}_c L_l(\theta) g_c(\theta)}{p_l}\right) d\theta. \tag{11.29}$$

For solution by quadrature over fixed points, define q points $X_k, k = 1, 2, \ldots, q$. Associated with each individual point X_k are $c = 1, \ldots, t$ weights, one for each component in the mixture:

$$A_c(X_k) = \frac{1}{\sqrt{2\pi\sigma_c^2}} \exp\left(-\frac{(X_k - \mu_c)^2}{2\sigma_c^2}\right). \tag{11.30}$$

Replacing integration with summation, denote the conditional probability that an examinee responding in pattern x_l has an ability X_k and belongs to component c as:

$$W_{cl}(X_k) = \frac{\pi_c L(x_l|X_k) A_c(X_k)}{\sum_{c=1}^{t} \pi_c \sum_{k=1}^{q} L(x_l|X_k) A_c(X_k)}, \tag{11.31}$$

where

$$L(x_l|X_k) = \prod_{j=1}^{n} \Phi_j(X_k)^{x_{lj}} (1 - \Phi_j(X_k))^{1-x_{lj}},$$

$$\Phi_j(X_k) = \int_{-\infty}^{c_j + a_j X_k} \frac{1}{\sqrt{2\pi}} \exp\left(-\frac{t^2}{2}\right) dt,$$

$$\phi_j(X_k) = \frac{1}{\sqrt{2\pi}} \exp\left(-\frac{(c_j + a_j X_k)^2}{2}\right).$$

The expressions (11.26)–(11.29) are approximated as:

$$\hat{\pi}_c = \frac{1}{N}\sum_{l=1}^{s} r_l \sum_{k=1}^{q} W_{cl}(X_k), \tag{11.32}$$

$$\hat{\mu}_c = \frac{\sum_{l=1}^{s} r_l \sum_{k=1}^{q} X_k W_{cl}(X_k)}{\sum_{l=1}^{s} r_l \sum_{k=1}^{q} W_{cl}(X_k)}, \tag{11.33}$$

$$\hat{\sigma}_c^2 = \frac{\sum_{l=1}^{s} r_l \sum_{k=1}^{q} (X_k - \hat{\mu}_c)^2 W_{cl}(X_k)}{\sum_{l=1}^{s} r_l \sum_{k=1}^{q} W_{cl}(X_k)}, \tag{11.34}$$

for $c = 1, 2, \ldots, t$. If $\sigma_c^2 = \sigma^2, c = 1, 2, \ldots, t$, then

$$\hat{\sigma}^2 = \frac{1}{N}\sum_{l=1}^{s}\sum_{c=1}^{t} r_l \sum_{k=1}^{q} (X_k - \hat{\mu}_c)^2 W_{cl}(X_k). \tag{11.35}$$

Iterating the optimization process with (11.32)–(11.34) (or (11.35)) may lead to slow convergence. To speed convergence, we may use the Berndt–Hall-Hall–Hausman Berndt et al. (1974) method which requires the computation of the first partial derivatives of the log likelihood function only. For a two component normal mixture distribution with a common variance, for example, the iteration formula is given by:

$$\Delta^{t+1} = \Delta^t + \left(\sum_{l=1}^{s} \left(\frac{r_l \partial \log p_l}{\partial \Delta} \cdot \frac{r_l \partial \log p_l}{\partial \Delta'} \right) \right)^{-1} \times \begin{pmatrix} \sum_{l=1}^{S} \frac{r_l}{p_l} \int L_l(\theta) g_1(\theta) d\theta - N \\ \sum_{l=1}^{S} \frac{r_l}{p_l} \pi_1 \int L_l(\theta) g_1(\theta) \frac{\partial \log g_1(\theta)}{\partial \mu_1} d\theta \\ \sum_{l=1}^{S} \frac{r_l}{p_l} \pi_2 \int L_l(\theta) g_2(\theta) \frac{\partial \log g_2(\theta)}{\partial \mu_2} d\theta \\ \sum_{l=1}^{S} \frac{r_l}{p_l} \sum_{c=1}^{2} \pi_c \int L_l(\theta) g_c(\theta) \frac{\partial \log g_c(\theta)}{\partial \sigma^2} d\theta \end{pmatrix}, \tag{11.36}$$

where $\Delta = (\pi_1, \mu_1, \mu_2, \sigma^2)'$.

11.3.2 The Likelihood Component

Assume that the parameters of the mixture distribution are fixed and maximize $\log L'$ with respect to the IRT parameters. Taking the derivative of the $\log L'$ yields:

$$\frac{\partial \log L'}{\partial u_j} = \sum_{l=1}^{s}\sum_{c=1}^{t} \int \frac{r_l}{p_l} L_l(\theta) \frac{x_{lj} - \Phi_j(\theta)}{\Phi_j(\theta)(1 - \Phi_j(\theta))} \frac{\partial \Phi_j(\theta)}{\partial u_j} g_c(\theta) d\theta, \tag{11.37}$$

where $u_j = c_j$ or a_j, $j = 1, \ldots, n$. For solution by quadrature over fixed points, replace integration with summation, and we get the likelihood equations:

$$c_j : \frac{\partial \log L'}{\partial c_j} = \sum_{k=1}^{q} \frac{\overline{r}_{jk} - \overline{N}_k \Phi_j(X_k)}{\Phi_j(X_k)(1 - \Phi_j(X_k))} \phi_j(X_k) = 0, \tag{11.38}$$

$$a_j : \frac{\partial \log L'}{\partial a_j} = \sum_{k=1}^{q} \frac{\overline{r}_{jk} - \overline{N}_k \Phi_j(X_k)}{\Phi_j(X_k)(1 - \Phi_j(X_k))} \phi_j(X_k) X_k = 0, \tag{11.39}$$

where

$$\overline{r}_{jk} = \sum_{l=1}^{s}\sum_{c=1}^{t} \frac{r_l x_{lj} L_l(X_k) A_c(X_k)}{p_l},$$

$$\overline{N}_k = \sum_{l=1}^{s}\sum_{c=1}^{t} \frac{r_l L_l(X_k) A_c(X_k)}{p_l},$$

$$p_l = \sum_{c=1}^{t} \pi_c \sum_{k=1}^{q} L(x_l|X_k) A_c(X_k).$$

To solve (11.38) and (11.39), we follow Bock and Jones (chapter 3, Bock and Jones (1968)) and Bock and Aitkin (1981) as described in Chapter 6. They used the Newton–Raphson method to estimate c_j and a_j separately for each item by conventional probit analysis. Taking the second derivatives of $\log L'$ with respect to c_j and a_j yields the elements of the two by two matrix:

$$\frac{\partial^2 \log L'}{\partial c_j^2} = \sum_{k=1}^{q} (F_{jk} - G_k), \tag{11.40}$$

$$\frac{\partial^2 \log L'}{\partial c_j \partial a_j} = \sum_{k=1}^{q} (F_{jk} - G_k) X_k, \tag{11.41}$$

$$\frac{\partial^2 \log L'}{\partial a_j^2} = \sum_{k=1}^{q} (F_{jk} - G_k) X_k^2, \tag{11.42}$$

where

$$F_{jk} = \frac{\left(\overline{r}_{jk} - \overline{N}_k \Phi_j(X_k)\right) \phi_j(X_k)}{\Phi_j(X_k)(1 - \Phi_j(X_k))} \left[-(c_j + a_j X_k) - \frac{\phi_j(X_k)}{\Phi_j(X_k)} + \frac{\phi_j(X_k)}{1 - \Phi_j(X_k)} \right],$$

$$G_k = \frac{\overline{N}_k \phi_j(X_k)^2}{\Phi_j(X_k)(1 - \Phi_j(X_k))}.$$

Substituting (11.38)–(11.42) into the following Newton–Raphson equations (11.43), we obtain new estimates for c_j and a_j:

$$\begin{pmatrix} c_j \\ a_j \end{pmatrix}^{t+1} = \begin{pmatrix} c_j \\ a_j \end{pmatrix}^{t} - \begin{pmatrix} \frac{\partial^2 \log L'}{\partial c_j^2} & \frac{\partial^2 \log L'}{\partial c_j \partial a_j} \\ \frac{\partial^2 \log L'}{\partial a_j \partial c_j} & \frac{\partial^2 \log L'}{\partial a_j^2} \end{pmatrix}^{-1} \begin{pmatrix} \frac{\partial \log L'}{\partial c_j} \\ \frac{\partial \log L'}{\partial a_j} \end{pmatrix}. \tag{11.43}$$

The information matrix for c_j and a_j is given by

$$I\begin{pmatrix} \hat{c}_j \\ \hat{a}_j \end{pmatrix} = N \sum_{k}^{q} \frac{\phi_j(X_k)}{\Phi_j(X_k)(1 - \Phi_j(X_k))} \begin{pmatrix} 1 & X_k \\ X_k & X_k^2 \end{pmatrix} \tag{11.44}$$

for $j = 1, \ldots, n$, and variances c_j and a_j are approximated by

$$\begin{pmatrix} \mathrm{var}(\hat{c}_j) \\ \mathrm{var}(\hat{a}_j) \end{pmatrix} \approx \mathrm{diag}\left(I \begin{pmatrix} \hat{c}_j \\ \hat{a}_j \end{pmatrix} \right)^{-1} \tag{11.45}$$

for $j = 1, \ldots, n$.

11.3.3 Algorithm

Assuming $\sigma_c^2 = \sigma^2, c = 1, 2, \ldots, t$, we can implement the following computational algorithm.

A Select plausible starting values for the model parameters $\hat{c}_j^0$, $\hat{a}_j^0$, $j = 1, \ldots, n$, and $\hat{\pi}_c^0, \hat{\mu}_c^0, c = 1, \ldots, t$ and $(\hat{\sigma}^2)^0$.
B Compute $W_{cl}(X_k)$ for all c, l, k with (11.31).
C Compute $\hat{\pi}_c^1, \hat{\mu}_c^1, c = 1, \ldots, t$ and $(\hat{\sigma}^2)^1$ with (11.32), (11.33), (11.35), where the BHHH algorithm, using (11.36), may optionally be used.
D Using the initial values given in (A), evaluate the first and second derivatives of $\log L'$, corresponding to (11.38)–(11.42).
E Solve the Newton–Raphson equations (11.43) for c_j^1, and a_j^1, $j = i, \ldots, n$.
F With new parameter values $\hat{c}_j^1, \hat{a}_j^1$, $j = 1, \ldots, n$, and $\hat{\pi}_c^1, \hat{\mu}_c^1$, $c = 1, \ldots, t$ and $(\hat{\sigma}^2)^1$, go back to (A), and repeat the same steps until convergence.

11.3.4 Unequal Variances

In some applications, we may want to allow the variances of the components of the normal mixture distributions to be different. Day (1969) noted that maximum likelihood worked well for the case of equal variances. In the case of unequal variances, however, Day (1969) argued that any small set of sufficiently close data points would generate a spurious maximizer of the likelihood function. Maximum likelihood estimation (MLE) could be successfully used whenever the relationship between component standard errors is known, and appropriate constraints of the form $\sigma_i = c_{ij}\sigma_j$ for $i, j = 1, \ldots, N$ are added. But the coefficient c_{ij} is typically unknown. Hathaway (1985) showed that adding simple constraints on component standard deviations results in a well-conditioned optimization problem. The constraints suggested by Hathaway (1985) are:

$$\pi_c \geq k_1 > 0 \text{ and } \sigma_c \geq k_2\sigma_{c+1} \quad \text{for } c = 1, 2, \ldots, t.$$

11.4 Multivariate Probit Analysis

Probit analysis (Bliss, 1935) is classically used to describe a dosage response relation between a continuous exposure variable and a binary "quantal" response (e.g. alive or dead). In multivariate probit analysis (Ashford and Sowden 1970), the dosage variable is assumed to be related to the joint response on n binary outcome variables. Typically, the n variables represent different outcomes jointly measured in the same subjects, but they may also represent a pattern of univariate observations made repeatedly over time in the same individual.

In terms of the sampling side of the model, classical probit analysis (Bliss 1935, Finney 1964) assumes that subjects can be placed into homogeneous groups that define linearly increasing or decreasing levels of a continuous variable (e.g. age). However, this is an artificial restriction, and there is no reason why the predictors cannot represent any combination of grouping variables and/or continuous predictors and their interactions (see (Gibbons and Wilcox-Gök, 1998)). The advantage of the probit formulation over logistic alternatives is in the ability to accommodate general correlated errors of measurement.

Ashford and Sowden (1970) generalized probit analysis to the bivariate case by assuming that the joint response of two biological systems followed a bivariate normal distribution with correlation coefficient r. This generalization provides a full information solution for the bivariate case, but is computationally intractable for more than two binary outcomes. Generalization to the case of three quantal variables has been considered by Hausman and Wise (1978). An alternate approach was proposed by Kolakowski and Bock (1981). Following Lawley (1943), they proposed an n-variate probit model for a single continuous latent variable. Their generalization represents a special case in that it assumes that the quantal responses are independent within subjects conditional on the latent variable (conditional independence: (Lord and Novick 1968; Bishop et al. 1975)).

Muthén (1979) introduced a more general probit model for n dichotomous indicators of m continuous latent variables. Muthen proposed an IRT model to relate a p-dimensional vector of dichotomous response variables to an m dimensional vector of continuous latent variables. Structural relations (e.g. dose responses) are then estimated in terms of the m latent variables and not n manifest binary response variables. As in Ashford and Sowden's model, maximum likelihood estimation was found to "involve too heavy computations in the general case."

Bock and Gibbons (1996) developed a full-information multivariate probit model that generalizes (Ashford and Sowden 1970) results to an unlimited number of binary response variables. For each binary outcome variable, a series of structural parameters are estimated (e.g. intercept and slope of a dose–response relation) and not the m latent variables as in Muthén (1979). In addition, the n-binary variates are assumed to be related to m latent variables which reproduce the residual correlation structure among the n manifest binary responses.

11.4.1 The Model

Following Bock and Gibbons (1996), we assume that each experimental or observational unit is assigned to one of n mutually exclusive groups. Gibbons and Wilcox-Gök (1998) extend the model to the general regression setting with main effects and interactions of any mixture of discrete and continuous predictors. We assume that the p binary variables for subject i are accounted for by a p vector of latent response strengths,

$$\boldsymbol{y}_i = \boldsymbol{B}\boldsymbol{x}_i + \boldsymbol{\Lambda}\boldsymbol{\theta} + \boldsymbol{\zeta}, \tag{11.46}$$

where $\boldsymbol{x}_i$ is the q-vector of covariates associated with subject i and $\boldsymbol{B}$ is a $p \times q$ matrix of coefficients of the regression of $\boldsymbol{y}$ on $\boldsymbol{x}$. The m-vector $\boldsymbol{\theta}$, $m < p$, contains values of the underlying factors that account for the correlation among the p outcome variables through a $p \times m$ matrix of factor coefficients, $\boldsymbol{\Lambda}$. Ultimately, these correlations are expressed in the within-group associations among the binary variables. Finally, $\boldsymbol{\zeta}$ is a p-vector of uncorrelated residuals.

Under multivariate normal assumptions

$$\boldsymbol{\theta} \sim N(\boldsymbol{0}, \boldsymbol{I}) \quad \text{and} \quad \boldsymbol{\zeta} \sim N\left(\boldsymbol{0}, d_j^2 I_p\right), \tag{11.47}$$

where $\boldsymbol{\theta}$ and $\boldsymbol{\zeta}$ are mutually independent and d_j^2 is the unique variance (i.e. uniqueness in IRT terms) for binary variable j, $d_j^2 = 1 - \sum_{k=1}^{m} \lambda_{jk}^2$. The joint distribution of $\boldsymbol{y}_i$ and $\boldsymbol{\theta}$ is $(p+m)$-variate normal:

$$\begin{bmatrix} \boldsymbol{y}_i \\ \boldsymbol{\theta} \end{bmatrix} = N\left(\begin{bmatrix} \boldsymbol{B}\boldsymbol{x}_i \\ \boldsymbol{0} \end{bmatrix}, \begin{bmatrix} \boldsymbol{\Lambda}\boldsymbol{\Lambda}' + d_j^2 I_p & \boldsymbol{\Lambda} \\ \boldsymbol{\Lambda}' & \boldsymbol{I} \end{bmatrix}\right). \tag{11.48}$$

The conditional distribution of $\boldsymbol{y}_i$, given $\boldsymbol{\theta}$, is therefore

$$\boldsymbol{y}_i | \boldsymbol{\theta} \sim N\left(\boldsymbol{B}\boldsymbol{x}_i + \boldsymbol{\Lambda}\boldsymbol{\theta}, d_j^2 I_p\right). \tag{11.49}$$

Letting $u_{ij} = (1, 0)$ denote a value of the jth binary response variable for subject i, we write the corresponding response probability, conditional on $\boldsymbol{\theta}$, as

$$\begin{aligned} P(u_{ij} = 1 | \boldsymbol{\theta}) &= \frac{1}{\sqrt{2\pi}} \int_0^{\infty} \exp\left\{-\frac{1}{2}\left[y_{ij} - (\boldsymbol{\beta}_j \boldsymbol{x}_i + \boldsymbol{\lambda}_j \boldsymbol{\theta})/d_j\right]^2\right\} dy \\ &= \Phi(z_{ij}), \end{aligned} \tag{11.50}$$

the standard univariate normal distribution function with argument

$$z_{ij} = (\boldsymbol{\beta}_j \boldsymbol{x}_i + \boldsymbol{\lambda}_j \boldsymbol{\theta})/d_j, \tag{11.51}$$

where $\boldsymbol{\beta}_j$ and $\boldsymbol{\lambda}_j$ are the jth rows of $\boldsymbol{B}$ and $\boldsymbol{\Lambda}$, respectively.

On these assumptions, the p binary responses, given $\boldsymbol{\theta}$, are independent, and the conditional probability of observing the p-vector response pattern $\boldsymbol{\mu}_i = [\mu_{ij}]$, for subject i is

$$L_i(\boldsymbol{\theta}) = \prod_j^p \left[\Phi(z_{ij})\right]^{u_{ij}} \left[1 - \Phi(z_{ij})\right]^{1-u_{ij}}. \tag{11.52}$$

Moreover, because the components of $\boldsymbol{\theta}$ are uncorrelated, the *unconditional* probability of observing the pattern $\boldsymbol{\mu}_i = [1, 1, \dots, 1]$ is

$$P(\boldsymbol{\mu}_i) = \int_{-\infty}^{\infty} \cdots \int_{-\infty}^{\infty} \int_{-\infty}^{\infty} L_i(\theta) \prod_k^m \phi(\theta_k) d\theta_1 d\theta_2 \cdots d\theta_m, \tag{11.53}$$

where $\phi(\theta_k)$ is the standard normal ordinate at θ_k. This definite integral is the probability of positive orthant of the p-variate tolerance distribution for subject i. To obtain any other orthant probability, we reverse the direction of integration for those variables with $u_{ij} = 0$. When these probabilities are needed in the estimation, we approximate the integral by m-fold Gauss–Hermite quadrature with five or more points per dimension (Stroud and Secrest 1966).

11.4.2 Identification

Identifiability of elements in $\boldsymbol{B}$, $\boldsymbol{\Lambda}$, and d_j is discussed in Muthén (1979) and Bock and Gibbons (1996). In the term $\boldsymbol{BX}$ (the mean tolerances within groups), the $q \times n$ matrix $\boldsymbol{X}$ is assumed to be of full rank, $q \leq n$. This will generally be the case when $\boldsymbol{X}$ is the matrix of a regression model or a basis of an experimental design model. Thus, all pq elements of $\boldsymbol{B}$ are identified.

For $\boldsymbol{\Lambda}$, the well-known rotational indeterminacy of the common factor model (see (Jöreskog 1979)) implies the nonidentifiability of $m(m-1)/2$ elements. In addition, the invariance of the orthant probabilities with respect to the scale of the latent variables (tolerances) implies the nonidentifiability of both d_j and λ_{jk}, for $k = j$. Thus, the number of identifiable parameters in the model is $pq + pm - m(m+1)/2$ for $q \leq n$ and $m < p$. The maximum number of parameters that can be estimated in the tolerance factor model for multivariate probit analysis therefore equals the number of coefficients in the saturated full-rank model for the mean tolerances plus the number of pairwise correlations between the p tolerances. The tolerance factor model as given is therefore over-parameterized even in the bivariate case.

11.4.3 Estimation

If a first-order method, such as the EM algorithm, is employed in solving the likelihood equations, identifiability in $\boldsymbol{\Lambda}$ can be ignored (at the expense of uniqueness of the solution). But this method does not provide standard errors for the estimated parameters, and the large number of iterations typically required is computationally burdensome. A second-order method, such as Newton–Raphson or Fisher scoring, which requires as many independent restrictions on the model as there are unidentified parameters, is a much better choice on both counts. In the present application, these restrictions must be placed on the elements of $\mathbf{\Lambda}$. There are several possible approaches; however, Bock and Gibbons (1996) found that the most computationally tractable approach is to eliminate the over-parameterization by fixing $m(m+1)/2$ elements of $\mathbf{\Lambda}$ and estimating the remaining parameters by Fisher scoring. Specifically, they determine the scale of the elements of θ by beginning the iterative solution of the likelihood equations with a number of EM cycles, initially using $\lambda_{jk} = 1$ when $k = j$, and zero otherwise, while fixing the d_j parameters at 1. This results in a suitable values of λ_{jk} $(k \geq j)$ that then remain fixed in the subsequent Fisher-scoring iterations.

Maximum likelihood estimation of the free parameters of the model is straightforward (see (Bock and Gibbons 1996) – Appendix and (Gibbons and Wilcox-Gök 1998)): with random sampling of units within groups, the distribution of the observed numbers of distinct response patterns in the data is product-multinomial and the standard asymptotics of MLE apply as the group sample sizes become large (see (Rao 1973)). If the number of binary response variables is not too large, the analysis can include all 2^p patterns, even those with zero observed frequencies. In that case, the usual form of the Fisher information matrix for multinomial data, which is the sum of terms over all patterns in all groups, is feasible for computation.

If p is large, however, the number of patterns becomes so large that computation with all possible patterns cannot be considered. Fortunately, many of these patterns will have zero frequencies in practical size samples, and the corresponding terms drop out of the likelihood equations and need not be included in the calculations. Because their counterparts in the Fisher information matrix do not drop out, we resort in these cases to the "empirical" information matrix of Louis (1982), which excludes terms with zero frequencies (see (Bock and Gibbons 1996) – Appendix).

11.4.4 Tests of Fit

When the sample sizes are sufficiently large that relatively few of the expected pattern frequencies are less than 1, the Pearson and likelihood-ratio chi-square tests of the fit of the probit model against the general multinomial model are available. In situations where the number of binary variables is moderate to large ($p > 10$, say), the expected values in typical applications will often be too small to justify the chi-square approximation of the null distribution. Even in those cases, however, the approximation is generally satisfactory for the change of log likelihood (deviance) as additional parameters are included in the model (Haberman 1977). This provides a useful statistical criterion not only for setting the rank of the probit regression model but also for choosing the number of factors to be included in the tolerance variables.

11.4.5 Illustration

Gibbons and Lavigne (1998) applied the multivariate probit model of Bock and Gibbons (1996) to the problem of predicting the emergence of multivariate patterns of four childhood psychiatric disorders: disruptive disorders (DD – attention deficit disorders, oppositional defiant disorder, conduct disorder); adjustment disorders (ADJ); emotional disorders (ED – all anxiety disorders, depression); and other DSM-III-R Axis I disorders (OTHER). They screened 3800 children in 68 pediatric practices, yielding a sample of 1164 children for analysis over three waves of data. Subjects ranged in age from two through seven years. Response patterns and associated observed frequencies for each age group are displayed in Table 11.3.

The first step of the analysis was to determine the number of factors required to model the underlying tolerance correlations. Beginning with a cubic model for the relationship between age and joint diagnostic status, no significant improvement in fit of a model with two factors over the fit of a unidimensional model was

Table 11.3 Response pattern frequencies DD, ADJ, ED, and OTHER.

	Age in years					
Pattern	2	3	4	5	6	7
0000	106	100	193	194	121	56
0001	1	2	5	12	7	6
0010	0	3	10	10	1	22
0100	0	0	3	2	0	5
1000	24	39	53	36	22	6
0011	0	1	2	3	2	7
0101	0	0	1	0	1	0
0110	0	0	1	1	0	2
1001	1	4	5	11	8	3
1010	1	6	13	12	13	15
1100	0	1	1	1	1	0
0111	0	0	0	0	0	0
1101	0	0	0	0	0	0
1011	0	0	0	1	3	4
1110	0	0	0	0	0	0
1111	0	0	0	0	0	0

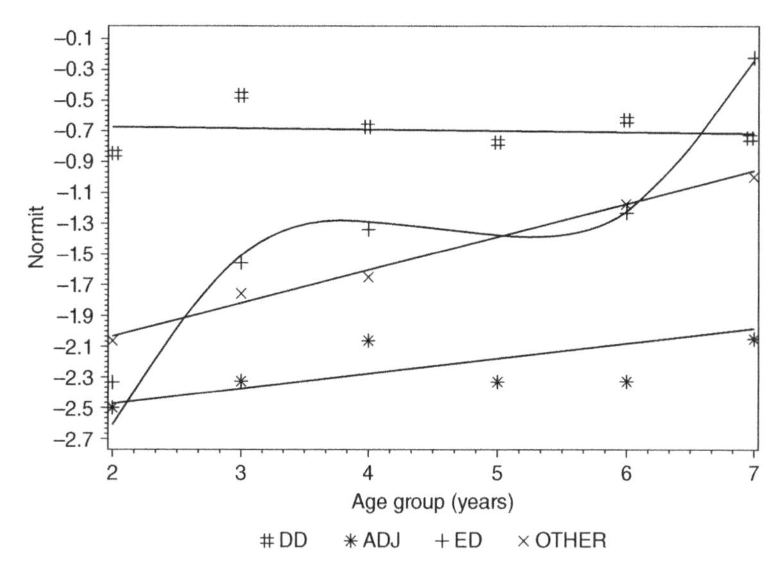

Figure 11.1 Normit regression lines of diagnostic prevalence.

found ($\chi_2^2 = 1.86, p < 0.40$). Based on this result, a one-factor model for the latent tolerance distribution was used in all subsequent models.

With respect to the relationship between age and diagnostic status, significant linear and curvilinear trends were observed for OTHER and ED, respectively (see Figure 11.1).

Inspection of Figure 11.1 reveals that there is an excellent fit between observed and predicted diagnostic category frequencies. The underlying correlation matrix and standard errors are given in Table 11.4. The estimated correlation matrix reveals moderate correlation between diagnoses of DD, ED, and OTHER but not

Table 11.4 Correlation matrix and standard errors.

	DD	ADJ	ED	OTHER
DD	1.0000			
ADJ	0.1247	1.0000		
ED	0.4902	0.1105	1.0000	
OTHER	0.3166	0.0713	0.2804	1.0000
Standard	0.04259			
errors	0.05376	0.07381		
	0.00973	0.00971	0.01320	

Table 11.5 Marginal maximum likelihood parameter estimates.

	λ_j	Constant	Linear	Quadratic	Cubic
DD	0.7439	*– 0.7009*	−0.0007	−0.0068	0.0129
ADJ	0.1676	*– 2.5209*	0.1617	−0.1043	0.0569
ED	0.6589	*– 1.3713*	*0.1806*	−0.0104	*0.0563*
OTHER	0.4255	*– 1.5049*	*0.1119*	−0.0081	0.0019
		0.0429	0.0143	0.0119	0.0078
Standard		0.2967	0.1135	0.0830	0.0377
errors		0.0744	0.0266	0.0211	0.0121
		0.0682	0.0237	0.0194	0.0118

Emphasized entries MLE/SE ≥ 1.96 $p < 0.05$

with ADJ. Marginal maximum likelihood estimates (MMLEs) of model parameters and corresponding standard errors are provided in Table 11.5. MMLEs for the regression of diagnostic prevalence on age in Table 11.5 indicate a significant linear increase from ages two through seven years for OTHER disorders, and significant linear and cubic terms for anxiety/depressive disorders. Estimated proportions of each individual diagnosis and estimated comorbidity prevalence rates for each age are provided in Table 11.6.

Table 11.6 Estimated prevalence and comorbidity rates.

	Age (yr)					
	2	3	4	5	6	7
DD	0.2127	0.2681	0.2550	0.2320	0.2143	0.2412
ADJ	0.0000	0.0018	0.0179	0.0123	0.0023	0.0141
ED	0.0031	0.0602	0.0878	0.0832	0.1108	0.3980
OTHER	0.0169	0.0293	0.0470	0.0839	0.1192	0.1590
DD + ADJ	0.0000	0.0018	0.0050	0.0032	0.0017	0.0000
DD + ED	0.0031	0.0376	0.0486	0.0474	0.0574	0.1596
DD + OTHER	0.0082	0.0136	0.0187	0.0366	0.0464	0.0660
ADJ + ED	0.0000	0.0000	0.0009	0.0006	0.0000	0.0052
ADJ + OTHER	0.0000	0.0000	0.0006	0.0000	0.0007	0.0000
ED + OTHER	0.0000	0.0013	0.0032	0.0155	0.0260	0.0910

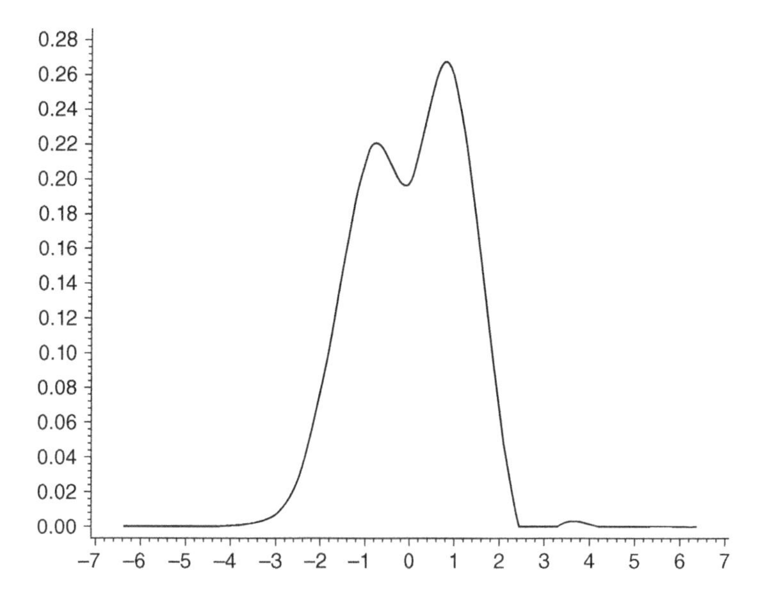

Figure 11.2 Empirical latent variable distribution.

Inspection of Table 11.6 reveals increasing risk for diagnoses of ED and OTHER with age as well as appreciable increases in comorbidity of the two or either ED or OTHER comorbid with DD. Highest comorbidity is between DD and ED which is most pronounced at age seven (15.96%).

Gibbons and Lavigne relaxed the assumption of normality of the latent tolerance distribution and empirically estimated the corresponding weights of $g(\boldsymbol{X})$ (see Figure 11.2), using the general methodology described in Section 10.2.7.

Inspection of Figure 11.2 reveals a bimodal tolerance distribution of the modes of which correspond to those children with one or more diagnoses versus those children who did not meet criteria for a diagnosis. Since the prior distribution was expressed on 10 points, there were nine degrees of freedom used and significant improvement over a model with a normal prior was found ($\chi^2_9 = 19.53, p < 0.02$).

11.5 Multilevel IRT Models

There are many similarities between generalized mixed-effects regression models (see (Hedeker and Gibbons 2006) for an overview) and IRT models. Both specify one or more random-effects that are related to unit-level response probabilities on multiple within-unit (e.g. subject) variables. In the following, we describe results for the Rasch model and the two-parameter logistic model for binary outcomes; however, the ordinal logistic regression model and polytomous IRT models share the same relation. Rijmen et al. (2003) explore these similarities for an even greater range of IRT models, including a two-dimensional MIRT model.

11.5.1 The Rasch Model

A random intercept model for repeated assessment of a binary variable on five occasions is formally equivalent to a Rasch model (Rijmen et al. 2003; Hedeker et al. 2006). To see this, denote a positive response $u_{ij} = 1$ else $u_{ij} = 0$. Under the Rasch model, the probability of a positive (e.g. correct) response to item j for subject i, $u_{ij} = 1$ conditional on a single latent variable (e.g. ability) θ_i is

$$P(u_{ij} = 1 \mid \theta_i) = \frac{1}{1 + \exp[-a(\theta_i - b_j)]}, \tag{11.54}$$

where $\theta \sim N(0, 1)$. The parameter b_j is the item difficulty, which determines of the logistic curve on the latent ability dimension, and $(1 + \exp[-(z)])^{-1}$ is the logistic cdf. Intuitively, if θ_i exceeds b_j then the probability of a positive response, $u_{ij} = 1$, is greater than 0.5 (see Figure 3.5). Note that to draw the equivalence to a mixed-effects model we add a common item discrimination parameter a, which when set to 1.0, yields the Rasch model, which does not typically include this parameter.

Following Hedeker et al. (2006), the equivalent mixed-model uses a logistic link function (see (Hedeker and Gibbons 2006), chapter 9) and can be written in terms of the n_i x 1 vector of logits $\boldsymbol{\lambda}_i$ for subject i as

$$\boldsymbol{\lambda}_i = \boldsymbol{X}_i\boldsymbol{\beta} + \boldsymbol{1}\sigma_\alpha\theta_i, \tag{11.55}$$

where $\boldsymbol{X}_i$ is an $n_i \times n$ item indicator matrix obtained from I_n, which is an $n \times n$ identity matrix, $\boldsymbol{\beta}$ is an $n \times 1$ vector of item difficulties (the IRT b parameters), $\boldsymbol{1}_i$ is an $n_i \times 1$ vector of ones, and θ_i is the latent trait (random effect) for subject i. The parameter σ_α is the common slope parameter (the a parameter in IRT), which denotes variability in the random subject effects. More typically, we write the mixed-effects logistic random intercept model as

$$\boldsymbol{\lambda}_i = \boldsymbol{X}_i\boldsymbol{\beta} + \boldsymbol{1}\alpha_i, \tag{11.56}$$

where the random subject effects α_i are distributed $N(0, \sigma_\alpha^2)$, which is equivalent to (11.55) since $\theta_i = \alpha_i/\sigma_\alpha$. Intuitively, the random intercept model can be thought of as a multilevel model in which level 1 reflects the item responses which are nested within the level 2 subjects. As such, the Rasch model is simply a random intercept logistic regression model where item indicators represent the level 1 predictors. In the longitudinal case, this is equivalent to a mixed-effects model with time-specific effects and no fixed intercept for a single repeatedly measured binary outcome.

Note that both models permit missing data in terms of the number of items administered and/or responded to under quite general Missing at Random (MAR) assumptions (Rubin 1976). In this context, MAR assumes that the missing items are ignorable conditional on the other items included in the model for IRT, and in addition any covariates included in a generalized mixed model. This is related to the conditional independence assumption where the item responses are assumed to be conditionally independent given θ in the IRT framework or conditional on the random-effects in the mixed-effects model. To implement this in a mixed model we simply reduce the rank of $\boldsymbol{X}$. For example, if there are $n_i = 3$ items and only the first and third are answered, then

$$\boldsymbol{X} = \begin{bmatrix} 1 & 0 & 0 \\ 0 & 0 & 1 \end{bmatrix}. \tag{11.57}$$

The number of rows in $\boldsymbol{X}_i$, λ_i, and $\boldsymbol{1}_i$ are the number of items responded to by subject i. By contrast $\boldsymbol{\beta}$ has dimension equal to the total number of items.

A final step to equate the two models is required. In IRT, we assume that

$$\sum_{j=1}^{n} b_j = 0 \quad \text{and} \quad \prod_{j=1}^{n} a_j = 1. \tag{11.58}$$

For the Rasch model, this requires the following transformation of the mixed-model estimates to place them in the same scale,

$$b_j = -\beta_j - \frac{1}{n}\sum_{j'=1}^{n} - \beta_{j'}. \tag{11.59}$$

11.5.2 The Two-Parameter Logistic Model

A similar equivalence exists for the two-parameter model:

$$\lambda_i = \boldsymbol{X}_i\boldsymbol{\beta} + \boldsymbol{X}_i'\boldsymbol{T}\theta_i, \tag{11.60}$$

where $\boldsymbol{T}$ is a vector of standard deviations,

$$\boldsymbol{T}' = \left[\sigma_{\alpha 1}, \sigma_{\alpha 2}, \ldots, \sigma_{\alpha n}\right]. \tag{11.61}$$

As noted by Hedeker et al. (2006), the standard deviations correspond to the discrimination parameters of the two-parameter IRT model. The two-parameter model is a logistic mixed-effects regression model that allows the random effect variance to vary across the level 1 units, in this case the items, nested within subjects. Several multilevel software programs can fit such models. As noted in (11.58) the estimated parameters from the mixed model must be rescaled so that they can be interpreted as item difficulty and discrimination parameters. This requires centering the b_j around zero and the product of the a_j around 1, which is the same as transforming the product of the σ_j to equal 1. The transformations are:

$$a_j = \exp\left[\log\,\sigma_j - \frac{1}{n}\sum_{j'=1}^{n}\log\,\sigma_{j'}\right], \tag{11.62}$$

and

$$b_j = -(\beta_j/a_j) - \frac{1}{n}\sum_{j'=1}^{n} - (\beta_{j'}/a_{j'}). \tag{11.63}$$

11.5.3 Estimation

Complete details regarding estimation of mixed-effects logistic regression models can be found in Hedeker and Gibbons (2006). We sketch the general ideas here, which parallel those described for IRT models presented previously. Denoting $\boldsymbol{\mu}_i$ as the vector of responses from subject i, the probability of the response pattern $\boldsymbol{\mu}_i$ of size n_i conditional on $\boldsymbol{\alpha}_i$ is equal to the product of the level 1 responses:

$$\ell(\boldsymbol{\mu}_i \mid \boldsymbol{\alpha}_i) = \prod_{j=1}^{n_i} P(u_{ij} = 1 \mid \boldsymbol{\alpha}_i). \tag{11.64}$$

Assuming conditional independence of the item responses conditional on the random effect(s), the marginal density of $\boldsymbol{\mu}_i$ is

$$h(\boldsymbol{\mu}_i) = \int_{\alpha} \ell(\boldsymbol{\mu}_i \mid \boldsymbol{\alpha}_i) f(\boldsymbol{\alpha}) d\boldsymbol{\alpha}, \tag{11.65}$$

where $f(\boldsymbol{\alpha})$ represents the distribution of the random effect(s). The marginal log-likelihood is

$$\log\,L = \sum_{i=1}^{N} \log\,h(\boldsymbol{\mu}_i). \tag{11.66}$$

MMLEs of the regression coefficients $\boldsymbol{\beta}$ and the variance–covariance matrix of the random effects Σ_α or its Cholesky factor $\boldsymbol{T}$ are obtained by maximizing the log-likelihood with respect to the model parameters. For non-identity link functions, we can use Gauss–Hermite quadrature to numerically evaluate the likelihood using the same general approach as described here for unidimensional and multidimensional IRT models. Alternatives include partial likelihood approaches such as a Laplace approximation or full-likelihood approaches based on Markov chain Monte Carlo (MCMC, (Gilks et al. 1998)).

Estimation of the random-effects is based on empirical Bayes methods and is similar to the EAP estimators used for IRT models described in Chapter 10. For a single random effect, the posterior mean and variance are

$$\hat{\alpha}_i = E(\alpha_i \mid \boldsymbol{\mu}_i) = h_i^{-1} \int_{\alpha} \alpha_i \ell(\boldsymbol{\mu}_i \mid \alpha_i) f(\alpha) d\alpha, \tag{11.67}$$

and

$$V(\hat{\alpha}_i \mid \boldsymbol{\mu}_i) = h_i^{-1} \int_{\alpha} (\alpha_i - \hat{\alpha}_i)^2 \ell(\boldsymbol{\mu}_i \mid \alpha_i) f(\alpha) d\alpha. \tag{11.68}$$

For the random intercept model, the random-effect estimate in (11.67) and its uncertainty in (11.68) correspond to the ability estimate and its variance in an IRT model.

11.5.4 Illustration

Hedeker et al. (2006) illustrate use of this approach using the LSAT section 6 dataset originally published by Bock and Lieberman ((Bock and Lieberman 1970), Table 1). Table 11.7 presents Rasch model estimates originally published by Thissen (1982). Hedeker and colleagues reanalyzed the data using MIXOR (Hedeker and Gibbons 1996) and SAS PROC NLMIXED, which gave identical results.

Table 11.8 reports the results of a similar comparison for a two-parameter normal–normal model (i.e. a probit response function and normally distributed ability random effects). In both cases, the IRT and transformed mixed-effects model estimates are in close agreement.

Beyond providing a better understanding between the correspondence between IRT and mixed models, the mixed model formulation of the IRT model has several additional benefits. First, it provides a direct approach to adding covariates and grouping structures to the IRT model, similar, but even more general than the traditional multiple group IRT model introduced in Section 11.2. The main effects of these covariates can adjust for differences in the latent ability distribution, whereas interactions can accommodate variation in item parameters as a function of person-level characteristics. Second, additional levels of clustering (e.g. the nesting of students within classrooms) can be added to the IRT model when

Table 11.7 Rasch model estimates for the LSAT-6 data.

		NLMIXED	
Item	IRT ($\hat{b}_j$)	Raw ($\hat{h}_j$)	Transformed ($\hat{b}_j$)
1	−1.255	2.730	−1.255
2	0.476	0.999	0.476
3	1.234	0.240	1.235
4	0.168	1.306	0.168
5	−0.624	2.099	−0.625

Table 11.8 Two-parameter probit model estimates for the LSAT-6 data.

	IRT		MIXOR Raw		MIXOR Transformed	
Item	$\hat{b}_j$	$\hat{a}_j$	$\hat{\beta}_j$	$\hat{\sigma}_j$	$\hat{b}_j$	$\hat{a}_j$
1	−0.6787	0.9788	1.5520	0.4169	−0.6804	0.9779
2	0.3161	1.0149	0.5999	0.4333	0.3165	1.0164
3	0.7878	1.2652	0.1512	0.5373	0.7867	1.2603
4	0.0923	0.9476	0.7723	0.4044	0.0926	0.9487
5	−0.5174	0.8397	1.1966	0.3587	−0.5154	0.8415

estimated as a generalized mixed model. This would represent a 3-level model, where students (level 2) are assessed on n_i items (level 1) and are nested within k classrooms (level 3).

12

Test and Scale Development and Maintenance

The statistician who supposes that his main contribution to the planning of an experiment will involve statistical theory, finds repeatedly that he makes his most valuable contribution simply by persuading the investigator to explain why he wishes to do the experiment.

(Source: Gertrude Cox)

12.1 Introduction

This chapter covers a variety of topics related to the practical aspects of the use of item response theory (IRT) in the development and maintenance of tests and scales. We begin with an overview of item banking and calibration. Next, we discuss matrix item sampling using incomplete block designs. We follow with a discussion of test equating and conclude with a discussion of item parameter drift. Although there are large literatures related to many of these topics in the classical test theory (CTT) literature (particularly equating), our focus throughout is based on IRT (and various generalizations) to these problems because of their numerous advantages over CTT.

12.2 Item Banking

An item bank is a collection of items stored in an electronic format; i.e., a database. The item bank contains the text of the item, the answer categories, details regarding the development of the item, and various IRT parameters that describe the association between the item and the latent variable(s) of interest, and the item's difficulty or associated severity. There are several approaches to the development

Item Response Theory, First Edition. R. Darrell Bock and Robert D. Gibbons.

of an item bank. Most item banks contain an item identifier with keywords for additional description, the item's text, the response categories, including information on reverse scoring for ordinal items that describe level of agreement with the question posed (e.g. do you feel sad versus do you feel happy in a mental health assessment, with answer categories of never, occasionally, often and always), item information including number of response categories, item type (binary, ordinal, nominal), category numbering, author, source, and statistical information (e.g. item parameters including discrimination parameters, factor loadings, item difficulties, intercepts, thresholds, and guessing parameters as appropriate). Graphical information can also be included showing the item response function and item information function.

Test developers can use the item bank in a number of ways to construct a test. Items can be selected based on a minimum threshold for item discrimination or factor loadings to ensure that the selected items are good discriminators of high and low levels on the underlying construct(s) of interest. For a bifactor model (see Chapter 6), we may select the item based on its ability to discriminate high and low levels of the underlying multidimensional construct based on its factor loading on the primary dimension of interest that cuts across all of the relevant subdomains. Test developers may also wish to develop a fixed-length short form test that provides good performance across the entire range of the latent variable, by selecting items with difficulty/severity across the range of the latent variable of interest. The selected items can be evaluated in terms of the resulting test information function. This will allow the test developer to determine the extent that the test provides or does not provide sufficient information across the range of the underlying ability or severity continuum. For example, traditional tests of depression often provide good information for patients with higher levels of psychopathology, but very little information at sub-clinical levels. While one may argue that this is acceptable, it is not, because one of the goals of mental health measurement is to provide measurement-based care. This requires constant level of precision at both high and low levels of severity so that longitudinal assessments can demonstrate that a patient has transitioned from high severity to normal levels of depressive symptomatology. As an illustration, Figure 12.1 presents the test information function for the anxiety example from Chapter 3, calibrated using a graded response model. As described earlier, the test is informative at levels of severe psychopathology, but has limited information at lower levels of psychopathology. The test is useful for case finding, but has limited utility for the measurement of change.

One of the most important uses of item banking is for adaptive testing. Here we are not encumbered by the size of the item bank in terms of respondent burden because each subject will receive items adaptively drawn from the bank based on their ability or severity. The number of items adaptively selected is proportional to

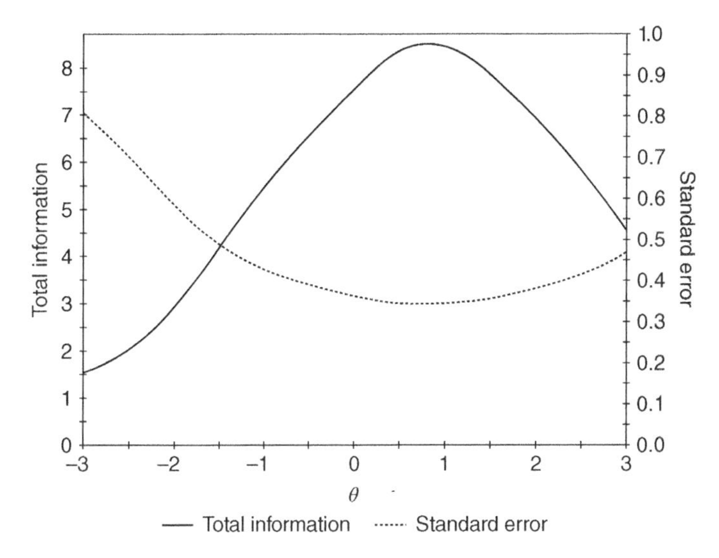

Figure 12.1 Test information function – Anxiety example.

the required precision of measurement. The process is fully automated so that the test developer does not have to preselect the items to be administered. Even if there are items with poor discrimination in the bank, they will typically not be selected for adaptive administration because they do not maximize test information.

A useful example of item banking is for the development of a computerized adaptive depression test, the CAT-DI (Gibbons et al. 2012). Here a bank of 452 depression items was developed by extracting items from over 100 existing depression and anxiety scales. The items were organized into conceptually meaningful categories using a hierarchical approach informed by previous empirical work. The hierarchy included the primary domain, depression, subdomains (mood, cognition, somatic complaints, and suicidality), factors (e.g. within depressed mood, factors included increased negative affect and decreased positive affect), and facets (e.g. within increased negative affect, facets included sadness, lowered threshold for negative affect, irritability, moodiness, and poor mood distinct from grief or loss). The total number of facets was 46 for depression ((Gibbons et al. 2012), Supplement Table 1). We note that the inherent multidimensionality of depression in which items are sampled from subdomains, factors, and facets precludes calibration using traditional IRT models. However, the hierarchical structure of these items fits naturally within the framework of a bifactor model, where the primary dimension is depression and the secondary dimensions could be the subdomains, factors, or facets, depending on the level of detail desired. As noted in Chapter 6, a major advantage of the bifactor model is its computational advantage over unrestricted full-information item factor analysis for models of high dimensionality. Under the bifactor restriction, the likelihood reduces to a two-dimensional integration problem regardless of the number of dimensions. As such, there is no difference computationally for a model with four subdomains as secondary factors or a model with all 46 facets as secondary factors. Gibbons et al. used a bifactor model to calibrate the CAT-DI using the four subdomains as the secondary factors, and developed a corresponding CAT algorithm to score the primary depression dimension in the presence of conditional dependencies produced by the subdomains (see Chapter 13). Of course there is no reason why this same strategy could not be used in educational measurement, where for example, a mathematics item bank could be hierarchically arranged into subdomains based on different branches (e.g. algebra, trigonometry, calculus, statistics and probability). A bifactor model could then be used to score the primary dimension of mathematical ability, but also to provide proficiency estimates for the different subdomains.

12.3 Item Calibration

While this entire book is devoted to item calibration, there are features of the item calibration process that are relevant to test development, maintenance, and item banking. First, to be of value, all items in the bank must be jointly calibrated. As the item banks become larger with widening of the applications of IRT beyond educational testing, and extending to complex multidimensional constructs, the computational burden of fitting IRT and multidimensional item response theory (MIRT) models to such data increases. Second, as new items are added to the item bank, they must also be co-calibrated with items in the existing bank, both for the purpose of scoring and for ensuring that the new items measure the same latent construct as the original items. Third, perhaps most pronounced for educational measurement is the need to replenish the items in the bank with new items to replace older items that may have become obsolete or have become too familiar to respondents who may be re-taking the test. Fourth, as the population sampled from widens or changes over time, methods for determining differential item functioning in special subpopulations and testing for item parameter drift (Bock et al. 1988) become important steps in the calibration process. Fifth, as new items are added or the item bank is replenished, data for these items are required in order to calibrate and link them to the existing test. Several of these topics are explored in Sections 12.3.1–12.3.4.

A special calibration problem related to item bank maintenance arises when the item bank is to be administered via computerized adaptive testing (CAT). This is referred to as online calibration (Stocking 1988). We consider two pools of items, the existing operational items and a new pool of pretest items that are being evaluated to determine if they can be used to replenish the item bank. A challenge of adaptive testing is that the number of items administered (both operational and pretest) is typically small, leading to sparse coverage of the item bank when item administration is based on CAT. This lowers the precision of the item parameter estimates. A further complication is that different examinees are administered different pretest items, further adding to the spareness of the data.

There are several approaches that have been proposed for online calibration of new pre-test items that we review in Sections 12.3.1–12.3.4.

12.3.1 The OEM Method

Wainer and Mislevy (1990) proposed a marginal maximum likelihood (MML) estimator in which a single *E*-step is performed to obtain the posterior distribution of ability based solely on the responses from the operational CAT items using the originally estimated item parameters (one *EM* cycle – OEM). The *M*-step is used to estimate the pretest item parameters, using only the item responses for those items. Only a single *M*-step of the *EM* cycle is used to estimate the pretest item parameters. The *M*-step of the EM algorithm is done item by item, so a poor fitting pretest item(s) cannot contaminate other pretest items. The pretest items are on the same scale as the operational items because the posterior ability distribution information used in the *E*-step is restricted to the operational items. The obvious question is whether the item parameters for the pretest items have converged to their MMLEs.

12.3.2 The MEM Method

Ban et al. (2006) extended the OEM method by increasing the number of *EM* cycles until convergence. The multiple EM (MEM) cycles method begins with the OEM estimator, but adds additional *EM* cycles. In subsequent *E*-steps all of the item responses (operational and pretest) are used to estimate the posterior ability distribution. In subsequent *M*-steps, item parameters for the operational items remain fixed, but pretest item parameters are updated until convergence. This prevents contamination of the operational items by the new pretest items, but cross-contamination of the pretest items is not prevented using the MEM method. Ban et al. (2002) suggest using a Bayes Modal estimation method by multiplying the MML equations by a prior distribution for the pretest item parameters, to reduce the effect of extreme parameter estimates.

12.3.3 Stocking's Method A

Stocking (1988) developed two approaches to online calibration of new items added to an existing item bank. The first method, Method A, involves the computation of abilities based on the operational items already in the item bank using previously estimated item parameters. These ability estimates are then fixed in a second IRT calibration restricted to the newly developed pretest items. Because the ability estimates are fixed in the estimation of the item parameters of the new items, no rescaling is required (i.e. the item parameters from the pretest items are anchored to the ability estimates for the operational items). The limitation of Method A is that it assumes the abilities as known when in fact they are estimates and only true abilities in expectation. Stocking's Method B attempts to overcome this limitation.

12.3.4 Stocking's Method B

In an attempt to improve upon Method A, Stocking (1988) developed Method B where a set of anchor items from the operational items in the existing bank are added to the pretest items. As in Method A, ability is fixed to the existing operational items in the bank. However, in Method B, the newly estimated item parameters for the anchor items are used to create a scaling transformation (Stocking and Lord 1983) which is then applied to the newly estimated pretest item parameters. The Stocking and Lord (1983) method produces a scaling transformation to align the item characteristic curves. In the following, we provide a sketch of their method.

Let b_{j1} be the estimated item difficulty for anchor item j from the operational item bank, and b_{j2} the estimated item difficulty for anchor item j in the pretest. We seek the transformation coefficients A and B such that

$$b^*_{j2} = Ab_{j2} + B, \tag{12.1}$$

$$\theta^*_{j2} = A\theta_{j2} + B, \tag{12.2}$$

$$a^*_{j2} = a_{j2}/A + B\,. \tag{12.3}$$

b^*_{j2} is the value of b_{j2} transformed to the scale of the operational item bank.

These transformations do not change $a_{j2}(\hat{\theta}_{i2} - b_{j2})$ so $P_j(\hat{\theta}_{i2} \mid a_{j2}, b_{j2}) = P_j(\hat{\theta}^*_{i2} \mid a^*_{j2}, b^*_{j2})$. These are simple linear transformations, which could be easily computed if we knew the true values of these parameters, which of course we do not. There have been several scale transformations proposed to accommodate the uncertainty in the item parameters. The distinction among them is that some are estimated using only the item difficulties and then applied to the ability and discrimination parameters, whereas others estimate the scale factors from the entire item characteristic curve. The former are termed "mean and sigma" transformations and the latter "characteristic curve methods." Stocking and Lord (1983) describe one of each.

Their mean and sigma transformation begins with the a weighted estimator originally described by Linn et al. (1980), which gives small weight to extreme deviations from the fitted linear transformation function, and iterates until convergence. The initial linear transformation begins by computing the weights as:

$$A = S_{b_1}/S_{b_2}, \tag{12.4}$$

and

$$B = \overline{X}_{b_1} - A\overline{X}_{b_2}, \tag{12.5}$$

where $\overline{X}$ and S are the mean and standard deviation for the difficulty parameters across the anchor items in the operational and pretest calibration samples. The weights were originally described by Mosteller and Tukey (1977) as

$$T_j = \begin{cases} (1 - (D_j/CM)^2)^2 & \text{when}(D_j/CM)^2 < 1 \\ 0 & \text{otherwise,} \end{cases} \tag{12.6}$$

where D_j is the absolute value of the deviation between b_{j2} and b^*_{j2}, M is the median of the D_j, and $C = 6$. The iterative algorithm is as follows:

1. Let W_j be the larger of the two inverse variances of the difficulty parameter for item j in the two samples.
2. Rescale the weights so that they sum to one $W'_j = W_j/(\sum_{j=1}^{n} W_j)$.
3. Compute the mean and sigma transformation line using the weighted item difficulties (i.e. compute A and B).
4. Compute the absolute perpendicular distance between each item difficulty and the transformation line D_j.
5. Compute the Tukey weights T_j.
6. Reweight the item difficulties using the combined weight $U_j = (W_jT_j/(\sum_{j=1}^{n} W_jT_j)$.
7. Recompute the weighted transformation line using these new combined weights.
8. Repeat steps 4, 5, and 6 until the maximum change in the perpendicular distances is less than 0.01.

Once the final transformation parameters A and B are determined, they can be applied to a_j and θ.

Stocking and Lord's Stocking and Lord (1983) characteristic curve method, which they describe as their "new method," is based on equating estimated true scores for the two test administrations. For examinee i with ability θ_i, the true scores for the anchored items in the operational and pretest samples are δ_{i1} and δ_{i2},

$$\delta_{i1} = \sum_{j=1}^{n} P_{j1}(\hat{\theta}_{i1} \mid a_{j1}, b_{j1}, c_{j2}), \tag{12.7}$$

and

$$\delta_{i2} = \sum_{j=1}^{n} P_{j2}(\hat{\theta}_{i2} \mid a_{j2}, b_{j2}, c_{j2}) . \tag{12.8}$$

The idea is to select A and B such that the average squared distance between δ_{i1} and δ_{i2} is small. To do this, we minimize

$$F = \frac{1}{n} \sum_{i=1}^{N} (\delta_{i1} - \delta_{i2})^2, \tag{12.9}$$

where N is the number of examinees. Stocking and Lord (1983) provide computational details for the minimization of F, which varies with the specific IRT model used.

Ban et al. (2002) compared OEM, MEM, and Stocking Method B using sixteen 60 item ACT mathematics test forms. The estimated item parameters were used as true parameters for the simulated CATs. Of the 940 items used in the study, there were 240 pretest items, 100 anchor items, and 600 operational items. Overall, the best performance was found for the MEM method, which has the added advantage of not requiring anchor items and thereby decreasing respondent burden. The OEM method did not perform well.

12.4 IRT Equating

Score equating is important for high-stakes testing where alternate forms of a test are used over time to reduce familiarity with the questions that might produce artificial inflation of ability estimates. In general, the different tests are designed from a common "blueprint" to help ensure that they measure the same construct; however, differences in item composition across the test forms can lead to differences in the difficulty of the tests and produce ability estimates that are no longer comparable. This is of course a much larger problem for CTT than IRT because in IRT, we have estimates of each item's difficulty, which inform the ability estimation. The terminology used to describe equating is somewhat confusing. In many cases, score linking and equating are used synonymously. Holland et al. (2006) define linkage as the transformation from a score on one test to another test. They describe three types of linkage: predicting, scale aligning, and equating.

12.4.1 Linking, Scale Aligning and Equating

Predicting is the oldest form of score linking (Dorans et al. 2010). It involves a regression of the scores of one test on another test, resulting in a prediction equation that provides an estimated operational test score based on the score on the new or candidate test. There is nothing to limit the prediction equation from containing other auxiliary information such as demographic information or other test scores. The advantage of prediction is that there is no need for the two tests to share any items in common.

Scale aligning transforms the scores from two (or more) tests onto a common scale. Scale aligning and equating are often confused because they often use the same or similar statistical procedures. There are many scale aligning methods that have been described in detail by Holland et al. (2006) and Kolen (2006).

Equating is the strongest form of linkage. It can be viewed as a form of scale alignment that ensures that the scores from each test form can be used as if they came from the same test (Dorans et al. 2010). It requires that the tests measure the same construct at approximately the same level of difficulty.

12.4.2 Experimental Designs for Equating

There are several types of equating that are exemplified by their different experimental designs, described by Dorans et al. (2010).

12.4.2.1 Single Group (SG) Design

In a single group design, all subjects from a well-defined population take both (all) tests. This eliminates any confounding with ability that could be produced by a multi-group design. We assume that there is no carryover effect on the second test produced by practice that is gained from the first test. While statistically attractive due to its within-subject nature, it is generally impractical to have examinees take both tests unless they are quite short.

12.4.2.2 Equivalent Groups (EG) Design

In the EG design, different, but equivalent samples from a given population take different test forms, also designed to be equivalent from the same blueprint. Randomly assigning subjects drawn from the same population will help minimize differences in the distribution of ability between the groups. Stratified random sampling or "spiraling" can also be used where tests are alternated to examinees within strata (e.g. classrooms) In the EG design, there is no requirement that the tests have any items in common.

12.4.2.3 Counterbalanced (CB) Design

The CB design attempts to eliminate order effects introduced by the SG design by randomizing order of test presentation between two groups sampled from the same population.

12.4.2.4 The Anchor Test or Nonequivalent Groups with Anchor Test (NEAT) Design

In anchor test designs, there are two populations with samples of subjects from one population taking test X and the other population taking test Y. In addition, both samples take an anchor test consisting of the same items. The anchor test is used to quantify differences between the two populations sampled that may affect performance on the two tests to be equated. It is desirable to have the anchor test measure the same construct as tests X and Y; however, the anchor test is usually much shorter than the target tests. The NEAT test consists of two parallel SG designs, linked through the anchor test. A limitation of the NEAT design is that tests X and Y arc confounded with population, and it is only the anchor test that allows them to be equated. In some ways, the anchor test is used as an instrumental variable, that will potentially eliminate confounds between the two populations. We note that an anchor test can be either internal or external. An external anchor test is a mini-version of the two tests being equated and is administered as a separate test. An internal anchor test contains items that are common between the two tests and are embedded within each test. Typically, the internal anchor items are included in the scoring of the tests.

The reader should now realize that many of the IRT-based methods for on-line calibration are actually IRT equating methods. They typically involve an SG design with internal anchors. Popular approaches include mean and sigma equating (Stocking 1988) and characteristic curve equating (Stocking and Lord 1983) that we have reviewed in Section 12.3.

12.5 Harmonization

Gibbons et al. (2014) introduced an approach to align scales that measure the same construct but have different items and item formats. Their example is based on the need to compute quality-adjusted life years (QALYs) for health services research applications to facilitate health economic risk benefit determinations. QALYs are calculated by combining life years gained with a measure of societal preferences or "utilities" over different levels of health functioning. To obtain a measure of societal preferences over different levels of health functioning, health status needs to be measured. Over the years, the EuroQol five-dimension (EQ-5D) questionnaire has become the standard way of measuring health status in economic evaluations. The EQ-5D descriptive system is a generic, non-disease specific instrument that describes general health functioning. The EQ-5D descriptive system is transformed into a measure of preferences by applying country-specific scoring algorithms derived from large-scale valuation studies. Even though the EQ-5D is widely-used in economic evaluations, it is not yet routinely employed in clinical trials and other datasets that could potentially be used to conduct economic evaluations. On the other hand, these studies do have generic measures of health functioning, like the Short Form 12 (SF-12) survey, which measures physical and mental functioning. The SF-12 instrument, however, has not been extensively used in valuation studies and therefore cannot be transformed into preferences, a key component of many economic evaluations.

Gibbons et al. (2014) use the bifactor model to provide harmonization between the two measures. Here each measure represents a subdomain and all items load on the primary dimension as well. This is an approach to scale alignment rather than equating because the two scales are being transformed into a common scale, the primary dimension, which provides cross-walk between the two measures. Regression methods can then be used to re-express estimates of θ_{1i} onto the original score metric for the target test. A limitation of this approach is that theta now represents an admixture of the two scales being harmonized. An alternative approach is to estimate the bifactor model that combines the two tests and to impute the missing item responses for the target scale that was not administered. To do this note that for the bifactor model

$$z_j = \sum_{v=1}^{d} a_{jv}\theta_v, \quad P_{jh}(\theta) = \Phi(z_j + c_{jh}) - \Phi(z_j + c_{jh-1}), \tag{12.10}$$

where $\phi(z_j + c_{j0}) = 0$ and $\phi(z_j + c_{jm_j}) = 1 - \phi(z_j + c_{jm_{j-1}})$. Here c_{jh} represents a category threshold for a graded bifactor model for an item with m ordered categories. We can then select the category for a missing scale item that has maximum probability based on the theta estimate(s) for the available scale(s). Noting that $\hat{\theta}_v$ is an estimate of θ_v, a somewhat more sophisticated approach is to sample n random draws from the distribution of θ_1, which is normal with mean $\overline{\theta}_1$ and variance $V(\Theta_1)$, compute the estimated category probabilities, and select the category with maximum average probability.

Harmonization and test linking share some features in common. Harmonization is a type of scale alignment in which two tests are expressed in a common metric, which is defined by the primary dimension that they share. The primary dimension of the bifactor model allows us to express the total combined test in a common metric, whereas the secondary dimensions partition out the test-specific differences. This is different from equating, where the goal is to express one test in the metric of the other test, thereby making them exchangeable. If the two tests being harmonized are substantively different, the primary dimension may tap a different construct than what is actually intended by the target test. As noted by Gibbons et al. (2014), "Using an IRT model that has been jointly calibrated for two or more instruments, it is possible to estimate the true score of interest when only a subset of the instruments has been administered." An advantage of this approach is that there is no limit on the number of tests that can be harmonized. The second approach in which the item responses on the first scale are imputed from the estimated primary dimension for the second test item responses, is a form of equating. The imputed item responses can then be used for ability estimation based on item parameters for the target test only, and are therefore in exactly the same scale, assuming that both tests have similar loadings on the primary dimension. This is true even if the item difficulties are different for the two tests. As designed, we have assumed that there are at least some subjects that have taken both tests, so that the loadings on the primary dimension are anchored. If this is not the case and the tests are never co-administered, the loadings on the primary dimension are not interpretable without additional strong assumptions.

12.6 Item Parameter Drift

Item parameter drift refers to the differential change in item parameters over time (Goldstein 1983; Bock et al. 1988). The difficulty levels of some items in an item bank may change over time while others remain constant. Changes in educational technology, school curricula, cultural change, and substance familiarity may all lead to item parameter drift. For example changes in technology introduced by the Internet may make knowledge of traditional topics in library science superfluous. Bock et al. (1988) describe a time-dependent IRT model to accommodate changes in item difficulty over time. They note that item parameter drift is similar to differential item functioning designed to detect item bias between groups, but here the differences are typically restricted to the item difficulty and the bias can be parameterized as smooth changes in difficulty of an item over time. To estimate item parameter drift, item by time (e.g. year) terms are added to the 3-PL model. This assumes that guessing parameters and discrimination parameters are unaffected and that overall trends in item difficulty or ability are absorbed in the estimate of yearly means.

The linear location drift model is defined as

$$P(x_{jk} = 1 \mid \theta) = g_{jk} + (1 - g_{jk})\Psi[z_{jk}(\theta)], \tag{12.11}$$

where g_{jk} is the lower asymptote or guessing parameter of the 3-PL model, Ψ is the logistic response function, and z_{jk} is the logistic deviate. Assuming $g_{jk} = g_j$ and $a_{jk} = a_j$ for all k,

$$\begin{aligned} z_{jk}(\theta) &= a_j(\theta - b_j - \delta_j t_k) \\ &= c_j + a_j\theta + d_j t_k. \end{aligned}$$

It is convenient to scale time from an arbitrary origin and to add the constraint

$$\sum_{j=1}^{n} \delta_j = 0. \tag{12.12}$$

It should be immediately apparent that this model is a special case of the much more general multivariate probit regression model described in Chapter 11.

12.7 Summary

The methods described in this chapter for scale development and maintenance overlap in many ways. There is also overlap with methods for multiple group IRT in Chapter 11 and with differential item functioning in Chapter 9. Linking and equating share many common elements, but differ in terms of the degree to which the resulting test scores can be used exchangeably. Item parameter drift is an extension of multiple group IRT where linear constraints on the time trends are added to the model. More general "item bias" comparisons can also be made using multivariate probit regression models, which have the added advantage of accommodating multidimensionality.

13

Some Interesting Applications

Statistics ... is the most important science in the whole world, for upon it depends the practical application of every other (science) and of every art.

(Source: Florence Nightingale)

13.1 Introduction

While the majority of applications of item response theory (IRT) have been in educational measurement, with the advent of multidimensional item response theory (MIRT; (Bock and Aitkin 1981)) and computerized adaptive testing (CAT) (Kingsbury and Weiss 1980; Weiss 1985), there are many novel applications of IRT for the measurement of complex traits that go beyond the traditional application of IRT for unidimensional constructs that are the focus of application to educational measurement problems. In Sections 13.2–13.4, we consider several examples with the hope that these will stimulate much wider applications of the general theory.

13.2 Biobehavioral Synthesis

Stan et al. (2020) created a crosswalk between 51 clinician-rated symptoms of psychosis and the cortical thickness of 68 regions of the brain, using a full-information item bifactor model. While reduced cortical thickness has been found in psychotic disorders, prior to this, the relationship with cortical thickness in general and in specific regions of the brain, with specific psychotic symptoms had not been established. Specifically, they used MIRT to investigate whether there are specific regions of the brain in which cortical thinning is associated with a specific pattern of psychotic symptom ratings.

Item Response Theory, First Edition. R. Darrell Bock and Robert D. Gibbons.

The analysis was performed using the National Institute of Mental Health B-SNIP study sample (Tamminga et al. 2013). Of the 2450 subjects, 560 persons who lacked both complete symptom score ratings and cortical thickness measurements were excluded from the analysis. The remaining sample of 1890 participants included (i) 865 probands with a psychotic disorder, including schizophrenia (SZ), schizoaffective disorder (SZA), and psychotic bipolar I disorder (BP), (ii) 678 of their first-degree relatives, and (iii) 347 healthy controls. First-degree relatives with and without a psychiatric diagnosis were included. All probands were clinically stable and receiving consistent psychiatric medication treatment in the community during the previous month. Cortical thinness was established using magnetic resonance imaging using a 3-T magnet across five research sites. Cortical thinness was transformed into quintiles of the distribution of each brain area to create ordinal measures for all variables and facilitate the analysis.

For the bifactor model, five subdomains for the symptom data were selected based on the domains from which the 51 items were drawn. The five domains were depression (10 items), positive psychosis (7 items), general psychosis (16 items), negative psychosis (7 items), and mania (11 items). For the 68 regional cortical thickness variables, a maximum likelihood exploratory factor analysis model was used to identify nine biological subdomains to accommodate the clustering of brain regions. The bifactor model therefore had $5 + 9 + 1 = 15$ dimensions. Despite the high dimensionality, the bifactor model is computationally tractable due to its dimension reduction feature, such that the degree of the integration is reduced from 15 to 2. The goal of the bifactor model was to identify a subset of biological and behavioral variables that were associated based on loading on the primary dimension, while accommodating their 14 subdomain-specific associations.

Results of the analysis revealed a subset of 16 out of 68 regions of the brain in which cortical thinning is associated with a subset of 7 out of the 51 psychopathological symptoms (see Figure 13.1). The newly synthesized psychosis measure was characterized by the following symptoms: (i) delusions, (ii) hallucinatory behavior, (iii) suspiciousness, persecution, (iv) passive, apathetic social withdrawal, (v) depression, (vi) unusual thought content, and (vii) active social avoidance. This cluster of psychosis symptoms is unique in itself, including the reality distortion characteristics of psychosis, along with negative symptoms, depression, and social avoidance. The cortical thickness measures contributing

Figure 13.1 Schematic representation of the bifactor model, showing the effect of the overarching construct, the latent biobehavioral integration variable (left column) and the subdomains. The symptoms and the biological variables with high loading factors are shown in highlighted characters.

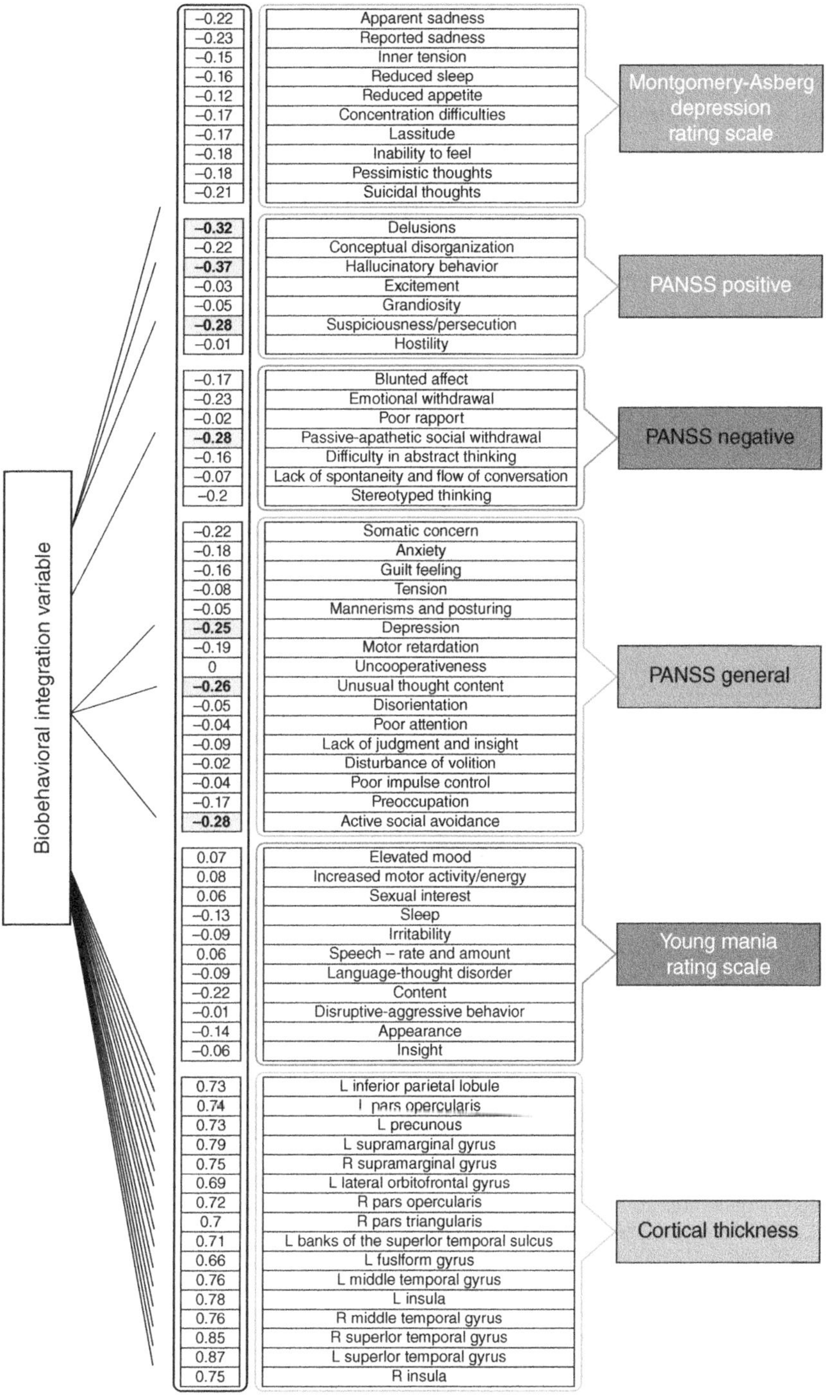

Biobehavioral integration variable

Loading	Item	Scale
−0.22	Apparent sadness	Montgomery-Asberg depression rating scale
−0.23	Reported sadness	
−0.15	Inner tension	
−0.16	Reduced sleep	
−0.12	Reduced appetite	
−0.17	Concentration difficulties	
−0.17	Lassitude	
−0.18	Inability to feel	
−0.18	Pessimistic thoughts	
−0.21	Suicidal thoughts	
−0.32	Delusions	PANSS positive
−0.22	Conceptual disorganization	
−0.37	Hallucinatory behavior	
−0.03	Excitement	
−0.05	Grandiosity	
−0.28	Suspiciousness/persecution	
−0.01	Hostility	
−0.17	Blunted affect	PANSS negative
−0.23	Emotional withdrawal	
−0.02	Poor rapport	
−0.28	Passive-apathetic social withdrawal	
−0.16	Difficulty in abstract thinking	
−0.07	Lack of spontaneity and flow of conversation	
−0.2	Stereotyped thinking	
−0.22	Somatic concern	PANSS general
−0.18	Anxiety	
−0.16	Guilt feeling	
−0.08	Tension	
−0.05	Mannerisms and posturing	
−0.25	Depression	
−0.19	Motor retardation	
0	Uncooperativeness	
−0.26	Unusual thought content	
−0.05	Disorientation	
−0.04	Poor attention	
−0.09	Lack of judgment and insight	
−0.02	Disturbance of volition	
−0.04	Poor impulse control	
−0.17	Preoccupation	
−0.28	Active social avoidance	
0.07	Elevated mood	Young mania rating scale
0.08	Increased motor activity/energy	
0.06	Sexual interest	
−0.13	Sleep	
−0.09	Irritability	
0.06	Speech – rate and amount	
−0.09	Language-thought disorder	
−0.22	Content	
−0.01	Disruptive-aggressive behavior	
−0.14	Appearance	
−0.06	Insight	
0.73	L inferior parietal lobule	Cortical thickness
0.74	L pars opercularis	
0.73	L precunous	
0.79	L supramarginal gyrus	
0.75	R supramarginal gyrus	
0.69	L lateral orbitofrontal gyrus	
0.72	R pars opercularis	
0.7	R pars triangularis	
0.71	L banks of the superior temporal sulcus	
0.66	L fusiform gyrus	
0.76	L middle temporal gyrus	
0.78	L insula	
0.76	R middle temporal gyrus	
0.85	R superior temporal gyrus	
0.87	L superior temporal gyrus	
0.75	R insula	

to the measure include contiguous regions of middle temporal gyrus, superior temporal gyrus, insula, supramarginal gyrus, and operculum bilaterally, and on the left side, the fusiform gyrus, banks of the superior temporal sulcus, inferior parietal lobule, precuneus, and lateral orbitofrontal gyrus, and on the right, the pars triangularis (see Figure 13.2). The core set of cortical regions identified in the analysis was validated by calculating the sum of the thicknesses of all 16 of the regional components for each subject, then correlating the composite thickness with clinical, demographic, and cognitive measures of psychosis severity (duration of illness, number of hospitalizations, number of unique psychotropic medications, and total dose of antipsychotic in chlorpromazine equivalents), social and functional impairment (Global Assessment of Functioning, Social Functioning Scale, and its subscales) and cognition (Wide Range Achievement Test and the Brief Assessment of Cognition in Schizophrenia and its component tests). All of these measures significantly correlated with the composite cortical thickness measure.

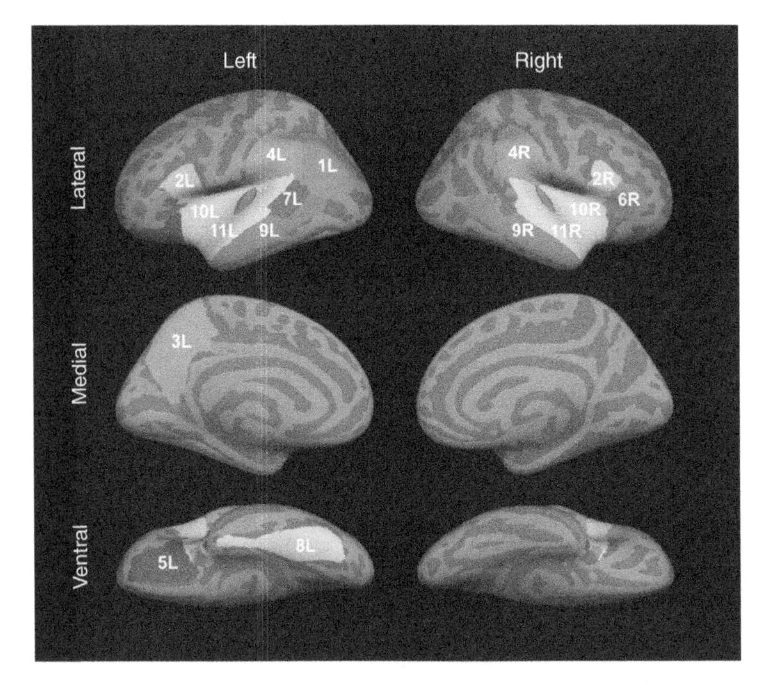

Figure 13.2 Inflated representation (lateral, medial, and ventral views) of the cortical regions generated by the bifactor model analysis. Cortical regions are labeled as follows: 1L: left inferior parietal lobule; 2L, 2R: left and right pars opercularis; 3L: left precuneus; 4L, 4R: left and right supramarginal gyrus; 5L: left lateral orbitofrontal gyrus; 6R: right pars triangularis; 7L: left bank of the superior temporal sulcus; 8L: left fusiform gyrus; 9L, 9R: left and right middle temporal gyrus; 10L, 10R: left and right insula; 11L, 11R: left and right superior temporal gyrus.

As a statistical comparison to the IRT approach, they conducted a canonical correlation analysis to relate the biological and symptomalogical variables. None of the eigenvalues exceeded 1.0, and the canonical variates had only a single clinical and biological variable with loadings in excess of 0.25 and were therefore not interpretable. The authors concluded that "These results support the increased sensitivity of the MIRT analysis approach for investigating and defining the relationships between clinical symptoms and neocortical anatomic alterations associated in psychosis."

Figure 13.2 shows that the identified cortical regions clustered around the frontotemporal axis, and included contiguous temporal, parietal, and frontal cortical areas. The regions appear to be both spatially and functionally connected, adding further to the validity of the biobehavioral integration. The key finding was that a large continuous region of heteromodal cortex was associated with a unique set of severe psychosis symptoms that included reality distortions (hallucinations, delusions, and paranoia), along with symptoms of depression and social withdrawal that are often considered secondary to psychosis.

13.3 Mental Health Measurement

In Chapter 8, we introduced examples of MIRT-based CAT used for the assessment of complex multidimensional mental health constructs. Here we expand further on some of the novel applications of these developments.

13.3.1 The CAT-Depression Inventory

Gibbons et al. (2012) introduced the first MIRT-based CAT for the measurement of depression, the CAT-Depression Inventory (CAT-DI). An item bank consisting of 452 depression items was developed based on the review of over 100 existing depression or depression-related rating scales. Items were recoded to refer to symptomatology experienced during the past two weeks. Previous work showed that these items could be categorized into mood, cognition, and somatic subdomains, suggesting that a bifactor model would be a good choice to accommodate their multidimensional structure. Items were further categorized into factors (e.g. within depressed mood, factors included negative affect and decreased positive affect), and facets (e.g. within negative affect, facets included sadness, irritability, and moodiness). Most items were rated on a 5-point ordinal scale with increasing frequency or severity of the symptom. Study participants included treatment-seeking psychiatric outpatients from two clinics serving populations of different severity and community controls to ensure coverage of the entire severity continuum. A sample of 798 individuals were used to calibrate the item-bank. Three-hundred and eight participants took the entire bank of 452 items, and 490 calibration participants took a 252 item subset based on a balanced incomplete block design (Cochran and Cox, 1957), designed to maximize the two-way pairing of the items. Simulated adaptive testing was used to select CAT tuning parameters in the 308 subjects with complete item bank data, permitting computation of the correlation between CAT and total item bank administration. CAT tuning parameters included stopping rules based on final posterior standard deviation (PSD) and minimal item information conditional on estimated severity score, and the probability of selecting the maximally informative item and the second maximally informative item (to expand use of items across the bank). To establish validity, 816 new patients received the live version of the CAT-DI, and convergent validity was assessed using extant clinician and self-rating scales including the Patient Health Questionnaire 9 (PHQ-9, (Kroenke et al. 2001)), 24-item Hamilton Rating Scale for Depression (HAM-D, (Hamilton 1960)), and the Center for Epidemiologic Studies Depression Scale (CES-D, (Radloff 1977)). The HAM-D was administered by a trained clinician, and the PHQ-9 and CES-D were self-reports. Major depression, minor depression, and dysthymia were defined according to DSM-IV criteria, based on the Structured Clinical Interview for DSM-IV (SCID, (First et al. 1996)).

Results of the calibration revealed the bifactor model with five subdomains (mood, cognition, behavior, somatic, and suicide) significantly improved fit over a unidimensional model ($\chi^2_{389} = 6825$, $p < 0.0001$). A total of 389 items were retained in the model that had a factor loading on the primary dimension of 0.3 or greater (96% > 0.4 and 79% > 0.5). Simulated CAT found that for a PSD of 0.3 on an underlying normal severity distribution, or approximately 5 points on a 100 point scale, an average of 12 items was required to adaptively administer the test, while maintaining a correlation of $r = 0.95$ with the 389 item bank score. The test was completed in an average of 2.29 minutes (interquartile range of 1.72–2.97 minutes). The distribution of the CAT-DI severity scores was resolved into a mixture of two normals (see Figure 8.1), with a threshold of −0.61 on the underlying normal distribution. Using this threshold provided sensitivity of 0.92 and specificity of 0.88 for the hour-long SCID DSM-IV diagnosis of major depressive disorder (MDD). In terms of the continuous CAT-DI score, there was a 24-fold increase in the probability of a positive MDD diagnosis across the range of the scale (area under the curve (AUC) = 0.92, 95% confidence interval (CI) = 0.89, 0.94). Convergent validity was also demonstrated against extant depression scales, both clinician-rated (HAM-D, $r = 0.75$) and self-rated (PHQ-9, $r = 0.81$ and CES-D $r = 0.84$). The CAT-DI is now in widespread use throughout the world.

13.3.2 The CAT-Anxiety Scale

Gibbons et al. (2014) report on the development of a MIRT-based CAT for the measurement of anxiety (the CAT-ANX). Using an item bank of 467 anxiety items, they calibrated a bifactor model using the same sample of 798 patients used for the CAT-DI, and validated the final CAT using the same sample of 816 used for the CAT-DI. To assess validity, a logistic regression was used to relate the CAT-ANX to a diagnosis of generalized anxiety disorder (GAD). In addition, a multinomial logistic regression was used to simultaneously estimate the relation between the CAT-DI and CAT-ANX to MDD, GAD and comorbid GAD and MDD.

Results of the calibration revealed the bifactor model with four subdomains (mood, cognition, behavior, somatization) significantly improved fit over a unidimensional model ($\chi^2_{431} = 7304$, $p < 0.0001$). A total of 431 items were retained in the model that had a factor loading on the primary dimension of 0.3 or greater. Simulated CAT found that for a PSD of 0.3 on an underlying normal severity distribution, or approximately 5 points on a 100 point scale, an average of 12 items (range 6–24) was required to adaptively administer the test, while maintaining a correlation of $r = 0.94$ with the 431 item bank score. The test was completed in an average of 2.48 minutes. In terms of the continuous CAT-DI score, there was a 12-fold increase in the probability of a positive GAD diagnosis across the range of the scale. Depending on the application, a threshold of −0.50 produced specificity of 0.93 and sensitivity of 0.65 for a SCID diagnosis of GAD. Alternatively, a threshold of −0.85 produced a classifier with sensitivity and specificity both equal to 0.86 for a SCID diagnosis of GAD. The multinomial logistic regression using both the CAT-DI and CAT-ANX to predict GAD, MDD, or comorbid GAD and MDD had overall classification accuracy of 84.3%. Depression and anxiety are highly correlated. The correlation between the CAT-DI and CAT-ANX was $r = 0.82$.

13.3.3 The Measurement of Suicidality and the Prediction of Future Suicidal Attempt

Gibbons et al. (2017) developed a MIRT-based CAT for the measurement of suicidality, the underlying construct that is related to suicidal ideation, attempt, and death by suicide (the CAT-SS). Using the 1008 item bank used to develop the CAT-DI, CAT-ANX, and a corresponding CAT for mania/hypomania, they used the 11 suicide subdomain items in the depression bank to identify depression, anxiety, and mania symptom items that were strongly related to the 11 suicidality items (ideation and behavior). The net result was the development of a 111 item bank that provides a crosswalk between psychopathology and suicidality.

Example items are anhedonia (loss of interest and pleasure), hopelessness, and helplessness, in addition to traditional suicide items. Using a similar methodology to that described above, they demonstrated that adaptive administration of an average of 10 items (range 5–20) maintained a correlation of $r = 0.96$ with the 111 item bank and could be administered in an average of 110 seconds. Validation against structured clinical interviews revealed a 52-fold increase in suicidal ideation or worse across the range of the CAT-SS. Sensitivity was 1.0 and specificity was 0.95 for current suicidal ideation or worse, sensitivity of 1.0 and specificity of 0.92 for active suicidal ideation or worse, sensitivity of 1.0 and specificity of 0.89 for suicide plan or plan and intent, and sensitivity of 0.58 and specificity of 0.88 for lifetime suicide attempt. The last result for lifetime suicide attempt is remarkable because the CAT-SS time-frame is the past two weeks. Studies of the ability of the CAT-SS to predict suicide attempts and intentional self-harm in the next three months, six months, and one year are now being completed in the Veterans Administration and as a part of several National Institute of Mental Health (NIMH) funded grants. Data are available from a DIF study of the CAT-SS in sexual and gender minority population, which found an AUC of 0.70 for prior suicide attempt predicting suicidal ideation, plan, or attempt at six-month follow-up and an AUC of 0.85 for the model that added the CAT-SS score and prior suicide attempt to the prediction equation (Mustanski et al. 2021). Similarly, Berona (2021), conducted a separate analysis of these data and showed predictive validity of the baseline CAT-SS in predicting time to suicide attempt during 6 months follow-up (HR = 1.34, 95% CI (1.03, 1.74)) overall and HR = 1.51, 95% CI (1.06, 2.15) for the transition from suicidal ideation to suicide attempt, for each 10% increment on the CAT-SS.

Gibbons et al. (2019) developed a similar adaptive suicidality scale for children and adolescents ages 7–17 as a part of the NIMH-funded Y-CAT study (the K-CAT-SS). Beginning with a bank of 75 items, the final item bank contained 64 items and can be adaptively administered using an average of 7 items in approximately one minute. Using a similar structured clinical interview for suicidality as a validator, an AUC of 0.966 was found for the K-CAT-SS scores and clinician-rated suicidal ideation or worse, and an AUC of 0.83 for lifetime suicide attempt.

King et al. (2021) developed a similar MIRT-based CAT for suicidality in children (the CASSY). The sample consisted of 6536 adolescents (ages 12–17) recruited from 13 emergency departments affiliated with the Pediatric Emergency Care Network (PECARN). A subset of 2845 adolescents was randomly assigned to a three-month follow-up to assess emergent suicidal ideation, attempts, and nonsuicidal self-injury. The initial item bank consisted of 92 items providing a crosswalk between suicidal ideation and behavior, psychopathology, posttraumatic stress disorder (PTSD), social functioning, sleep disturbance, anger/aggression, and substance use, which formed the subdomains in the bifactor model. The final item bank contained 72 items which were adaptively administered using an average of 11 items in less than a minute and one-half

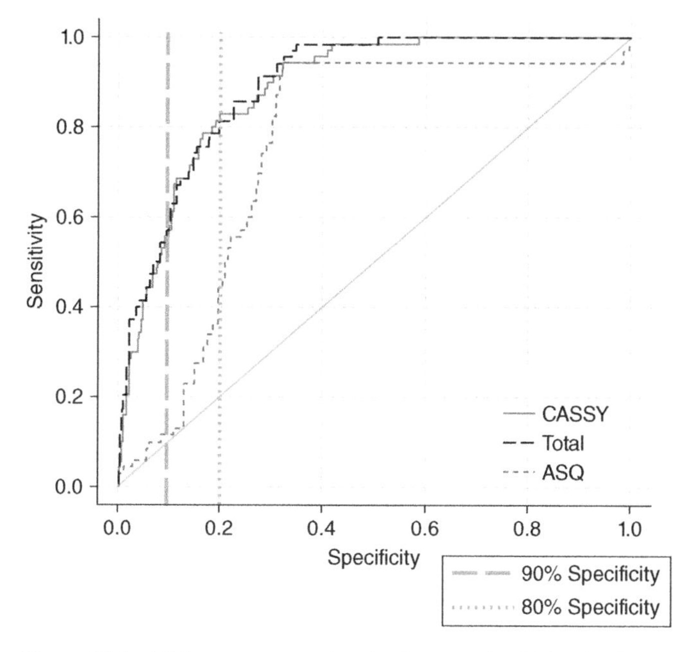

Figure 13.3 ROC curves comparing the Computerized Adaptive Screen for Suicidal Youth (CASSY) to full 72 item bank and Ask Suicide-Screening Questions (ASQ).

and maintaining a correlation of $r = 0.94$ with the total bank score. AUC for the prediction of suicide attempt within three months of the initial CASSY administration was 0.89. Figure 13.3 displays the receiver operator curve (ROC) for the CASSY, the total bank score, and the most commonly used traditional fixed length suicidality test used in emergency departments, the ASQ (National Institute of Mental Health 2017). Interestingly, the 11 item CAT performed almost identically to the 72 item total bank score and both dramatically outperformed the ASQ. The traditional measure can achieve high sensitivity but at the expense of unacceptably low specificity. Kalb et al. (2019) report that there were approximately 15 million adolescent emergency department visits annually between 2011 and 2015. Figure 13.3 reveals that for specificity of 0.90, the CASSY achieves sensitivity of 0.61 for a future suicide attempt. This would be viable for a national roll out, with a low false positive rate (10%) and identification of 61% of three-month suicide attempts based on results of a 1.5 minute test.

13.3.4 Clinician and Self-Rated Psychosis Measurement

Traditionally, the measurement of psychosis and the diagnosis of related disorders such as schizophrenia, schizoaffective disorder, psychotic disorders, have been exclusively based on clinician ratings and structured clinical interviews. Guinart et al. (2020) developed the first MIRT-based CAT for the adaptive measurement of psychosis, both for clinician-rated and self-rated assessments. They developed an item bank based on commonly used clinician rating scales: Schedule for Affective Disorders and Schizophrenia (SADS, (Endicott and Spitzer 1978)), the Scale for the Assessment of Positive Symptoms (SAPS, (Andreasen 1984)), the Brief Psychiatry Rating Scale (BPRS, (Overall and Gorham 1988)), and the Scale for the Assessment of Negative Symptoms (SANS, (Andreasen 1984)). Items were classified into the positive, negative, cognitive, and manic subdomains for analysis using a bifactor model. This yielded 144 items for which existing clinician ratings were available for 649 subjects (535 schizophrenia, 43 schizoaffective, 54 depression, and 17 mania). Items were reworded to make them appropriate for patient self-report in addition to clinician administration.

Items were calibrated in the full 649 patient sample, and simulated adaptive testing was performed from the complete set of item parameters and CAT tuning parameters selected based on the method described previously. Validation was then conducted in a separate sample of 200 subjects based on a SCID DSM-5 diagnostic interview to assess psychotic disorders. A total of 160 subjects were affected by affective and nonaffective psychoses (76 schizophrenia, 35 schizoaffective disorder, 26 bipolar disorder, 11 MDD, 7 psychosis not otherwise specified and 5 schizophreniform disorder), and 40 were healthy controls.

Results of the calibration revealed that adaptive testing required an average of 12 items and maintained a correlation of $r = 0.92$ with the total bank score. Convergent validity against the clinician-rated BPRS was demonstrated for both clinician ($r = 0.69$) and self-report $r = 0.69$ versions. For the clinician version, inter-rater reliability was strong, with an intra-class correlation (ICC) = 0.73. Test–retest reliability was also strong for both the clinician version $r = 0.86$ and for the self-report version $r = 0.82$. In terms of psychosis disorder diagnostic prediction, the five-minute clinician version demonstrated near-perfect accuracy for the hour-long SCID diagnostic interview AUC = 0.97. Perhaps more importantly, the self-report version also accurately tracked the results of the structured clinical interviews with AUC = 0.85. This indicates that we can, in fact, rely on patient self-reports of psychosis that can be completed in less than two minutes in or out of the field, to do diagnostic screening and to provide longitudinal measurement-based care. The self-report psychosis CAT as well as several of the other adaptive tests that are a part of the CAT-Mental Health (CAT-MH, (Gibbons et al. 2016)) are being used as first-stage screeners in the SAMHSA Mental and Substance Use Disorders Prevalence Study (MDPS).

13.3.5 Substance Use Disorder

Substance use disorders (SUDs) represent a public health emergency (Haffajee and Frank 2018), with urgent need for screening and treatment. However, most people with SUD do not receive treatment for their behavioral health conditions (Creedon and Lê Cook 2016). A major barrier to receiving treatment is the fast and effective identification of those in need (Priester et al. 2016). SUD prevention requires accurate initial risk detection, monitoring changes in risk over time and effective, timely intervention delivery (SAMHSA 2016). Instruments are needed that not only assist with identification of those at risk but also provide efficient, accurate quantification of nonnegligible risk and severity of illness to assist clinical decision-making and resource allocation across diverse health-care settings (i.e. emergency departments, in-patient units, out-patient primary care, and behavioral health settings).

Gibbons et al. (2020) developed the first adaptive scale for the screening and measurement of SUD. Using MIRT-based CAT, they provide a psychometric harmonization between SUD, depression, anxiety, trauma, social isolation, functional impairment, and risk-taking behavior symptom domains, providing a more balanced view of SUD. The item bank contained 252 items. A bifactor model was used to calibrate the item bank in a sample of 513 participants from primary care, community clinics, emergency departments, and patient-to-patient referrals in Spain (Barcelona and Madrid) and the United States (Boston and Los Angeles). Interestingly, the initial calibration was conducted in Spanish. Subdomains for the bifactor model included (i) SUD, (ii) psychological disorders, (iii) risky behavior, (iv) functional impairment, and (v) social support.

Results of the calibration revealed that 168 items loaded strongly on the primary SUD dimension, the majority with factor loadings in excess of 0.6. An average of 11 items was required for adaptive administration and maintained a correlation of $r = 0.91$ with the 168 item bank score. Validation against the Composite International Diagnostic Interview (CIDI) revealed an AUC = 0.85 with a 20-fold increase in the likelihood of an SUD diagnosis across the CAT-SUD scale. All sessions include specific abuse questions involving frequency of use during the past month of alcohol, sedatives/hypnotics, opioids/analgesics, heroin/methadone, and cocaine/amphetamines. This ensures that specific substances of abuse are identified in each adaptive testing session, in addition to the overall severity of the SUD. The CAT-SUD provides a solution to many of the barriers identified above and can be used in clinic or remotely to screen populations for risk, identify those in need of treatment, and monitor the effectiveness of treatment and to prevent relapse.

13.3.6 Special Populations and Differential Item Functioning

In Chapter 9, we covered differential item functioning (DIF) to determine the extent to which items that are good discriminators of high and low ability or severity in one population may not be in another. DIF provides an alternative to recalibrating and revalidating a mental health CAT in each population for which the test is to be used. These populations may be defined by language/culture (e.g. Latinos taking tests in Spanish), different patient populations (e.g. perinatal women), different medical contexts (primary care or emergency medicine), different sexual orientations (LGBTQ populations), and different settings (e.g. criminal justice), all of which may impact the way in which specific symptom items contribute to the estimated severity score on one or more adaptive tests.

13.3.6.1 Perinatal

Kim et al. (2016) examined DIF in a sample for 419 perinatal women relative to the original psychiatric calibration sample of 1614 adult psychiatric outpatients. Specifically, they examined DIF for depression, anxiety, and mania/hypomania constructs in women during pregnancy and postpartum. DIF was observed for one depression item, one anxiety item, and 15 mania/hypomania items. These items were related to fatigue and engagement in risk taking behaviors. Removal of these items resulted in correlations between the original psychiatric population and perinatal calibrations of $r = 0.98$ for depression, $r = 0.99$ for anxiety, and $r = 0.93$ for mania/hypomania indicating that the originally calibrated item bank could be used following removal of these aberrant items. The authors also administered the extant perinatal depression Edinburgh Postnatal Depression Scale (EPDS) and found that 91.3% of the perinatal women were concordant for either at-risk or low-risk between the EPDS and the perinatal version of the CAT-MH. The EPDS and perinatal CAT-MH scores were correlated $r = 0.82$ for depression, $r = 0.78$ for anxiety and $r = 0.31$ for mania/hypomania, indicating the limitation of traditional measures that are designed to assess risk of a single disorder (depression) missing other disorders (mania and possible bipolar disorder).

13.3.6.2 Emergency Medicine

Beiser et al. (2019) conducted a study of 999 emergency department patients who sought emergency treatment for a nonpsychiatric indication. Patients were screened for depression and suicidality. The presence of MDD conveyed a significant 61% increased risk for ED visits in the next year and a 49% increased rate of hospitalizations in the following year. For each 10 point increase in the CAT-DI score, there was a 10% increase in both future ED visits and hospitalizations or a 2.5-fold increase of future ED visits and hospitalizations in the following year, across the range of the scale. In terms of suicide risk, 3% of the 1000 (ED) patients generated a serious suicide alert, based on ideation plus intent, plan, or recent behavior. None of these suicidal patients were identified by medical attendings or staff. Using the same methodology as described in Chapter 9 and developed by Kim et al. (2016), DIF was evaluated in these data (see Beiser et al. 2019). Of the 69 items that are routinely adaptively administered ($n \geq 50$), only

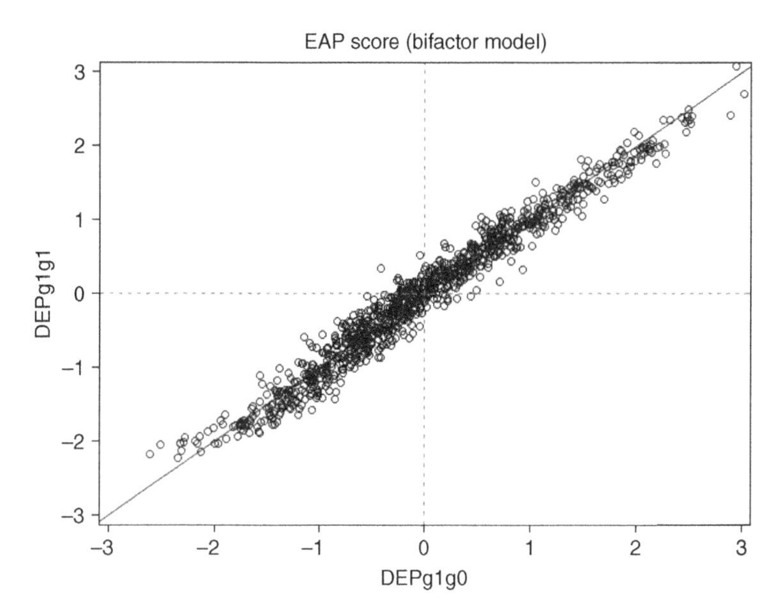

Figure 13.4 Correlation between Expected *A Posteriori* (EAP) severity scores for the ED group (g1) based on the original calibration (DEPg1g0) and the ED calibration (DEPg1g1).

a single item exhibited DIF, *In the past 2 weeks, I felt that everything I did was an effort*. Removal of this item yielded a correlation of $r = 0.98$ between the ED-specific calibration and the original outpatient psychiatric calibration (see Figure 13.4). The authors conclude that "We found little evidence of item-level DIF within the CAT-DI item bank in our ED population. This demonstrates the validity of the CAT-DI as an appropriate diagnostic screening instrument in the ED."

13.3.6.3 Latinos Taking Tests in Spanish

Gibbons et al. (2018) examined DIF in a Latino population taking the CAT-MH (depression, anxiety, mania/hypomania) in Spanish. A sample of $N = 1276$ Latino subjects from Spain and the United States were used to study DIF relative to the original psychiatric calibration sample. The Latino subjects were recruited from clinic waiting rooms of mental health, substance use, primary care, and HIV clinics in Boston Massachusetts, and Madrid and Barcelona Spain. First-generation Latino immigrants (born in a country other than the interview site) constituted 99.5% of the Spain sample and 95.0% of the Massachusetts sample. Massachusetts participants were born in Central America (49.8%), Puerto Rico (18.5%), the Caribbean (12.6%), South America (13.5%), and the continental United States (5.0%). The majority of participants in Spain were of South American origin

(80.2% in Madrid and 86.8 in Barcelona), followed by Caribbean (12.7%) in Madrid and Central American (7.6%) in Barcelona.

The Spanish translation followed the well-known methodology described by Matías-Carrelo et al. (2003) to attain a Spanish version with semantic, content, and technical equivalence to that of the original English version. All the CAT-MH items were first sent to a professional translation company to have them translated from English to Spanish. A separate team of bilingual investigators reviewed the translation and identified that some of the terms were not identical to the English terms in describing the symptoms or questions. Two research investigators who were fully bilingual and had worked on translations of diagnostic and symptoms measures revised the professional translation of the 1008 items to change some literal translations to ensure better content equivalence. After this step was completed, the modified Spanish translation was sent to two bilingual investigators to translate all the 1008 CAT-MH items from Spanish to English (i.e. a reverse translation). Four investigators reviewed the back translation. For those items where there were differences between the back translation from Spanish to English and the original English version, the team of investigators reviewed the discrepancies and determined how to make sure that the Spanish item was consistent with the English version. A multinational bilingual committee comprising four researchers and four clinicians that included Spanish speakers from six diverse countries or territories (Puerto Rico, Mexico, Panama, Columbia, Spain, Peru) adjudicated the suggested changes and produced the final Spanish translation of the 1008 CAT-MH item bank. The Spanish version of the CAT-MH was then piloted with a small group of Latino migrant workers to see whether the wording was understood and to ensure necessary adjustments, with some items modified from their original version to ensure that respondents with a lower education level understood the wording. Cognitive debriefing was used to identify items that were difficult to answer or confusing. Based on their responses, some words were changed or the sequence of words was changed to better capture the meaning of the item. In all cases, the items remained comparable, if not identical, to the original items.

In this population, the rate of MDD was 25% with 9% in the moderate or severe CAT-DI categories. For depression, four items exhibited DIF, these items were related to cheerfulness, life satisfaction, concentration, and fatigue. The correlation between the original psychiatric calibration and the Latino Spanish calibration after eliminating these items was $r = 0.99$. For anxiety, no items with DIF were identified. The correlation between the original psychiatric and Spanish calibrations was $r = 0.99$. For mania/hypomania, four items with DIF were identified related to risk-taking, self-assurance, and sexual activity. The correlation between the original and new calibration was $r = 0.96$. The authors conclude that "These findings reveal that the CAT-MH can be reliably used to measure depression, anxiety, and mania in Latinos taking these tests in Spanish."

13.3.6.4 Criminal Justice

On any given day, 300 000 to 400 000 people with mental illness are incarcerated in jails and prisons across the United States, and an additional 500 000 are under correctional supervision in the community (National Leadership Forum on Behavioral Health/Criminal Justice Services 2009). Reported rates of severe mental illness were 14.5% for men and 31% for women (Steadman et al. 2009). Rates of less-severe mental illness (e.g. some anxiety disorders) were 35% for men and 27% for women (Black et al. 2010). The prevalence of PTSD and suicide are at least three times higher in jails and prisons (Goff et al. 2007) than in the community, and the rate of SUD is seven times higher (Tsai 1958). Gibbons et al. (2019) examined DIF for the CAT-MH in a criminal justice setting. They studied MIRT-based CATs for depression, anxiety, mania/hypomania, suicidality, and SUD in English and Spanish speakers in 475 defendants in the Cook County Bond Court. The bond court is connected to the Cook County Jail, the largest single-site jail in the United States. Every person arrested and detained (either in the Cook County Jail or a police precinct) for a felony charge in the City of Chicago goes through Bond Court, typically within 48 hours of arrest. At Bond Court, a judge determines whether the person may be released on bond and, if so, the dollar amount of that bond. If the person is not released on bond, they go to the Cook County Jail. On 12 December 2017, 95% of the people in the Cook County Jail were incarcerated pretrial, meaning they were either not given a bond or could not pay the bond amount (chicagodatacollaborative.org).

The median time to complete the battery of five CATs was 10 minutes with an interquartile range of 8–12 minutes. Across the five measures, there were a total of nine items that exhibited DIF. These items included symptoms related to feeling everything was an effort, difficulty concentrating, difficulty falling asleep, sexual activity, feeling fearful, and hopeless. These items are likely reflecting the experience of incarceration or fear of imminent incarceration rather than an underlying mental health or SUD. Elimination of these suspect items yielded correlations ranging from $r = 0.96$ to $r = 0.99$ between the criminal justice calibration scores and the original psychiatric calibration scores based on the criminal justice response patterns. The authors concluded that "Our results show that the revised version of the CAT-MH can be used to screen and assess a variety of mental health conditions in the criminal justice population."

13.3.7 Intensive Longitudinal Data

MIRT-based CAT has special advantages for the collection of longitudinal data in general and intensive longitudinal data in particular. Because the same items are not repeatedly administered, response bias produced by answering the same questions over and over again is eliminated. Sections 13.3.1–13.3.6 have demonstrated that CAT for mental health applications can be administered in two minutes or less per construct, so collection of even daily high-quality measurements is possible. As an example, La Porte et al. (2020) have demonstrated the feasibility of perinatal mood screening on patients' personal smartphones, using remote administration of CAT-MH depression, anxiety, mania, and suicidality tests with a median test time of 5.8 minutes. No differences in completion rates were observed between pregnant and postpartum women. The CAT-MH was completed by 67% of women (137/203) with no financial incentive to participate. This group is now screening pregnant and postpartum women on a weekly basis for depression.

Sani et al. (2017) illustrated intensive longitudinal data collection in a case study of a treatment resistant depressed patient who received deep brain stimulation. Daily postsurgical e-mail prompts sent for a period of six months resulted in 93 administrations of the CAT-DI on the patient's smartphone. There was an average of 3.37 weekly measurements with an average separation of 2.12 days. No additional incentive was provided to the patient for completing the adaptive tests. The patient is a 55-year-old female with six psychiatric hospitalizations for depression, two suicide attempts, marginal response to eight electroconvulsive therapy (ECT) treatments, and 35 psychotropic medications. The authors demonstrated feasibility of daily depressive severity measurement at high levels of precision and compliance. Clinician ratings confirmed the general pattern of treatment benefit, but mask the marked variability in mood and more marked periods of benefit and decline revealed by the daily adaptive depression measurement.

From these two studies, it appears that remote high-frequency mental health measurement is highly feasible. Taking this a step further would be the use of MIRT-based CAT for ecological momentary assessments, where several assessments are made during the course of a day (see Figure 13.5).

13.4 IRT in Machine Learning

Martínez-Plumed et al. (2016) explore the use of IRT in connection with evaluating properties of different machine learning algorithms. A general limitation of machine learning methods is that they are evaluated empirically by comparing different methods against a set of problems. Their relative performance is aggregated across a set of problems and is a measure of average predictive quality. In this context, "instances" (i.e. a single row of data for a set of features) correspond to items and classifiers (e.g. different machine learning algorithms such as random forests, support vector machines, and neural networks) correspond to the subjects for which we are interested in their "ability." In the supervised learning setting (where we have the true state that we are trying to predict), for instance j, we have a binary indicator of whether classifier i produced a correct $x_{ij} = 1$

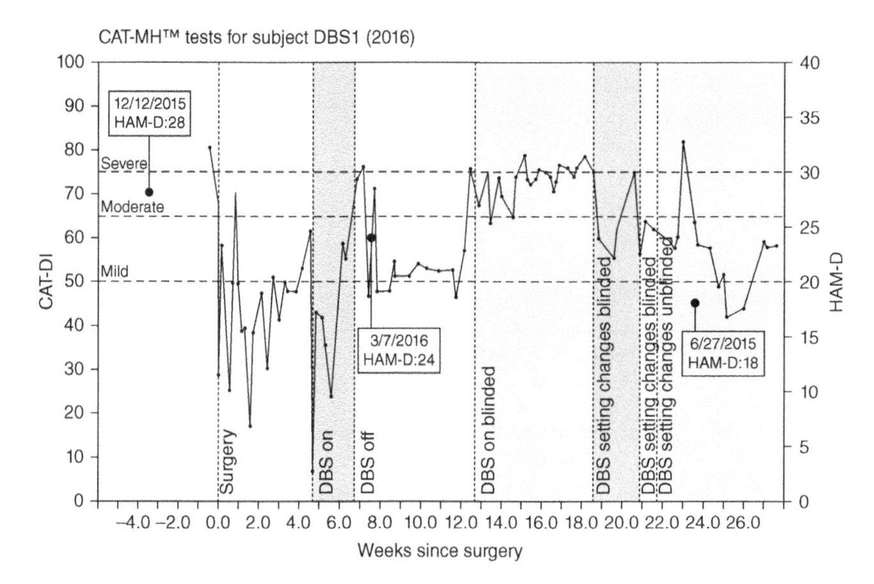

Figure 13.5 Intensive longitudinal depression severity measurement.

or incorrect $x_{ij} = 0$ classification. The authors show that instance difficulty, or "hardness" is an important feature of an instance and that recently less proficient means of estimating instance hardness have been used in machine learning to increase classification accuracy (Smith et al. 2014). However, the IRT discrimination parameter appears to be of even much greater value in that for machine learning applications, the discrimination parameter of an instance is a measure of how well the instance differentiates strong and weak classifiers for a given dataset.

To test the applicability of IRT to machine learning, the authors analyzed a dataset with 200 instances (feature patterns) and 128 different classifiers. To a psychometrician, this may seem counterintuitive in that the instances are likely derived from different subjects, but are considered the items and the classifiers are the subjects, but seem more like items. Although not considered by the authors, it would be possible to treat the classifiers as the items and the instances as subjects. In this case, the discrimination parameter for a classifier would indicate how well the classifier differentiates easy and more difficult instances. However, returning to their formulation, here the ability measure describes the accuracy of the classifier over the n instances. The authors export the idea of a person characteristic curve to propose a classifier characteristic curve (CCC). The CCC is a plot of the response probability (accuracy) of a particular classifier as a function of either the instance difficulty or the instance discrimination and allows for the comparison of different classifiers for different levels of instance difficulty or

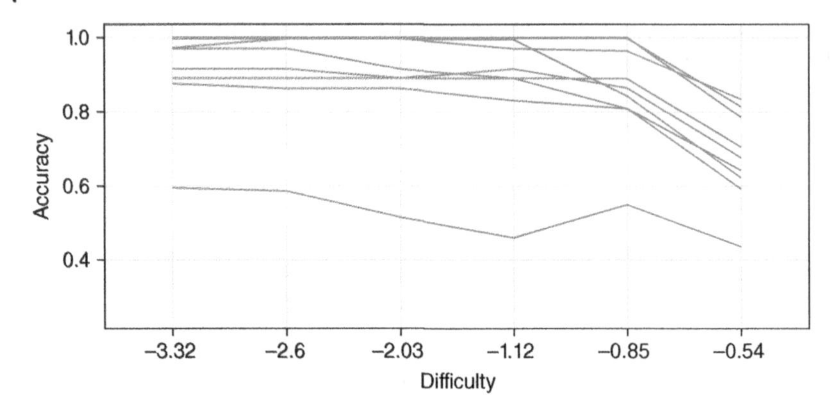

Figure 13.6 CCC plots across bins on the difficulty parameter for nine classifiers.

instance discrimination. Ideally, we seek a classifier that is robust to difficulty and discriminability of the instance.

Figure 13.6 provides an example of a CCC as a function of instance difficulty. The lowest performing classifier in Figure 13.6 is the random classifier which is expected to do poorly. Apart from the random classifier, all other classifiers do well for the first two difficulty bins (easy instances). Once instance of difficulty reaches −2.0 and the classifiers begin to differentiate in terms of their accuracy, with neural networks and support vector machines doing the best in terms of classification accuracy. In the fifth bin (instance difficulty = −0.85, 17% of the instances), flexible discriminant analysis and recursive partitioning do the worst with classification accuracy of 81% and neural networks and support vector machines still have classification accuracy of 1.0. In the final bin (instance difficulty = −0.54, 17% of the instances), random forest becomes the best classifier (accuracy = 83%) with neural networks (accuracy = 81%) and support vector machines (accuracy = 78%) close behind. Interestingly, the two nearest neighbor classifier shifts from being the best classifier for instances with low or medium difficulty, to the second worst classifier for high-difficulty instances. This type of nuanced view of the performance of classifiers as a function of properties of the instances is a major advantage of the IRT approach to determining differential classifier accuracy.

Bibliography

Achtyes, E.D., Halstead, S., Smart, L.A. et al. (2015). Validation of computerized adaptive testing in an outpatient nonacademic setting: the VOCATIONS trial. *Psychiatric Services* 66 (10): 1091–1096.

Ackerman, T.A. (1994). Using multidimensional item response theory to understand what items and tests are measuring. *Applied Measurement in Education* 7 (4): 255–278.

Ackerman, T.A. (1996). Graphical representation of multidimensional item response theory analysis. *Applied Psychological Measurement* 20: 311–329.

Aitchison, J. and Silvey, S.D. (1958). Maximum-likelihood estimation of parameters subject to restraints. *The Annals of Mathematical Statistics* 29 (3): 813–828.

Alegría, M., Alvarez, K., Ishikawa, R.Z. et al. (2016). Removing obstacles to eliminating racial and ethnic disparities in behavioral health care. *Health Affairs* 35 (6): 991–999.

Andersen, E.B. (1977). Sufficient statistics and latent trait models. *Psychometrika* 42 (1): 69–81.

Andersen, E.B. (1980). *Discrete Statistical Models with Social Science Applications*. Amsterdam: North Holland.

Andersen, E. and Madsen, M. (1977). Estimating the parameters of the latent population distribution. *Psychometrika* 42 (3): 357–374.

Anderson, T.W. (1984). *An Introduction to Multivariate Statistical Analysis*, 2e. New York: Wiley.

Andreasen, NC. (1984). The Scale for the Assessment of Positive Symptoms (SAPS). Iowa City, IA: University of Iowa.

Andrich, D. (1978). Application of a psychometric rating model to ordered categories which are scored with successive integers. *Applied Psychological Measurement* 2 (4): 581–594.

Andrich, D. (1988). A general form of Rasch's extended logistic model for partial credit scoring. *Applied Measurement in Education* 1 (4): 363–378.

Item Response Theory, First Edition. R. Darrell Bock and Robert D. Gibbons.

Anscombe, F.J. (1956). On estimating binomial response relations. *Biometrika* 43 (3/4): 461.

Ashford, J. and Sowden, R.R. (1970). Multi-variate probit analysis. *Biometrics* 26 (3): 535–546.

Baek, S.-G. (1997). Computerized adaptive testing using the partial credit model for attitude measurement. In: *Objective Measurement: Theory Into Practice*. (ed. M. Wilson, G. Engelhard and K. Draney), 37–55.

Baker, F.B. (1992). *Item Response Theory: Parameter Estimation Techniques*. New York: Marcel Dekker.

Ban, J.C., Hanson, B.A., Yi, Q., and Harris, D.J. (2002). Data sparseness and on-line pretest item calibration-scaling methods in CAT. *Journal of Educational Measurement* 39 (3): 207–218.

Ban, J.C., Hanson, B.A., Wang, T. et al. (2006). A comparative study of online pretest item calibration/scaling methods in computerized adaptive testing. *American Educational Research Association* 38 (3): 191–212.

Bartholomew, D.J. and Tzamourani, P. (1999). The goodness of fit of latent trait models in attitude measurement. *Sociological Methods and Research* 27 (4): 525–546.

Beiser, D., Vu, M., and Gibbons, R. (2016). Test-retest reliability of a computerized adaptive depression screener. *Psychiatric Services* 67 (9): 1039–1041.

Beiser, D.G., Ward, C.E., Vu, M. et al. (2019). Depression in emergency department patients and association with health care utilization. *Academic Emergency Medicine* 26 (8): 878–888.

Berkson, J. (1956). Estimation by least squares and by maximum likelihood. *Proceedings of the Third Berkeley Symposium* 1: 1–11.

Berndt, E.R., Hall, B.H., Hall, R.E., and Hausman, J.A. (1974). Estimation and inference in nonlinear structural models. *Annals of Economic and Social Measurement* 3 (4): 653–665.

Berona, J., Whitton, S., Newcomb, M.E. et al. Prospective risk and protective factors for the transition from suicide ideation to attempt among sexual and gender minority youth. *Psychiatric Services*, in press.

Birnbaum, A. (1957). Probability and Statistics in Item Analysis and Classification Problems: Efficient Design and Use of Tests of Mental Ability for Various Decision-making. *Technical report, Ser. Rep. No. 15*. Randolph Air Force Base, TX: USAF School of Aviation Medicine.

Birnbaum, A. (1958a). Further Considerations of Efficiency in Tests of a Mental Ability. *Technical report, Ser. Rep. No. 17*. Randolph Air Force Base, TX: USAF School of Aviation Medicine.

Birnbaum, A. (1958b). On the Estimation of Mental Ability. *Technical report, Ser. Rep. No. 17*. Randolph Air Force Base, TX: USAF School of Aviation Medicine.

Birnbaum, A. (1968). Some latent trait models and their use in inferring an examinee's ability. In: *Statistical Theories of Mental Test Scores* (ed. F.M. Lord and M.R. Novick), 397–479. Reading, MA: Addison-Wesley.

Bishop, Y.M., Holland, P.W., and Fienberg, S.E. (1975). *Discrete Multivariate Analysis Theory and Practice*. Cambridge, MA: Massachusetts Institute of Technology Press.

Black, D.W., Gunter, T., Loveless, P. et al. (2010). Antisocial personality disorder in incarcerated offenders: psychiatric comorbidity and quality of life. *Annals of Clinical Psychiatry* 22 (2): 113–120.

Bliss, C.I. (1935). The calculation of the dosage-mortality curve. *Annals of Applied Biology* 22 (1): 134–167.

Bock, R. (1972). Estimating item parameters and latent ability when responses are scored in two or more nominal categories. *Psychometrika* 37 (1): 29–51.

Bock, R. (1975). *Multivariate Statistical Methods in Behavioral Research*. New York: McGraw-Hill.

Bock, R.D. and Moore, E.G.J. (1986). *Advantage and Disadvantage: A Profile of American Youth*. Hillsdale, NJ: Erlbaum.

Bock, R. (1989a). Addendum: measurement of human variation: a two-stage model. R. Darrell Bock. In: *Multilevel Analysis of Educational Data*, 319–342. Academic Press.

Bock, R. (1989b). *Measurement of Human Variation: A Two-Stage Model*. Academic Press.

Bock, R.D. (1997). The nominal categories model. W.J. van der Linden; R.K. Hambleton. In: *Handbook of Modern Item Response Theory*, 33–49. New York: Springer.

Bock, R.D. and Aitkin, M. (1981). Marginal maximum likelihood estimation of item parameters: application of an EM algorithm. *Psychometrika* 46 (4): 443–459.

Bock, R.D. and Gibbons, R.D. (1996). High-dimensional multivariate probit analysis. *Biometrics* 52 (4): 1183–1194.

Bock, R. and Gibbons, R. (2010). Factor analysis of categorical item responses. In: *Handbook of Polytomous Item Response Theory Models* (ed. M.L. Nering and R. Ostini). Florence, KY: Lawrence Erlbaum. 155–184.

Bock, R.D. and Jones, L.V. (1968). *The Measurement and Prediction of Judgment and Choice*. San Francisco, CA: Holden-Day.

Bock, R.D. and Lieberman, M. (1970). Fitting a response model for n dichotomously scored items. *Psychometrika* 35 (2): 179–197.

Bock, R.D. and Mislevy, R.J. (1982). Adaptive EAP estimation of ability in a microcomputer environment. *Applied Psychological Measurement* 6 (4): 431–444.

Bock, R.D. and Schilling, S. (1997). High-dimensional full-information item factor analysis. M. Berkane. In: *Latent Variable Modeling and Applications to Causality*, 163–176. New York: Springer.

Bock, R.D. and Zimowski, M.F. (1997). Multiple group IRT. In: *Handbook of Modern Item Response Theory* (ed. W.J. van der Linden and R.K. Hambleton), 433–448. New York: Springer.

Bock, R.D., Mislevy, R., and Woodson, C. (1982). The next stage in educational assessment. *Educational Researcher* 11 (3): 4–16.

Bock, R.D., Muraki, E., and Pfeiffenberger, W. (1988). Item pool maintenance in the presence of item parameter drift. *Journal of Educational Measurement* 25 (4): 275–285.

Bock, R.D., Thissen, D., and Zimowski, M.F. (1997). IRT estimation of domain scores. *Journal of Educational Measurement* 34 (3): 197–211.

Böckenholt, U. (2001). Hierarchical modeling of paired comparison data. *Psychological Methods* 6 (1): 49–64.

Bradley, R.A. and Terry, M.E. (1952). Rank analysis of incomplete block designs: I. The method of paired comparisons. *Biometrika* 39 (3/4): 324.

Brennan, R. (2001). *Generalizability Theory*. New York: Springer.

Brown, J. and Weiss, D. (1977). An Adaptive Testing Strategy for Achievement Test Batteries. *Technical report (Research Rep. No. 77-6)*. Minneapolis, MN: University of Minnesota, Department of Psychology, Psychometric Methods Program, Computerized Adaptive Testing Laboratory.

Browne, M.W. and Cudeck, R. (1993). Alternative ways of assessing model fit. In: *Testing Structural Equation Models* (ed. K.A. Bollen and J.S. Long), pp. 136–162. Beverly Hills, CA: Sage.

Cai, L. (2010). A two-tier full-information item factor analysis model with applications. *Psychometrika* 75: 581–612.

Cai, L. and Hansen, M. (2013). Limited-information goodness-of-fit testing of hierarchical item factor models. *British Journal of Mathematical and Statistical Psychology* 66 (2): 245–276.

Cai, L., Maydeu-Olivares, A., Coffman, D.L., and Thissen, D. (2006). Limited-information goodness-of-fit testing of item response theory models for sparse 2P tables. *British Journal of Mathematical and Statistical Psychology* 59 (1): 173–194.

Cai, L., Thissen, D., and du Toit, S.H. (2011). *IRTPRO*. Lincolnwood, IL: Scientific Software International.

Camilli, G. and Shepard, L. (1994). *Methods for Identifying Biased Test Items*. Thousand Oaks, CA: Sage.

Chang, H.H. (2004). Understanding computerized adaptive testing: from Robbins–Monro to Lord and beyond. In: *The Sage Handbook of Quantitative Methodology for the Social Sciences* (ed. D. Kaplan), pp. 117–133. Thousand Oaks, CA: Sage.

Chang, H.-H. and Ying, Z. (1996). A global information approach to computerized adaptive testing. *Applied Psychological Measurement* 20: 213–229.

Chang, H.-H. and Ying, Z. (1999). A-stratified multistage computerized adaptive testing. *Applied Psychological Measurement* 23 (3): 211–222.

Chang, H.-H. and Ying, Z. (2009). Nonlinear sequential designs for logistic item response theory models with applications to computerized adaptive tests. *Annals of Statistics* 37 (3): 1466–1488.

Chang, H.-H., Qian, J., and Ying, Z. (2001). A-stratified multistage computerized adaptive testing with b blocking. *Applied Psychological Measurement* 25 (4): 333–341.

Chapman, L. and Bock, R.D. (1958). Components of variance due to acquiescence and content in the F scale measure of authoritarianism. *Psychological Bulletin* 55 (5): 328–333.

Chen, S.-Y., Ankenmann, R.D., and Chang, H.-H. (2000). A comparison of item selection rules at the early stages of computerized adaptive testing. *Applied Psychological Measurement* 24 (3): 241–255.

Cochran, W. and Cox, G. (1957). *Experimental Designs*. New York: Wiley.

Cooper, B.E. (1968). Algorithm AS 2: the normal integral. *Applied Statistics* 17 (2): 186.

Creedon, T.B. and Lê Cook, B. (2016). Datawatch: access to mental health care increased but not for substance use, while disparities remain. *Health Affairs* 35 (6): 1017–1021.

Cronbach, L. (1970). *Essentials of Psychological Testing*. New York: Harper & Row.

Cronbach, L.J., Gleser, G.C., Nanda, N., and Rajaratnam, N. (1972). *The Dependability of Behavioral Measurements: Theory of Generalizability for Scores and Profiles*. New York: Wiley.

Day, N.E. (1969). Estimating the components of a mixture of normal distributions. *Biometrika* 56 (3): 463.

Dempster, A.P., Laird, N.M., and Rubin, D.B. (1977). Maximum likelihood from incomplete data via the EM algorithm. *Journal of the Royal Statistical Society: Series B (Methodological)* 39 (1): 1–22.

Dempster, A.P., Rubin, D.B., and Tsutakawa, R.K. (1981). Estimation in covariance components models. *Journal of the American Statistical Association* 76 (374): 341–353.

Divgi, D.R. (1979a). Calculation of the tetrachoric correlation coefficient. *Psychometrika* 44 (2): 169–172.

Divgi, D.R. (1979b). Calculation of univariate and bivariate normal probability functions. *The Annals of Statistics* 7 (4): 903–910.

Dodd, B.G., de Ayala, R.J., and Koch, W.R. (1995). Computerized adaptive testing with polytomous items. *Applied Psychological Measurement* 19 (1): 5–22.

Dorans, N.J., Moses, T.P., and Eignor, D.R. (2010). Principles and practices of test score equating. *ETS Research Report Series* 2010 (2): i–41.

Dunnett, C. (1964). New tables for multiple comparisons with a control. *Biometrics* 20 (3): 482–491.

DuToit, M. (2003), *IRT from SSI: Bilog-MG, multilog, parscale, testfact*, Scientific Software International, Chicago, IL.

Edwards, A.L. and Thurstone, L.L. (1952). An internal consistency check for scale values determined by the method of successive intervals. *Psychometrika* 17 (2): 169–180.

Elderon, L., Smolderen, K.G., Na, B., and Whooley, M.A. (2011). Accuracy and prognostic value of american heart association-recommended depression screening in patients with coronary heart disease. *Circulation: Cardiovascular Quality and Outcomes* 4 (5): 533–540.

Embretson, S. and Reise, S. (2000). *Item Response Theory for Psychologists*. Mahway, NJ: Lawrence Erlbaum Associates.

Endicott, J. and Spitzer, R.L. (1978). A diagnostic interview: the schedule for affective disorders and schizophrenia. *Archives of General Psychiatry* 35 (7): 837–844.

Fechner, G.T. (1966). *Elements of Psychophysics* (ed. D.H. Howes and E.G. Boring). Leipzig: Breitkopf und Härtel. First published in 1860, translated by Adler, H.E.

Fechner, G.T. (1860). *Elemente der psychophysik*. Leipzig: Breitkopf und Härtel.

Fedorov, V.V. and Hackl, P. (1997). *Model-Oriented Design of Experiments, Lecture Notes in Statistics*. New York: Springer-Verlag.

de Finetti, B.D. (1972). *Probability, Induction and Statistics: The Art of Guessing*. New York: Wiley.

Finney, D.J. (1952). *Statistical Method in Biological Assay*. New York: Hafner Publishing Co.

Finney, D.J. (1964). *Probit Analysis: A Statistical Treatment of the Sigmoid Response Curve*. London: Cambridge University Press.

First, M., Gibbon, M., Spitzer, R., and Williams, J. B. W. (1996). *User's Guide for the Structured Clinical Interview for DSM-IV Axis I Disorders-Research Version*. New York: Biometrics Research Department, New York State Psychiatric Institute.

Fisher, R.A. (1922). On the mathematical foundations of theoretical statistics. *Philosophical Transactions of the Royal Society of London. Series A, Containing Papers of a Mathematical or Physical Character* 222 (594–604): 309–368.

Fisher, R.A. and Yates, F. (1938). *Statistical Tables for Biological, Agricultural and Medical Research*. London: Oliver and Boyd.

Fletcher, R. (1987). *Practical Methods of Optimization*, 2e. Chichester: Wiley.

Fliege, H., Becker, J., Walter, O.B. et al. (2005). Development of a computer-adaptive test for depression (D-CAT). *Quality of Life Research* 14 (10): 2277–2291.

Gardner, W., Shear, K., Kelleher, K.J. et al. (2004). Computerized adaptive measurement of depression: a simulation study. *BMC Psychiatry* 4.

Garwood, F. (1941). The application of maximum likelihood to dosage-mortality curves. *Biometrika* 32 (1): 46.

Gauss, C.F. (1809). *Theoria Motus Corporum Coelestium in Sectionibus Conicis Solem Ambientium*. Perthes et Besser.

Gibbons, R.D. and Amatya, A. (2015). *Statistical Methods for Drug Safety*. Boca Raton, FL: Chapman and Hall.

Gibbons, R.D. and Cai, L. (2017). *Dimensionality Analysis From: Handbook of Item Response Theory: Applications*, vol. 3. CRC Press.

Gibbons, R.D. and Hedeker, D.R. (1992). Full-information item bi-factor analysis. *Psychometrika* 57 (3): 423–436.

Gibbons, R.D. and Lavigne, J.V. (1998). Emergence of childhood psychiatric disorders: a multivariate probit analysis. *Statistics in Medicine* 17 (21): 2487–2499.

Gibbons, R.D. and Wilcox-Gök, V. (1998). Health service utilization and insurance coverage: a multivariate probit analysis. *Journal of the American Statistical Association* 93 (441): 63–72.

Gibbons, R.D., Bock, R.D., Hedeker, D. et al. (2007a). Full-information item bifactor analysis of graded response data. *Applied Psychological Measurement* 31 (1): 4–19.

Gibbons, R.R.D., Immekus, J.J.C., and Bock, R.D. (2007b). The added value of multidimensional IRT models. *Multidimensional and Hierarchical Modeling Monograph* 1 (312): 1–49.

Gibbons, R.D., Weiss, D.J., Kupfer, D.J. et al. (2008). Using computerized adaptive testing to reduce the burden of mental health assessment. *Psychiatric Services* 59 (4): 361–368.

Gibbons, R.D., Weiss, D.J., Pilkonis, P.A. et al. (2012). Development of a computerized adaptive test for depression. *Archives of General Psychiatry* 69 (11): 1104–1112.

Gibbons, R.D., Weiss, D.J., Pilkonis, P.A. et al. (2014). Development of the CAT-ANX: a computerized adaptive test for anxiety. *American Journal of Psychiatry* 171 (2): 187–194.

Gibbons, R.D., Weiss, D.J., Frank, E., and Kupfer, D. (2016). Computerized adaptive diagnosis and testing of mental health disorders. *Annual Review of Clinical Psychology* 12 (1): 83–104.

Gibbons, R.D., Kupfer, D., Frank, E. et al. (2017). Development of a computerized adaptive test suicide scale-The CAT-SS. *Journal of Clinical Psychiatry* 78 (9): 1376–1382.

Gibbons, R.D., Alegría, M., Cai, L. et al. (2018). Successful validation of the CAT-MH scales in a sample of Latin American migrants in the United States and Spain. *Psychological Assessment* 30 (10): 1267–1276.

Gibbons, R.D., Kupfer, D.J., Frank, E. et al. (2019). Computerized adaptive tests for rapid and accurate assessment of psychopathology dimensions in youth. *Journal of the American Academy of Child & Adolescent Psychiatry*. 1264–1273.

Gibbons, R.D., Alegria, M., Markle, S. et al. (2020). Development of a computerized adaptive substance use disorder scale for screening and measurement: the CAT-SUD. *Addiction* 115 (7): 1382–1394.

Gilks, W.R., Roberts, G.O., and Sahu, S.K. (1998). Adaptive markov chain monte carlo through regeneration. *Journal of the American Statistical Association* 93 (443): 1045–1054.

Gill, P. and Murray, W. (1974). *Numerical Methods for Constrained Optimization*. New York: Academic Press.

Glas, C.A. (1998). Detection of differential item functioning using lagrange multiplier tests. *Statistica Sinica* 8 (3): 647–667.

Goff, A., Rose, E., Rose, S., and Purves, D. (2007). Does PTSD occur in sentenced prison populations? A systematic literature review. *Criminal Behaviour and Mental Health* 17 (3): 152–162.

Goldstein, H. (1983). Measuring changes in educational attainment over time: problems and possibilities. *Journal of Educational Measurement* 20 (4): 369–377.

Goodman, L.A. (1968). The analysis of cross-classified data: independence, quasi-independence, and interactions in contingency tables with or without missing entries. *Journal of the American Statistical Association* 63 (324): 1091–1131.

Guilford, J. (1954). The constant methods. In: *Psychometric Methods*, 2e, 597. New York: McGraw-Hill.

Guinart, D., de Filippis, R., Rosson, S. et al. (2020). Development and validation of a computerized adaptive assessment tool for discrimination and measurement of psychotic symptoms. *Schizophrenia Bulletin* 9: sbaa168.

Gulliksen, H. (1950). *Theory of Mental Tests*. Wiley.

Gumbel, E.J. (1961). Bivariate logistic distributions. *Journal of the American Statistical Association* 56 (294): 335–349.

Gupta, S.S. (1963). Probability integrals of multivariate normal and multivariate t. *The Annals of Mathematical Statistics* 34 (3): 792–828.

Guttman, L. (1945). A basis for analyzing test-retest reliability. *Psychometrika* 10 (4): 255–282.

Haberman, S.J. (1977). Log-linear models and frequency tables with small expected cell counts. *Annals of Statistics* 5: 1148–1169.

Haberman, S. (1978). *Analysis of Qualitative Data: Introductory Topics*, vol. 1. New York: Academic Press, Incorporated.

Haberman, S. (1979). *Analysis of Qualitative Data: New Developments*, vol. 2. New York: Academic Press, Incorporated.

Haffajee, R.L. and Frank, R.G. (2018). Making the opioid public health emergency effective. *JAMA Psychiatry* 75 (8): 767–768.

Haggard, E. (1958). *Intraclass Correlation and the Analysis of Variance*. New York: Dryden Press.

Haley, D.C. (1952). Estimation of the Dosage Mortality Relationship When the Dose is Subject to Error. Technical report. *Technical Report No. 15 (Office of Naval*

Research Contract No 25140, NR 342-022). Stanford, CA: Stanford University Applied Mathematics and Statistics Labs.

Hambleton, R.K. and Swaminathan, H. (1985). *Item Response Theory: Principles and Applications.* Boston, MA: Kluwer-Nijhoff.

Hamilton, M. (1960). A rating scale for depression. *Journal of Neurology, Neurosurgery, and Psychiatry* 23: 56–62.

Han, K.T. (2012). SimulCAT: windows software for simulating computerized adaptive test administration. *Applied Psychological Measurement* 36 (1): 64–66.

Harman, H. (1967). *Modern Factor Analysis.* Chicago: University of Chicago Press.

Hastings, C. (1955). *Approximations for Digital Computers.* Princeton, NJ: Princeton University Press.

Hathaway, R.J. (1985). A constrained formulation of maximum-likelihood estimation for normal mixture distributions. *The Annals of Statistics* 13 (2): 795–800.

Hausman, J.A. and Wise, D.A. (1978). A conditional probit model for qualitative choice: discrete decisions recognizing interdependence and heterogeneous preferences. *Econometrica, Econometric Society* 46 (2): 403–426.

Hawton, K. and Fagg, J. (1992). Trends in deliberate self poisoning and self injury in Oxford, 1976–90. *British Medical Journal* 304 (6839): 1409–1411.

Hedeker, D. (1989). Random regression models with autocorrelated errors. Unpublished PhD dissertation. University of Chicago.

Hedeker, D. and Gibbons, R.D. (1996). MIXOR: a computer program for mixed-effects ordinal probit and logistic regression analysis. *Computer Methods and Programs in Biomedicine* 49: 157–176.

Hedeker, D. and Gibbons, R. (2006). *Longitudinal Data Analysis.* New York: Wiley.

Hedeker, D., Mermelstein, R.J., and Flay, B.R. (2006). Application of item response theory models for intensive longitudinal data. In: *Models for Intensive Longitudinal Data* (ed. T.A. Walls and J.L. Schafer), 84–108. Oxford: Oxford University Press.

Henderson, H.V. and Searle, S.R. (1979). Vec and vech operators for matrices, with some uses in Jacobians and multivariate statistics. *Canadian Journal of Statistics* 7 (1): 65–81.

Hendrickson, A.E. and White, P.O. (1964). Promax: a quick method for rotation to oblique simple structure. *British Journal of Statistical Psychology* 17 (1): 65–70.

Holland, P. and Wainer, H. (1993). *Differential Item Functioning.* Hillsdale, NJ: Lawrence Erlbaum Associates.

Holland, P.W., Dorans, N.J., and Petersen, N.S. (2006). Equating test scores. C. R. Rao and S. Sinharay. In: *Handbook of Statistics*, vol. 26, 169–203. Amsterdam, Netherlands: Elsevier.

Holzinger, K.J. and Swineford, F. (1937). The Bi-factor method. *Psychometrika* 2 (1): 41–54.

Householder, A.S. (1953). *Principles of Numerical Analysis.* New York: McGraw-Hill.

Householder, A.S. (1964). *Theory of Matrices in Numerical Analysis*. New York: Blaisdell.

Jeon, M., Rijmen, F., and Rabe-Hesketh, S. (2013). Modeling differential item functioning using a generalization of the multiple-group bifactor model. *Journal of Educational and Behavioral Statistics* 38 (1): 32–60.

Joe, H. and Maydeu-Olivares, A. (2010). A general family of limited information goodness-of-fit statistics for multinomial data. *Psychometrika* 75 (3): 393–419.

Joiner, T. (2010). Myths about suicide. *Choice Reviews Online* 48 (03): 48–1761.

Jöreskog, K.G. (1969). A general approach to confirmatory maximum likelihood factor analysis. *Psychometrika* 34 (2): 183–202.

Jöreskog, K.G. (1979). Basic ideas of factor and component analysis. In: *Advances in Factor Analysis and Structural Equation Models*, 5–20. Cambridge: Abt Books.

Jöreskog, K.G. (1994). On the estimation of polychoric correlations and their asymptotic covariance matrix. *Psychometrika* 59: 381–390.

Kaiser, H.F. (1958). The varimax criterion for analytic rotation in factor analysis. *Psychometrika* 23: 187–200.

Kalb, L.G., Stapp, E.K., Ballard, E.D. et al. (2019). Trends in psychiatric emergency department visits among youth and young adults in the US. *Pediatrics* 143 (4).

Kelley, T. (1947). *Fundamentals of Statistics*. Cambridge: Harvard University Press.

Kennedy, W.J. and Gentle, E.J. (1980). *Statistical Computing*. New York: Marcel Dekker.

Kiefer, J. and Wolfowitz, J. (1956). Consistency of the maximum likelihood estimator in the presence of infinitely many incidental parameters. *The Annals of Mathematical Statistics* 27 (4): 887–906.

Kim, J.J., Silver, R.K., Elue, R. et al. (2016). The experience of depression, anxiety, and mania among perinatal women. *Archives of Women's Mental Health* 19 (5): 883–890.

King, C.A., Brent, D., Grupp-Phelan, J. et al. (2021). The computerized adaptive screen for suicidal youth (CASSY) development and independent validation. *JAMA Psychiatry*, published online ahead of print.

Kingsbury, G.G. and Weiss, D.J. (1980). An Alternate-Forms Reliability and Concurrent Validity Comparison of Bayesian Adaptive and Conventional Ability Tests. *Research Report 80-5*, Computerized Adaptive Testing Laboratory. Minneapolis, MN: University of Minnesota.

Kingsbury, G.G. and Weiss, D.J. (1983). A comparison of IRT-based adaptive mastery testing and a sequential mastery testing procedure. D. J. Weiss. In: *New Horizons in Testing*, 257–283. New York: Academic Press.

Kolakowski, D. and Bock, R.D. (1981). A multivariate generalization of probit analysis. *Biometrics* 37: 541–551.

Kolen, M.J. (2006). Scaling and norming. In: *Educational Measurement* (ed. R.L. Brennan), 155–186. Westport, CT: American Council on Education/Prager.

Kroenke, K., Spitzer, R.L., and Williams, J.B. (2001). The PHQ-9: validity of a brief depression severity measure. *Journal of General Internal Medicine* 16 (9): 606–613.

La Porte, L.M., Kim, J.J., Adams, M.G. et al. (2020). Feasibility of perinatal mood screening and text messaging on patients' personal smartphones. *Archives of Women's Mental Health* 23 (2): 181–188.

Laird, N. (1978). Nonparametric maximum likelihood estimation of a mixing distribution. *Journal of the American Statistical Association* 73 (364): 805–811.

Lawley, D.N. (1943). XXIII.On Problems connected with item selection and test construction. *Proceedings of the Royal Society of Edinburgh. Section A. Mathematical and Physical Sciences* 61 (3): 273–287.

Lazarsfeld, P. (1958). Evidence and inference in social research. *Daedalus* 87 (4): 99–130.

Lazarsfeld, P. (1959). Latent structure analysis. In: *Psychology: A Study of Science*, S. Koch, 476–543. New York: McGraw-Hill.

Lehman, A.F. (1988). A quality of life interview for the chronically mentally ill. *Evaluation and Program Planning* 11 (1): 51–62.

Lehmann, E. and Casella, G. (1998). *Theory of Point Estimation*. Springer-Verlag.

Leung, C.-K., Chang, H.-H., and Hau, K.-T. (2003). Computerized adaptive testing: a comparison of three content balancing methods. *The Journal of Technology, Learning and Assessment* 2 (5): 1–15.

Likert, R. (1932). A technique for the measurement of attitudes. *Archives of Psychology* 22: 5–55.

Lima Passos, V., Berger, M.P.F., and Tan, F.E. (2007). Test design optimization in CAT early stage with the nominal response model. *Applied Psychological Measurement* 31 (3): 213–232.

Lindquist, E. (1953). *Design and Analysis of Experiments in Psychology and Education*. Boston, MA: Houghton Miffin.

Linn, R.L., Rock, D.A., and Cleary, T.A. (1969). The development and evaluation of several programmed testing methods. *Educational and Psychological Measurement* 29 (1): 129–146.

Linn, R., Levine, M., Hastings, C., and Wardrop, J. (1980). An Investigation of Item Bias in a Test of Reading Comprehension. *Technical report (Technical Report No. 163)*. Urbana, IL: Center for the Study of Reading, University of Illinois.

Longford, N. (1987). A fast scoring algorithm for maximum likelihood estimation in unbalanced mixed models with nested random effects. *Biometrika* 74 (4): 817–827.

Lord, F. (1952). A theory of test scores. *Psychometric Monographs* 7: 84.

Lord, F.M. (1953). The relation of test score to the trait underlying the test. *Educational and Psychological Measurement* 13 (4): 517–549.

Lord, F.M. (1962). Estimating norms by item-sampling. *Educational and Psychological Measurement* 22 (2): 259–267.

Lord, F.M. (1968). *Some Test Theory for Tailored Testing*. Research Bulletin, 69/4. Princeton, NJ: Educational Testing Service.

Lord, F.M. (1980). *Applications of Item Response Theory to Practical Testing Problems*. Mahwah, NJ: Lawrence Erlbaum Associates.

Lord, F.M. (1983). Unbiased estimators of ability parameters, of their variance, and of their parallel-forms reliability. *Psychometrika* 48: 233–245.

Lord, F.M. and Novick, M.R. (1968). *Statistical Theories of Mental Test Scores*. Reading, MA: Addison-Wesley Pub. Co.

Louis, T.A. (1982). Finding the observed information matrix when using the EM algorithm. *Journal of the Royal Statistical Society: Series B (Methodological)* 44 (2): 226–233.

Luce, R.D. (1959). On the possible psychophysical laws. *Psychological Review* 66 (2): 81–95.

Magnus, J. and Neudecker, H. (1988). *Matrix Differential Calculus with Applications in Statistics and Econometrics*. New York: Wiley.

Martínez-Plumed, F., Prudêncio, R.B., Martínez-Usó, A., and Hernández-Orallo, J. (2016). Making sense of item response theory in machine learning. *Frontiers in Artificial Intelligence and Applications* 285: 1140–1148.

Masters, G.N. (1982). A Rasch model for partial credit scoring. *Psychometrika* 47 (2): 149–174.

Mathers, C.D. and Loncar, D. (2006). Projections of global mortality and burden of disease from 2002 to 2030. *PLoS Medicine* 3 (11): 2011–2030.

Matías-Carrelo, L.E., Chávez, L.M., Negrón, G. et al. (2003). The Spanish translation and cultural adaptation of five mental health outcome measures. *Culture, Medicine and Psychiatry* 27 (3): 291–313.

Maydeu-Olivares, A. and Joe, H. (2005). Limited- and full-information estimation and goodness-of-fit testing in 2n contingency tables: a unified framework. *Journal of the American Statistical Association* 100 (471): 1009–1020.

McBride, J.R. and Martin, J.T. (1983). Reliability and validity of adaptive ability tests in a military setting. In: *New Horizons in Testing* (ed. D.J. Weiss), 223–236. New York: Academic Press.

McFadden, D. (1973). Conditional logit analysis of qualitative choice behavior. In: *Frontiers in Econometrics* (ed. P. Zarembka), 105–142. New York: Academic Press.

Meredith, W. (1993). Measurement invariance, factor analysis and factorial invariance. *Psychometrika* 58 (4): 525–543.

Mislevy, R.J. (1983). Item response models for grouped data. *Journal of Educational Statistics* 8 (4): 271–288.

Mislevy, R.J. (1984). Estimating latent distributions. *Psychometrika* 49 (3): 359–381.

Mislevy, R.J. and Verhelst, N. (1990). Modeling item responses when different subjects employ different solution strategies. *Psychometrika* 55 (2): 195–215.

Mitchell, A., Vaze, A., and Rao, S. (2009). Clinical diagnosis of depression in primary care: a meta-analysis. *The Lancet* 374 (9690): 609–619.

Mosteller, F. and Tukey, J.W. (1977). *Data Analysis and Regression: A Second Course in Statistics*. Reading, MA: Addison-Wesley Pub. Co.

Moustaki, I. (2000). A latent variable model for ordinal variables. *Applied Psychological Measurement* 24 (3): 211–223.

Moustaki, I., Jöreskog, K.G., and Mavridis, D. (2004). Factor models for ordinal variables with covariate effects on the manifest and latent variables: a comparison of LISREL and IRT approaches. *Structural Equation Modeling* 11 (4): 487–513.

Mulder, J. and van der Linden, W.J. (2009). Multidimensional adaptive testing with optimal design criteria for item selection. *Psychometrika* 74 (2): 273–296.

Muraki, E. (1990). Fitting a polytomous item response model to likert-type data. *Applied Psychological Measurement* 14 (1): 59–71.

Muraki, E. (1992). A generalized partial credit model: application of an EM algorithm. *ETS Research Report Series* 1992 (1): i–30.

Mustanski, B. and Espelage, D.L. (2020). Why are we not closing the gap in suicide disparities for sexual minority youth? *Pediatrics* 145 (3).

Mustanski, B., Whitton, S., Newcomb, M. et al. Predicting suicidality using a computer adaptive test: Two longitudinal studies of sexual and gender minority youth. *Journal of Clinical and Consulting Psychology*, in press.

Muthén, B. (1979). A structural probit model with latent variables. *Journal of the American Statistical Association* 74 (368): 807–811.

Muthén, B.O. (1989). Latent variable modeling in heterogeneous populations. *Psychometrika* 54 (4): 557–585.

Nandakumar, R. and Stout, W.F. (1993). Refinements of Stout's procedure for assessing latent trait unidimensionality. *Journal of Educational Statistics* 18: 41–68.

National Institute of Mental Health (2017). Ask Suicide-Screening Questions (ASQ) Toolkit. https://www.nimh.nih.gov/research/research-conducted-at-nimh/asq-toolkit-materials/index.shtml (accessed 28 October 2020).

National Leadership Forum on Behavioral Health/Criminal Justice Services (2009). Ending an American Tragedy: Addressing the Needs of Justice-Involved People with Mental Illnesses and Co-Occurring Disorders. *Technical report*. https://www.usf.edu/cbcs/mhlp/tac/documents/behavioral-healthcare/samh/ending-an-american-tragedy.pdf (accessed 28 October 2020).

Naylor, J.C. and Smith, A.F.M. (1982). Applications of a method for the efficient computation of posterior distributions. *Applied Statistics* 31 (3): 214.

Neyman, J. and Scott, E.L. (1948). Consistent estimates based on partially consistent observations. *Econometrica* 16 (1): 1.

Nishisato, S. and Nishisato, I. (1994). *Dual Scaling in a Nutshell*. Toronto: MicroStats.

Nunnally, J. (1967). *Psychometric Theory*. New York: McGraw Hill.

Overall, J.E. and Gorham, D.R. (1988). The brief psychiatric rating scale (BPRS): recent developments in ascertainment and scaling. *Psychopharmacology Bulletin* 24 (1): 97–99.

Pilkonis, P.A., Choi, S.W., Reise, S.P. et al. (2011). Item banks for measuring emotional distress from the patient-reported outcomes measurement information system (PROMIS®): depression, anxiety, and anger. *Assessment* 18 (3): 263–283.

Priester, M.A., Browne, T., Iachini, A. et al. (2016). Treatment access barriers and disparities among individuals with co-occurring mental health and substance use disorders: an integrative literature review. *Journal of Substance Abuse Treatment* 61: 47–59.

Radloff, L.S. (1977). The CES-D scale: a self-report depression scale for research in the general population. *Applied Psychological Measurement* 1 (3): 385–401.

Rao, C.R. (1973). *Linear Statistical Inference and Its Applications*, 2e. New York: Wiley.

Rasch, G. (1960). *Studies in Mathematical Psychology: I. Probabilistic Models for Some Intelligence and Attainment Tests*. Copenhagen: Nielsen & Lydiche.

Rasch, G. (1961). On general laws and the meaning of measurement in psychology. *Proceedings of the 4th Berkeley Symposium on Mathematical Statistics and Probability, Volume 4: Contributions to Biology and Problems of Medicine*, pp. 321–334.

Reckase, M.D. (2009). *Multidimensional Item Response Theory*. New York: Springer.

Reckase, M.D. and McKinley, R.L. (1991). The discriminating power of items that measure more than one dimension. *Applied Psychological Measurement* 15: 361–374.

Reise, S.P. and Waller, N.G. (1993). Traitedness and the assessment of response pattern scalability. *Journal of Personality and Social Psychology* 65 (1): 143–151.

Rijmen, F., Tuerlinckx, F., De Boeck, P., and Kuppens, P. (2003). A nonlinear mixed model framework for item response theory. *Psychological Methods* 8 (2): 185–205.

Roy, S.N. (1957). *Some Aspects of Multivariate Analysis*. Kolkata: Statistical Publishing Society.

Rubin, D.B. (1976). Inference and missing data. *Biometrika* 63 (3): 581–592.

Samejima, F. (1969). Estimation of latent ability using a response pattern of graded scores. *Psychometrika* 35 (1): 139.

Samejima, F. (1972). A general model for free-response data. *Psychometrika Monograph Supplement* 37 (1): 68.

Samejima, F. (1979). A New Family of Models for the Multiple Choice Item. *Technical report (Research Report No. 79-4)*. Tennessee University Knoxville Department of Psychology.

SAMHSA (2016). National Survey on Drug Use and Health. *Technical report*. Mental Health Services Administration. https://www.samhsa.gov/data/release/2016-national-survey-drug-use-and-health-nsduh-releases (accessed 28 October 2020).

Sanathanan, L. and Blumenthal, S. (1978). The logistic model and estimation of latent structure. *Journal of the American Statistical Association* 73 (364): 794–799.
Sani, S., Busnello, J., Kochanski, R. et al. (2017). High-frequency measurement of depressive severity in a patient treated for severe treatment-resistant depression with deep-brain stimulation. *Translational Psychiatry* 7 (8): e1207.
Savage, L.J. (1954). *The Foundations of Statistics*. New York: Dover Press.
Schwarz, Gideon (1978). Estimating the dimension of a model. *Annals of Statistics* 6 (2): 461–464.
Schilling, S. and Bock, R.D. (2005). High-dimensional maximum marginal likelihood item factor analysis by adaptive quadrature. *Psychometrika* 70 (3): 533–555.
Schott, J.R. (1997). *Matrix Analysis for Statistics*. Hoboken, NJ: Wiley.
Schroeder, E. (1945). *On Measurement of Motor Skills; An Approach Through a Statistical Analysis of Archery Scores*. Oxford: King's Crown Press.
Segall, D.O. (1996). Multidimensional adaptive testing. *Psychometrika* 61 (2): 331–354.
Segall, D.O. (2000). Principles of multidimensional adaptive testing. In: *Computerized Adaptive Testing: Theory and Practice* (ed. W.J. van der Linden and C.A.W. Glas), pp. 53–73. Boston, MA: Kluwer Academic.
Seo, D.G. and Weiss, D.J. (2015). Best design for multidimensional computerized adaptive testing with the bifactor model. *Educational and Psychological Measurement* 75 (6): 954–978.
Shenton, L.R. and Bowman, K.O. (1977). *Maximum Likelihood Estimation in Small Samples*. London: Griffin.
Silvey, S.D. (1980). *Optimal Design*. London: Chapman & Hall.
Skrondal, A. and Rabe-Hesketh, S. (2004). *Generalized Latent Variable Modeling: Multilevel, Longitudinal, and Structural Equation Models*. London: Chapman and Hall/CRC.
Smith, M.R., Martinez, T., and Giraud-Carrier, C. (2014). An instance level analysis of data complexity. *Machine Learning* 95 (2): 225–256.
Spearman, C. (1904). Measurement of association, Part II. Correction of 'systematic deviations'. *American Journal of Psychology* 15: 88–101.
Spearman, C. (1907). Demonstration of formulae for true measurement of correlation. *The American Journal of Psychology* 18 (2): 161.
Spielberger, C.D., Gorsuch, R.L., Lushene, R., et al. (1983). *Manual for the State-Trait Anxiety Inventory*. Palo Alto, CA: Consulting Psychologists Press.
Stan, A.D., Tamminga, C.A., Han, K. et al. (2020). Associating psychotic symptoms with altered brain anatomy in psychotic disorders using multidimensional item response theory models. *Cerebral Cortex* 30 (5): 2939–2947.
Steadman, H.J., Osher, F.C., Robbins, P.C. et al. (2009). Prevalence of serious mental illness among jail inmates. *Psychiatric Services* 60 (6): 761–765.
Stevens, S.S. (1961). To honor Fechner and repeal his law. *Science* 133 (3446): 80–86.

Stewart, W.F., Ricci, J.A., Chee, E. et al. (2003). Cost of lost productive work time among US workers with depression. *Journal of the American Medical Association* 289 (23): 3135–3144.

Stocking, M.L. (1988). Scale drift in on-line calibration. *ETS Research Report Series* 1988 (1): i–122.

Stocking, M.L. and Lord, F.M. (1983). Developing a common metric in item response theory. *Applied Psychological Measurement* 7 (2): 201–210.

Stout, W. (1987). A nonparametric approach for assessing latent trait unidimensionality. *Psychometrika* 52 (4): 589–617.

Stout, W. (1990). A new item response theory modeling approach with applications to unidimensional assessment and ability estimation. *Psychometrika* 55: 293–326.

Stroud, A.H. and Secrest, D. (1966). *Gaussian Quadrature Formulas*. Englewood Cliffs, NJ: Prentice Hall.

Stuart, A. (1958). Equally correlated variates and the multinormal integral. *Journal of the Royal Statistical Society: Series B (Methodological)* 20: 373–378.

Swaminathan, H. and Rogers, H. (1990). Detecting differential item functioning using logistic regression procedures. *Journal of Educational Measurement* 27 (4): 361–370.

Tamminga, C.A., Ivleva, E.I., Keshavan, M.S. et al. (2013). Clinical phenotypes of psychosis in the Bipolar-Schizophrenia network on intermediate phenotypes (B-SNIP). *American Journal of Psychiatry* 170 (11): 1263–1274.

Thissen, D. (1982). Marginal maximum likelihood estimation for the one-parameter logistic model. *Psychometrika* 47 (2): 175–186.

Thissen, D. (1991). *MULTILOG User's Guide: Multiple, Categorical Item Analysis and Test Scoring Using Item Response Theory*. Lincolnwood, IL: Scientific Software, Inc.

Thissen, D. and Steinberg, L. (1984). A response model for multiple choice items. *Psychometrika* 49 (4): 501–519.

Thissen, D. and Steinberg, L. (1986). A taxonomy of item response models. *Psychometrika* 51 (4): 567–577.

Thissen, D., Steinberg, L., and Wainer, H. (1993). Detection of differential item functioning using the parameters of item response models. In: *Differential Item Functioning*, 67–113. Hillsdale NJ: Lawrence Erlbaum Associates.

Thissen, D., Cai, L., and Bock, R.D. (2010). The nominal categories item response model. In: *Handbook of Polytomous Item Response Theory Models* (ed. M.L. Nering and R. Ostini), 43–75. Routledge/Taylor & Francis Group.

Thurstone, L.L. (1927). Psychophysical analysis. *The American Journal of Psychology* 38 (3): 368–389.

Thurstone, L.L. (1928). Attitudes can be measured. *American Journal of Sociology* 33 (4): 529–554.

Thurstone, L.L. (1929). Theory of attitude measurement. *Psychological Review* 36 (3): 222–241.

Thurstone, L.L. (1947). *Multiple Factor Analysis*. Chicago, IL: University of Chicago Press.

Thurstone, L.L. (1959). *The Measurement of Values*. Chicago, IL: University of Chicago Press.

Tsai, J.; Gu, X. (1958). Utilization of addiction treatment among U.S. adults with history of incarceration and substance use disorders. *Addiction Science & Clinical Practice* 14 (9).

Tucker, L.R. (1958). An inter-battery method of factor analysis. *Psychometrika* 23 (2): 111–136.

Vale, C.D. and Weiss, D.J. (1977). A Rapid Item-Search Procedure for Bayesian Adaptive Testing 77-4. *Technical report 77-4*. Minneapolis, MN: University of Minnesota, Department of Psychology, Psychometric Methods Program, Computerized Adaptive Testing Laboratory.

van der Linden, W.J. (1999). Adaptive Testing with Equated Number-Correct Scoring. *OMD research report; No. 99-02*. Enschede, Netherlands: University of Twente, Faculty Educational Science and Technology.

van der Linden, W.J. and Pashley, P.J. (2010). Item selection and ability estimation in adaptive testing. In: *Elements of Adaptive Testing* (ed. W.J. van der Linden and C.A.W. Glas), pp. 3–30. New York: Springer.

van der Linden, W.J. and Reese, L.M. (1998). A model for optimal constrained adaptive testing. *Applied Psychological Measurement* 22 (3): 259–270.

Veerkamp, W.J.J. and Berger, M.P.F. (1997). Some new item selection criteria for adaptive testing. *Journal of Educational and Behavioral Statistics* 22 (2): 203–226.

Wainer, H. and Mislevy, R. (1990). Item response theory, item calibration, and proficiency estimation. In: *Computerized Adaptive Testing: A Primer* (ed. H. Wainer), 81–99. Mahwah, NJ: Lawrence Erlbaum Associates.

Wainer, H., Sireci, S.G., and Thissen, D. (1991). Differential testlet functioning definitions and detection. *ETS Research Report Series* 1991 (1): i–42.

Walker, H.M. and Lev, J. (1953). *Statistical Inference*. New York: Henry Holt & Company.

Warm, T.A. (1989). Weighted likelihood estimation of ability in item response theory. *Psychometrika* 54 (3): 427–450.

Weiss, D.J. (1985). Adaptive testing by computer. *Journal of Consulting and Clinical Psychology* 53 (6): 774–789.

Weiss, D.J. and Kingsbury, G.G. (1984). Application of computerized adaptive testing to educational problems. *Journal of Educational Measurement* 21 (4): 361–375.

Weiss, D.J. and McBride, J.R. (1984). Bias and information of Bayesian adaptive testing. *Applied Psychological Measurement* 8 (3): 273–285.

Whooley, M.A. (2012). Diagnosis and treatment of depression in adults with comorbid medical conditions: a 52-year-old man with depression. *Journal of the American Medical Association* 307 (17): 1848–1857.

Woods, C.M., Cai, L., and Wang, M. (2013). The Langer-improved Wald test for DIF testing with multiple groups: evaluation and comparison to two-group IRT. *Educational and Psychological Measurement* 73 (3): 532–547.

Yao, L. (2012). Multidimensional CAT item selection methods for domain scores and composite scores: theory and applications. *Psychometrika* 77: 495–523.

Yao, L. (2013). Comparing the performance of five multidimensional CAT selection procedures with different stopping rules. *Applied Psychological Measurement* 37: 3–23.

Yi, Q. and Chang, H. (2003). A-stratified CAT design with content blocking. *British Journal of Mathematical and Statistical Psychology* 56: 359–378.

Zhang, J. and Stout, W. (1999). Conditional covariance structure of generalized compensatory multidimensional items. *Psychometrika* 64 (2): 129–152.

Zimowski, M.F., Muraki, E., Mislevy, R.J., and Bock, R.D. (1996). *BILOG-MG: Multiple-Group IRT Analysis and Test Maintenance for Binary Items*. Chicago, IL: Scientific Software International.

Index

Item Response Theory, First Edition. R. Darrell Bock and Robert D. Gibbons.

m

Printed and bound by CPI Group (UK) Ltd, Croydon, CR0 4YY

16/04/2025

14658369-0002